Urban Water-Related Problems

Urban Water-Related Problems

Editors

Akira Kawamura
Kei Nakagawa

MDPI

Basel • Beijing • Wuhan • Barcelona • Belgrade • Novi Sad • Cluj • Manchester

Editors
Akira Kawamura
Tokyo Metropolitan University
Hachioji, Japan

Kei Nakagawa
Nagasaki University
Nagasaki, Japan

Editorial Office
MDPI
St. Alban-Anlage 66
4052 Basel, Switzerland

This is a reprint of articles from the Special Issue published online in the open access journal *Water* (ISSN 2073-4441) (available at: https://www.mdpi.com/journal/water/special_issues/Urban_Water_Related_Problems).

For citation purposes, cite each article independently as indicated on the article page online and as indicated below:

Lastname, A.A.; Lastname, B.B. Article Title. *Journal Name* **Year**, *Volume Number*, Page Range.

ISBN 978-3-0365-9242-8 (Hbk)
ISBN 978-3-0365-9243-5 (PDF)
doi.org/10.3390/books978-3-0365-9243-5

Cover image courtesy of Akira Kawamura

Contents

About the Editors

Akira Kawamura

Akira Kawamura mandatorily retired from Tokyo Metropolitan University in 2021 after serving as a professor for 16 and half years. Afterward, he was awarded the title of Professor Emeritus. Akira Kawamura graduated with a Doctor of Engineering degree in Hydrology and Water Resources Engineering from Kyushu University in 1985. He was first appointed as a research associate in the Department of Civil Engineering Hydraulics at Kyushu University in 1985 and promoted to associate professor in 1993. He transferred to Tokyo Metropolitan University as a professor in 2004. His main fields of research interest include the real-time prediction of hydro-meteorological time series, urban rainfall–runoff models, integrated flood risk management, the analysis of chaotic hydrological phenomena, the sustainability assessment of water resources, X-band MP radar rainfall data applications, and deep learning models in hydrology, among others. To date, Prof. Kawamura has contributed to 202 peer-reviewed scientific publications, 48 of which are in international journals with high impact factors.

Kei Nakagawa

Kei Nakagawa is a professor of environmental groundwater science with 30 years of research experience. He was first appointed as an assistant professor in soil science at the Department of Agricultural Chemistry in 1999 at Kyushu University and then promoted to associate professor in water use engineering at the Department of Agricultural Engineering in 2002 at Kagoshima University. In April 2011, he was appointed as a full professor of the Graduate School of Fisheries and Environmental Sciences at Nagasaki University. His main fields of research interest include reactive transport in groundwater, physical and chemical hydrogeology and heterogeneity, saltwater intrusion, the performance evaluation of subsurface dams in the coastal aquifer, groundwater modeling, and the remediation of contaminated soils and groundwater (phytoremediation and electro-kinetic remediation of heavy metal contaminated soils). Since being appointed at Nagasaki University in 2011, his main research topic has been nitrate pollution resulting from agricultural activities.

water

MDPI

Editorial

Urban Water-Related Problems

Akira Kawamura [1,*] and Kei Nakagawa [2]

1 Department of Civil and Environmental Engineering, Tokyo Metropolitan University, Hachioji 192-0397, Japan
2 Institute of Integrated Science and Technology, Nagasaki University, Nagasaki 852-8521, Japan; kei-naka@nagasaki-u.ac.jp
* Correspondence: kawamura@tmu.ac.jp

1. Introduction

Urban areas are considered to be the most vulnerable to water-related problems, which involve a lack or excess of water problems from the perspectives of quantity and quality. These specific phenomena include flash floods and inundation, droughts and water shortages, surface and ground water pollution, tsunamis and storm surges, landslides and mudflows, the degradation of fluvial and aquatic ecosystems, and unsanitary conditions and epidemics, among others. In urban areas, water-related problems cause immense human losses and economic damage. Water-related problems frequently reoccur in urban areas, and are intricately linked with each other, posing major obstacles to the achievement of human security and the sustainable socio-economic development of cities. Thus, it is crucial that they are scientifically and comprehensively discussed, so that they can be better understood, in order to fight against and mitigate these problems.

The special session "Urban Water-Related Problems" has been held in the AOGS (Asia Oceania Geosciences Society) annual meeting since 2017. Thus far, we have discussed a very wide range of urban water-related problems. For this Special Issue published in *Water*, we intend to invite studies on these broad topics and encourage a collective perspective on urban water-related problems.

This Special Issue comprises four review papers and nine research articles, with contributions from 55 authors of six countries. All four of the review papers are related to the problems associated with flash flood phenomena, among the many broad topics mentioned above, but they focus on very different aspects of the phenomena, such as urban runoff modeling in Japan [1], real-time urban flood forecasting systems for Southeast Asia [2], frequency analysis of urban floods [3], and the reduction of non-point pollution as well as flood runoff by porous concrete infiltration [4].

On the other hand, among nine research articles, five articles are related to water quality problems, of which two articles investigate groundwater pollution in Japan and Sweden [5,6], one article deals with lake contamination by phosphorus in Sweden [7], another targets lakes or reservoirs from the perspective of the hydrodynamic response by wind [8], and the last one considers the water pollution problem for a riverbank area in Bangladesh [9]. Three research articles out of the nine are related to problems of urban flood phenomena, in which studies on urban flooding area characteristics in Japan [10], water level forecasting caused by urban floods in Japan [11], and barriers to IFRM (Integrated Flood Risk Management) adaptation in the Philippines [12] are carried out. The last article is related to water resource problems, investigating the urban domestic water supply system from the viewpoint of willingness to pay [13].

Unfortunately, studies dealing with the phenomena related to tsunamis and storm surges, landslides and mudflows, and the degradation of fluvial and aquatic ecosystems are not published in this Special Issue.

Citation: Kawamura, A.; Nakagawa, K. Urban Water-Related Problems. *Water* **2023**, *15*, 3280. https://doi.org/10.3390/w15183280

Received: 28 August 2023
Accepted: 12 September 2023
Published: 16 September 2023

2. Overview of the Contributions of This Special Issue

The review paper "Urban Flood Runoff Modeling in Japan: Recent Developments and Future Prospects" [1] summarizes, discusses, and shares key outputs from some of the main research directions in urban flood runoff modeling, significant parts of which have been uniquely developed in Japan and only published in Japanese. In the paper, after a general introduction to urban runoff modeling, the authors present key historical works in Japan, followed by a description of the situation in Japan with respect to observations of precipitation and water level. Then, the storage function model approach is reviewed, including an extension to urban basins, as well as recent experiments with AI-based emulation in Japanese basins. Subsequently, the authors review the prospects of detailed hydrodynamic modeling involving high-resolution, vector-based Geographical Information System (GIS) data for the optimal description of the urban environment with applications in Tokyo. The authors conclude the paper with some future prospects related to urban flood risk modeling and assessment in Japan.

In the paper "Real-Time Urban Flood Forecasting Systems for Southeast Asia—A Review of Present Modelling and Its Future Prospects" [2], the authors reviewed state-of-the-art models of real-time forecasting systems in countries in Southeast Asia, such as Thailand, for urban flash floods. A real-time system basically consists of rainfall forecasting, drainage system modeling, and inundation area mapping. The authors summarized the recent radar data utilization methods for rainfall forecasting, physical-process-based hydraulic models for flood inundation prediction, and data-driven artificial intelligence (AI) models for the real-time forecasting systems. The authors also dealt with available technologies for modeling digital surface models (DSMs) for the finer urban terrain of drainage systems. The review indicated that an obstacle to using process-based hydraulic models was the limited computational resources and shorter lead time for real-time forecasting in many urban areas of tropical Southeast Asia.

The review paper "Frequency Analysis of Hydrological Data for Urban Floods—Review of Traditional Methods and Recent Developments, Especially an Introduction of Japanese Proper Methods" [3] investigates in detail the frequency analysis of hydrological data for urban floods, targeting the Japanese methods in particular. First, the authors introduced well-used Japanese frequency analysis methods, because some techniques that are slightly different from the international standard have been used in Japan for many years. This review emphasized discussions of the parameter estimation of stochastic models and the selection of optimal statistical models, which include the evaluation of goodness-of-fit techniques of statistical models. Based on these results, the authors criticized the Japanese standard procedures recommended by the central government, and indicated that consistency between parameter estimation and the evaluation of goodness-of-fit was necessary. From this perspective, the authors recommended using the maximum likelihood method and AIC. The authors also recommended, when using SLSC method, to apply not only SLSC itself but also SLSC's non-exceedance probability.

In the review paper "Application of Porous Concrete Infiltration Techniques to Street Stormwater Inlets That Simultaneously Mitigate against Non-Point Heavy Metal Pollution and Stormwater Runoff Reduction in Urban Areas: Catchment-Scale Evaluation of the Potential of Discrete and Small-Scale Techniques" [4], first, the literature related to the expansion of pervious areas was introduced, because the expansion of pervious areas was an essential and common concept in mitigating nonpoint pollution runoff in urban areas. Then, the potential application of porous concrete as a medium for constructing the bottom and side walls of street stormwater inlets was investigated. The effectiveness of the medium in reducing (i) the stormwater runoff volume and (ii) the heavy metal pollution runoff loads was assessed. In the paper, sensitivity tests were performed by changing the exfiltration rates. The results indicate that porous concrete used at only the bottom and side walls of the street stormwater inlets was suitable for reducing the runoff volume and removing any heavy metals from stormwater at a catchment scale.

In the paper "Groundwater Quality and Potential Pollution in the Southern Shimabara Peninsula, Japan" [5], the authors aimed to assess the groundwater quality in the southern regions of the Shimabara Peninsula in Nagasaki Prefecture, Japan, for a comprehensive understanding of the overall situation against the background that the peninsula is known for serious groundwater contamination caused by nitrate pollution. To achieve this goal, groundwater samples were collected at 56 locations in Minami-Shimabara City from 28 July to 4 August 2021. The authors showed using a Piper-trilinear diagram that almost 80% of the samples were classified as the alkaline earth carbonate type. From the Stiff diagrams, they showed that most of the locations exhibited a Ca-HCO_3 water type, while the Na-HCO_3 and Mg-HCO_3 types were also observed in coastal areas. Finally, from the principal component analysis, the authors revealed that the first component corresponded to dissolved constituents and denitrification, and the second component indicated the effects of ion exchange and low nitrate pollution.

The paper "PFAS in the Drinking Water Source: Analysis of the Contamination Levels, Origin and Emission Rates" [6] delves into the prevailing groundwater pollution situation in southern Sweden against the background that waterborne pollution caused by polyfluoroalkyl substances (PFAS) has been reported in numerous countries globally, emerging as a novel concern that is capturing attention. The objective of this study was to enhance understanding regarding the historical contamination of water sources due to PFAS and to explore interconnected facets of PFAS-containing aqueous film-forming foam (PFAS-AFFF) application in fire training exercises and equipment trials. To achieve this goal, the authors encompassed the analysis of contamination extents and PFAS compositions, the evaluation of potential PFAS and PFAS-AFFF emission scenarios, and the assessment of contaminant transport conditions.

The paper "Decision Support for Lake Restoration: A Case Study in Swedish Freshwater Bodies" [7] presents techniques aimed at providing decision support for lake restoration through a combination of multi-criteria analysis and decision analysis. The main objective is to identify effective and efficient measures for restoring lakes. The study focuses on evaluating six commonly used lake restoration methods for reducing internal phosphorus loads in two selected lakes, using criteria such as cost, longevity, and effectiveness. The findings indicate that aluminum treatment emerges as the most favorable option due to its superior effectiveness and cost efficiency. The authors suggested that these methodologies could be utilized not only in specific countries but also in the restoration of lakes and ponds worldwide.

The paper "A Detailed Analysis on Hydrodynamic Response of a Highly Stratified Lake to Spatio-Temporally Varying Wind Field" [8] explores the hydrodynamic reactions of a highly stratified lake or reservoir to various inhomogeneous wind conditions through the application of numerical models and integrated analysis. This approach was taken because wind was commonly acknowledged as a vital factor driving transport and mixing processes in stratified, enclosed systems. The authors demonstrated that under the influence of non-uniform wind conditions, often characterized by weak-to-moderate wind with high spatial variances, the lake's hydrodynamic responses differed from those under uniform wind conditions. On the other hand, low spatial variances, often associated with strong wind, resulted in hydrodynamic responses identical to those of uniform wind. Thus, the authors recommended using non-uniform wind in model simulation for achieving realistic results, improving water management in lakes and reservoirs.

The paper "Spatial Distribution and Source Identification of Water Quality Parameters of an Industrial Seaport Riverbank Area in Bangladesh" [9] employed a range of water quality indices, including the Metal Index (MI), Comprehensive Pollution Index (CPI), and Weighted Arithmetic Water Quality Index (WQI), to enhance understanding of pollution dispersion and the underlying processes influencing river water quality. Additionally, multivariate statistical techniques were utilized to assess pollutant loads and sources within the Pasur River system in Bangladesh. The authors indicated that contaminant origins encompass both geogenic and anthropogenic factors, involving untreated or inadequately treated wastewater from industries and the discharge of urban domestic waste. The effectiveness of the water quality assessment and pollution source identification methodologies introduced by the authors was demonstrated, with potential applicability on a global scale.

In the paper "Topographical Characteristics of Frequent Urban Pluvial Flooding Areas in Osaka and Nagoya Cities, Japan" [10], frequent urban pluvial flooding areas using 20 years of urban pluvial flooding area records during 1993–2012 were identified and analyzed using the principal component analysis of their topographical characteristics in Osaka and Nagoya Cities, Japan. The authors showed that the topographical characteristics of the frequent urban pluvial flooding areas in both cities were different. In Osaka City, not only the topographical characteristics, but also the influence of anthropogenic factors and stormwater drainage improvements were influential. On the other hand, in Nagoya City, the mere presence of dominant structures dammed up the inundated water and caused urban pluvial flooding. The authors quantitatively showed the paradigm shift of urban pluvial flooding factors from topographical characteristics to anthropogenic characteristics by the statistical analysis of newly defined urban pluvial flooding frequency areas.

The paper "Study on a Water-Level-Forecast Method Based on a Time Series Analysis of Urban River Basins—A Case Study of Shibuya River Basin in Tokyo" [11] investigated a vector autoregressive model to develop a water level forecast system that uses observed rainfall and water level. The model was targeted to ensure information conducive to evacuation approximately 20 min in advance without the need to build a physical model. The authors showed that the method based on time series analysis achieved a stable forecast, and indicated that the method can be applied as a water level forecast method for basins with an extremely fast flood arrival times and limited observation data.

In the paper "An Expanded Interpretive Structural Modeling Analysis of the Barriers to Integrated Flood Risk Management Adaptation in Metro Manila" [12], the authors proposed an expanded ISM (Ex-ISM) method to comprehensively analyze the interrelationships between the barriers to integrated flood risk management (IFRM) adaptation in Metro Manila, Philippines. The Ex-ISM enhanced conventional ISM in that the symbolism was modified to explicitly show the contextual interrelationships; then, the hierarchy assignment step was simplified, and a diagram was used to show all of the inter-relationships that allowed a comprehensive analysis. The authors showed that the results obtained using the Ex-ISM method did not deviate from those yielded by the conventional ISM method, but the Ex-ISM method allowed an easy assignment of hierarchy. The authors also showed that the Ex-ISM method was able to draw a diagram incorporating not only the direct but also the indirect interrelationships to provide a comprehensive analysis of the relationships between barriers.

The paper "Willingness to Pay for Improved Urban Domestic Water Supply System: The Case of Hanoi, Vietnam" [13] investigated water users' willingness to pay (WTP) for the improvement of Hanoi's domestic water supply. This was because in Hanoi, the capital of Vietnam, the municipal government is facing a number of difficulties in providing sufficient water in a sustainable manner due to not only the increasing urban population and the serious pollution of water resources, but also a lack of resources to invest in the supply system. In this paper, a contingent valuation process based on a survey of 402 respondents was used to explore citizens' willingness to pay for the improvement of their urban water supply. The authors revealed that Hanoi's urban communities were generally satisfied with the quantity of their water supply, but tended to be dissatisfied with its quality. The developed WTP regression model based on the survey findings showed that the average WTP was 1.4% of the average household income, taken as the affordability level of monthly water payments.

Conflicts of Interest: The authors declare no conflict of interest.

References

1. Kawamura, A.; Amaguchi, H.; Olsson, J.; Tanouch, H. Urban Flood Runoff Modeling in Japan: Recent Developments and Future Prospects. *Water* **2023**, *15*, 2733. [CrossRef]
2. Chitwatkulsiri, D.; Miyamoto, H. Real-Time Urban Flood Forecasting Systems for Southeast Asia—A Review of Present Modelling and Its Future Prospects. *Water* **2023**, *15*, 178. [CrossRef]
3. Mizuki, C.; Kuzuha, Y. Frequency Analysis of Hydrological Data for Urban Floods—Review of Traditional Methods and Recent Developments, Especially an Introduction of Japanese Proper Methods. *Water* **2023**, *15*, 2490. [CrossRef]
4. Harada, S. Application of Porous Concrete Infiltration Techniques to Street Stormwater Inlets That Simultaneously Mitigate against Non-Point Heavy Metal Pollution and Stormwater Runoff Reduction in Urban Areas: Catchment-Scale Evaluation of the Potential of Discrete and Small-Scale Techniques. *Water* **2023**, *15*, 1998. [CrossRef]
5. Nakagawa, K.; Amano, H.; Yu, Z.; Berndtsson, R. Groundwater Quality and Potential Pollution in the Southern Shimabara Peninsula, Japan. *Water* **2022**, *14*, 4106. [CrossRef]
6. Mussabek, D.; Söderman, A.; Imura, T.; Persson, K.M.; Nakagawa, K.; Ahrens, L.; Berndtsson, R. PFAS in the Drinking Water Source: Analysis of the Contamination Levels, Origin and Emission Rates. *Water* **2023**, *15*, 137. [CrossRef]
7. Sellergren, M.; Li, J.; Drakare, S.; Thöns, S. Decision Support for Lake Restoration: A Case Study in Swedish Freshwater Bodies. *Water* **2023**, *15*, 668. [CrossRef]
8. Le, H.N.; Shintani, T.; Nakayama, K. A Detailed Analysis on Hydrodynamic Response of a Highly Stratified Lake to Spatio-Temporally Varying Wind Field. *Water* **2023**, *15*, 565. [CrossRef]
9. Islam, M.S.; Nakagawa, K.; Abdullah-Al-Mamun, M.; Khan, A.S.; Goni, M.A.; Berndtsson, R. Spatial Distribution and Source Identification of Water Quality Parameters of an Industrial Seaport Riverbank Area in Bangladesh. *Water* **2022**, *14*, 1356. [CrossRef]
10. Komori, D.; Nakaguchi, K.; Inomata, R.; Oyatsu, Y.; Tachikawa, R.; Kazama, S. Topographical Characteristics of Frequent Urban Pluvial Flooding Areas in Osaka and Nagoya Cities, Japan. *Water* **2022**, *14*, 2795. [CrossRef]
11. Koyama, N.; Sakai, M.; Yamada, T. Study on a Water-Level-Forecast Method Based on a Time Series Analysis of Urban River Basins—A Case Study of Shibuya River Basin in Tokyo. *Water* **2023**, *15*, 161. [CrossRef]
12. Mercado, J.M.R.; Kawamura, A.; Medina, R. An Expanded Interpretive Structural Modeling Analysis of the Barriers to Integrated Flood Risk Management Adaptation in Metro Manila. *Water* **2023**, *15*, 1029. [CrossRef]
13. Bui, N.T.; Darby, S.; Vu, T.Q.; Mercado, J.M.R.; Bui, T.T.P.; Kantamaneni, K.; Nguyen, T.T.H.; Truong, T.N.; Hoang, H.T.; Bui, D.D. Willingness to Pay for Improved Urban Domestic Water Supply System: The Case of Hanoi, Vietnam. *Water* **2022**, *14*, 2161. [CrossRef]

water

Review

Urban Flood Runoff Modeling in Japan: Recent Developments and Future Prospects

Akira Kawamura [1,*], Hideo Amaguchi [1], Jonas Olsson [2] and Hiroto Tanouchi [3]

[1] Department of Civil and Environmental Engineering, Tokyo Metropolitan University, Hachioji 192-0397, Japan; amaguchi@tmu.ac.jp
[2] Hydrology Research, Swedish Meteorological and Hydrological Institute, SE-601 76 Norrköping, Sweden; jonas.olsson@smhi.se
[3] Faculty of Systems Engineering, Wakayama University, Wakayama 640-8510, Japan; tanouchi@wakayama-u.ac.jp
* Correspondence: kawamura@tmu.ac.jp; Tel.: +81-42-677-1111 (ext. 4573)

Abstract: Since the 20th century, Japan has experienced a period of very rapid urbanization. Cities have experienced substantial densification and expansion, resulting in gradually elevated flood risk. Urban flooding has also occurred in most large cities in Japan, particularly in Tokyo. In response to this growing problem, much effort and resources have been spent on research and development aimed at understanding, simulating, and managing urban flood risk in Japan. The objective of this review is to summarize, discuss, and share key outputs from some of the main research directions in this field, significant parts of which have been uniquely developed in Japan and only published in Japanese. After a general introduction to urban runoff modeling, in the next section, key historical works in Japan are summarized, followed by a description of the situation in Japan with respect to observations of precipitation and water level. Then, the storage function model approach is reviewed, including an extension to urban basins, as well as recent experiments with AI-based emulation in Japanese basins. Subsequently, we review the prospects of detailed hydrodynamic modeling involving high-resolution, vector-based Geographical Information System (GIS) data for the optimal description of the urban environment with applications in Tokyo. We conclude the paper with some future prospects related to urban flood risk modeling and assessment in Japan.

Keywords: urban flood risk assessment; USF model; TSR model; urban landscape GIS delineation; rainfall data

Citation: Kawamura, A.; Amaguchi, H.; Olsson, J.; Tanouchi, H. Urban Flood Runoff Modeling in Japan: Recent Developments and Future Prospects. *Water* **2023**, *15*, 2733. https://doi.org/10.3390/w15152733

Academic Editor: Fernando António Leal Pacheco

Received: 25 June 2023
Revised: 26 July 2023
Accepted: 26 July 2023
Published: 28 July 2023

1. Introduction

Development in urban areas greatly increases surface runoff, owing to the increase in impervious surfaces, as well as the total runoff volume flowing to the receiving watershed. In addition, the construction of storm sewer systems and the culverting of rivers during the urbanization process increase the runoff rate. The peak velocity of runoff inevitably increases as more runoff is discharged at shorter intervals, resulting in an increased risk of overflow. In conjunction with the concentration of population and property in the watershed, urbanization thus generally makes flood potential higher [1–4].

Since the latter half of the 20th century, urbanization has progressed rapidly in both developed and developing countries. Especially in Japan, urbanization has progressed much more rapidly than in Europe and the United States [5]. Mountains and hills in watershed areas that had been forests and fields were developed one after another, and agricultural land, including paddy fields, was simultaneously converted into residential land [5]. As a result, the watersheds were transformed into urban areas with a high risk of inundation. In an urban watershed, there is a complex distribution of artificially developed impervious areas, such as houses, buildings, paved roads, and parking lots, where rainwater cannot infiltrate. Runoff pathways by storm drainage facilities, such as sewers and artificial drainage channels, as well as runoff control and flood control facilities (e.g., storage and infiltration facilities), form a very complex urban runoff system that is constantly changing [6]. In Japan, the 1958 Kano River Typhoon Event is considered the first case of urban flooding [7], which caused extensive inundation damage not only in the lowlands but also in previously flood-free plateau areas [7,8].

In terms of flood damage in Japan, there has been a decreasing trend in human losses, although the economic losses have not decreased. The damage per unit area has exhibited an upward trend since the 2000s. Recent floods in urban areas have become more complex than past urban floods. One reason is the increasing frequency of unpredictable localized torrential rains, commonly called "guerrilla downpours" [9–11], which might be considered to be an effect of urbanization [12]. Seino et al. [13] explored how urbanization affects precipitation in the Tokyo metropolitan area. The risk of pluvial flooding caused by surcharge from sewerage facilities has been shown to increase compared to fluvial flooding caused by overflow from river channels [14].

Urban flooding, especially associated with recent record-breaking rainfall events, has become one of the most pressing social issues. Thus, flood mitigation is a crucial water management strategy for reducing destructive damage. For this purpose, rainfall-runoff models are important tools that play a central role, especially in urban watersheds. Generally, the purpose of runoff analysis is to clarify the physical mechanism of the runoff process by applying the amount of precipitation (rainfall, snowfall, etc.) to predict hydrological impacts, such as floods and droughts, in the target watershed. The results are used for the planning and management of rivers and watersheds.

In terms of process description, runoff models can be classified into two main types: lumped and distributed models. The former considers the watershed as a homogeneous unit without spatial variability, and the runoff process is generally expressed as ordinary differential equations of time. Distributed models explicitly consider the spatial variability of watershed characteristics and are usually expressed as partial differential equations of time and place [15,16]. In addition, semi-distributed models also exist, where a watershed is divided into sub-watersheds, each of which is represented by a lumped model. Based on how the response of runoff as output to rainfall (precipitation) as input is described, models may be classified into "response models", which simply use a response relationship, "conceptual models", which conceptualize the response relationship, and "physical models", which model the runoff from the basic equations of physics.

The most commonly used lumped conceptual models in Japan today are the tank model [17] and the storage function (SF) model [18,19]. The tank model was developed by Sugawara in 1956 [17], and the SF model was proposed by Kimura in 1961 [18,20]. The SF model is a lumped model used to simulate individual flood events and has the ability to represent nonlinearity in the runoff process with a relatively simple structural equation and fast calculation. The SF model is commonly used in watersheds with relatively large catchment areas under the jurisdiction of the Ministry of Land, Infrastructure, and Transport (MLIT), and has become the standard runoff model for flood control planning in Japan [21–24].

In contrast to lumped models, distributed models explicitly incorporate the spatial variability of watershed characteristics. With the widespread use of Geographic Information System (GIS) technology, distributed physical models have been widely used in small-to-medium-sized urban watersheds, where the effects of land use change and flood control facility development on runoff characteristics have been evaluated [25–27]. The main advantages of commonly used grid-based distributed models are their simple structure and their use of spatially explicit watershed information, which is generally readily available. Because of these advantages, grid-based distributed models are widely used not only in rural areas but also in urban watersheds. In Japan, Sayama et al. [28] proposed the rainfall-runoff-inundation (RRI) model, which is a 2D model capable of simultaneously analyzing rainfall runoff and flood inundation. The RRI model has been integrated with modeling and visualization tools by the International Centre for Water Hazard and Risk

Management (ICHARM) of the Public Works Research Institute (PWRI) [29] and is used not only in Japan but also in other countries.

Globally, many different basin-scale hydrological models have been developed, including the TOPMODEL [30], Xinanjiang model [31,32], Systeme Hydrologique Europeen (SHE) [33], and the Institute of Hydrology Distributed model (IHDM) [34], and those are also commonly used in Japan. The Hydrologiska Byrån Vattenbalans (HBV) model was developed in Sweden in the 1970s and has been widely used worldwide [35]. More recently, the hydrological processes for the environment (HYPE) model [36,37] was developed, which has been set up from the basin to the global scale and is used for operational forecasting as well as climate change impact assessment. Tanouchi et al. [21] developed the semi-distributed HYPE model to improve the description of urban runoff.

The accurate characterization of impervious surface type and extent is required for the prediction of the corresponding impacts on watershed hydrology [38]. However, in general, the land use information used as watershed characteristics in grid-based models is simply the impervious area ratio and the runoff coefficient. Therefore, the ability to accurately represent the impermeable characteristics of urban watersheds in grid-based models is limited [39]. Many models use raster-type land use data generated from aerial photographs.

With recent technological advances in GIS, digital information has become readily available for urban watersheds. In addition to grid-shaped raster land use information, vector-type polygonal (polygon shape) land use data, which can accurately represent buildings, roads, and other land features, have also been created [40]. However, there are still only a few cases in which polygon-type land use information is directly used for runoff analysis. Sample et al. [41] used GIS to facilitate urban stormwater analysis using land use parcel boundaries. Rodriguez et al. [42] used a vector-based watershed description based on a so-called urban databank, which includes the categories cadastral parcel, building, street, sewer system, and river, to calculate an urban unit hydrograph. Lee et al. [43] developed a model to evaluate low-impact development (LID) by defining a watershed as a hydrological response unit (HRE), which is a sub-watershed for calculating stormwater inflow to the sewer system. However, the area of the HRE is very big at several thousand square meters, and land use information is averaged over the HRE.

In light of this background, we review flood runoff analysis and modeling with a focus on urban watersheds in Japan. Regarding previous research related to urban flood runoff modeling in Japan, this has already been reviewed in detail in the authorized Handbook [44] and Hydroscience Formulary [16,45–47] published by the Japan Society of Hydrology and Water Resources (JSHWR) and the Japan Society of Civil Engineers (JSCE), respectively. Therefore, in this review paper, we mainly focus on recent developments in urban flood runoff modeling. Following this introduction, in Section 2, we briefly describe the history of urban runoff modeling in Japan. Section 3 is devoted to observations of precipitation and water levels, which are required for accurate modeling and the current situation in Japan. Section 4 focuses on lumped modeling, the urban storage function (USF) model is outlined as an example, and the hydrological data required to build the model for urban areas are described together with the prediction accuracy obtained using artificial intelligence (AI). Distributed physical modeling is reviewed in Section 5, including different approaches to describing the urban structure, as well as the Tokyo storm runoff (TSR) model as an example of a distributed model using vector-type GIS data. This paper closes with some future prospects related to urban flood runoff modeling in Japan.

2. Brief History on Urban Runoff Modeling in Japan

In Japan, Morita [48] characterized an urban watershed as a so-called "small- to medium-sized urban river basin", where most of the watershed is flat land, such as alluvial plains, flood plains, alluvial fans, and hills. Normally, the entire upper, middle, and lower reaches are developed into urban areas, and the watershed area is generally smaller than 100 km^2.

As for urban runoff research in Japan, Kinoshita [1] was the first to comprehensively and quantitatively describe and discuss urban flood runoff mechanisms in 1967. This paper was followed by numerous studies on urban flood runoff mechanisms [2,3], which were summarized by Kadoya [4] in 1985. There, the urban flood runoff mechanisms are characterized by (1) an increased runoff rate during floods, (2) a reduced flood concentration time before rainwater flows into rivers, and (3) extremely large peak runoff rates due to high-intensity rainfall peaks with a short duration. Generally, most of the flood runoff models were mainly intended for mountainous and natural watersheds and were not developed with urban watersheds in mind, as pointed out in the Hydrology and Water Resources Handbook published by JSHWR in 1997 [49]. In addition, physical runoff processes in urban watersheds are not fundamentally different from those in natural watersheds. As a result, when applying a runoff model to an urban watershed, only the model parameters are usually adjusted [49].

The rational equation [50], which is used in many small-to-medium-sized river and sewage system plans in Japan, was originally proposed as a model for estimating peak flows as a basis for stormwater management in urban watersheds. Parameters, such as the runoff coefficient and flood concentration time, need to be calibrated for the target urban watershed. For example, in 1976, Kadoya and Fukushima [51] proposed an equation relating flood concentration time to flood magnitude based on the equivalent roughness method. The synthesized rational formula [52,53] has also been developed as a model for urban watersheds in 1998. This method creates a hydrograph by dividing an urban watershed into sub-watersheds, each characterized by area, runoff coefficient, and flood concentration time, and combining the peak flows by superimposing the results from the rational equation. The quasi-linear storage model [54] developed in 1977 and the modified RRL method [55] developed in 1972 are also used as flood runoff models for urban areas by setting flood concentration times and runoff coefficients for the urban watershed of interest [56,57].

In 1954, Iwagaki and Sueishi applied the kinematic wave theory to the analysis of hillslope runoff [58,59], and soon after, Lighthill and Whitham [60] developed the kinematic wave theory for river flow tracking. Since then, this theory has been accepted as a standard tool for modeling surface flows and various other elementary hydrological processes [61,62]. The kinematic wave model, in which the watershed is described in terms of a collection of rectangular slopes and river channels, and rainwater flow, which is represented by a hydraulic continuity equation and a kinematic equation, is a widely used distributed physical model in Japan (see, for example, [24,62]). As an example of applications of the kinematic wave model to urban watersheds, in 1991, Ando et al. [63] divided an urban watershed into rectangular sub-watersheds using spatially distributed rainfall and applied the kinematic wave model to each divided sub-watershed. The reproducibility of the hydrograph was improved by using the equivalent roughness value of the urban watershed as an input. In 1995, Fujimura et al. [64] showed that diffusion wave or dynamic wave models are more accurate than kinematic wave models for nonstationary and unsteady river channel flow calculations, although kinematic wave models are used for the slope flow, as well as surface flow, of low-lying watersheds. In 1996, Suzuki et al. [65] developed the Doken model as a distributed model for real-time flood forecasting. The model has been incorporated into the Integrated Flood Analysis System (IFAS) for flood forecasting in watersheds with inadequate water level observations and has been applied in Asian countries, such as Pakistan [66]. In 1997, Chikamori et al. [67] attempted to obtain model parameters, such as equivalent roughness, using GIS information when constructing a kinematic wave model for an urban watershed. In the Technical Standard for River Erosion Control, Survey Edition by MILT [68], the rational equation [50], the SF model [18], the kinematic wave model [24], the quasi-linear storage model [54], the tank model [17], and the Doken model [65] are introduced as runoff models.

Since then, along with the rapid development of computer and IT technology, the analysis of flood runoff in urban watersheds has been improved in several respects. Data collection by, e.g., satellites and radar is a significant advancement, as is the advancement of GIS technology [69]. Furthermore, a range of new methods based on, e.g., optimization methods, network theory, and AI, have been applied to flood runoff analysis and modeling. Shibuo and Furumai [70] recently reviewed urban stormwater management in Japan, focusing on rainfall observations, numerical modeling, and emerging technologies.

3. Rainfall and Water Level Observation System in Japan

A key component of flood runoff modeling is access to observational data. In Japan, information on rainfall and river levels is mainly provided by the Water Information and National Land Data Management Centre established by the Water Management and Land Conservation Bureau under the MLIT [71]. As of 2023, real-time river observations (water level and streamflow) and rainfall observations are provided online at 7580 and 10,619 stations, respectively [72]. In particular, the recently developed crisis water level gauges have improved river level monitoring during heavy rainfall, with approximately 6000 stations installed from 2018 to 2020. For rainfall, in addition to these, approximately 1300 precipitation monitoring stations have been installed by the Japan Meteorological Agency (JMA) through the Automated Meteorological Data Acquisition System (AMeDAS), with one station per 289 km^2 across the country [73].

The X-band multi-parameter (MP) network of weather radars, known as the eXtended RAdar Information Network (XRAIN), was developed in 2008 by MLIT. The network uses 16 high-performance C-band MP radars and 39 X-band MP radars that are specialized for rainfall observation [74]. XRAIN provides high-resolution (250 m), high-frequency (1 min), and near-real-time rainfall information for almost the entire land area of Japan without the need for correction by rain gauges at the ground surface. However, owing to its short wavelength, XRAIN has only a narrow observation area (80 km radius) because of attenuation caused by rainfall, and in some cases, it cannot fully penetrate heavy rainfall areas.

In urban areas, small-to-medium-sized rivers are at high risk of river overflow and flood damage due to flooding caused by short periods of high-intensity rainfall. Therefore, urban runoff analysis requires very high-resolution observations to accurately capture the spatial and temporal distributions of rainfall. Schilling [75] suggested that rainfall data of at least 1–5 min and 1 km resolutions should be used for urban drainage modeling. Fabry et.al. [76] recommended the use of higher-resolution data (1–5 min temporal and 100–500 m spatial) for urban hydrological applications to provide the necessary hydraulic details. In the case of localized torrential rainfall, there is often a delay of only tens of minutes from the start of the rainfall to the peak river water level, and flooding damage associated with such events is frequent in large cities in Japan [77]. For this reason, the validation of flood runoff analysis models for urban areas requires a time resolution of river observations similar to that of the rainfall observations.

In Tokyo, for example, the Tokyo Metropolitan Flood Control Integrated Information System (FCIS) plays a role in flood prevention activities by automatically collecting real-time observations of rainfall and river water levels in Tokyo and providing them to the relevant organizations [78]. The FCIS began observing 10 min rainfall in January 1978 and later increased the resolution to 1 min. Water level observations were initiated to inform residents of the danger of flooding due to the rapid increase in water levels caused by heavy rainfall. A total of 124 water level observation stations were established in Tokyo, and 15 stations (1.67 km^2/station) were established in the Kanda River basin. Rainfall observations were introduced to measure high-intensity rainfall with high resolution in time and space, as well as high precision. The number of rainfall observation stations has increased annually, and as of 2022, 93 stations (19 km^2/station) have been installed (Figure 1). This is a very large number, considering that there are only 10 AMeDAS stations operated by the JMA [73] in Tokyo (178 km^2/station). FCIS data are expected to accurately describe the spatial variability of heavy rainfall.

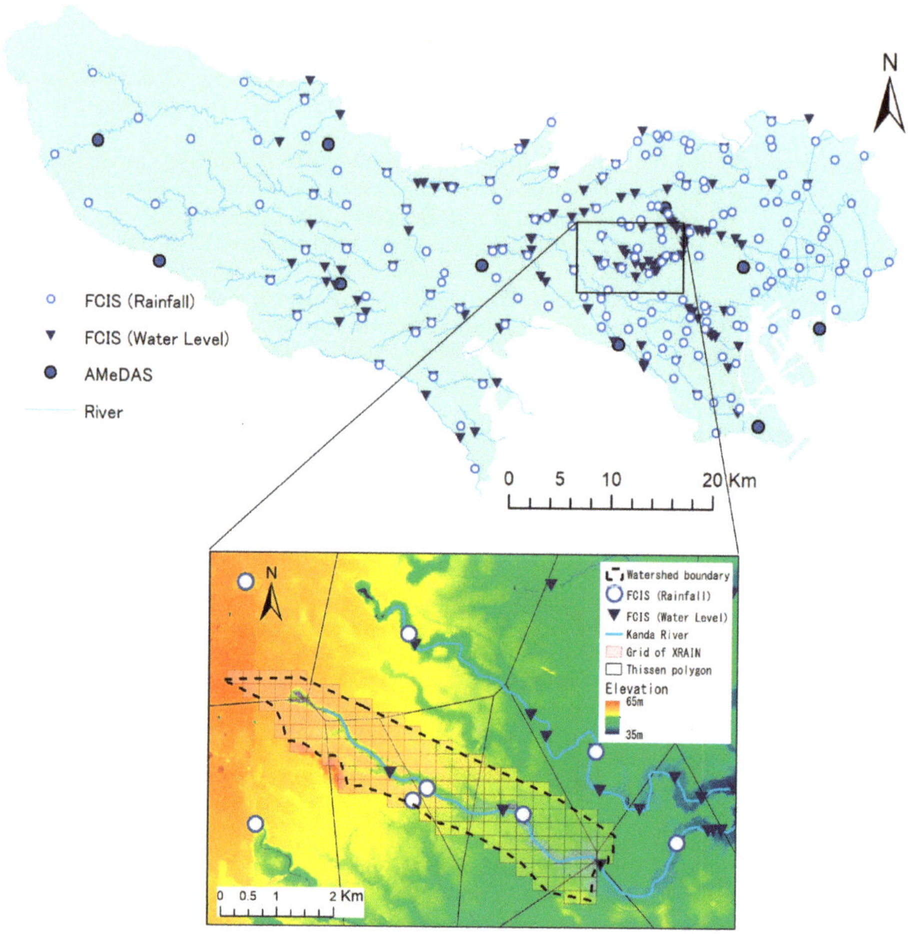

Figure 1. Map of FCIS and AMeDAS stations in Tokyo Metropolis (top) and a zoomed-in map of FCIS stations and XRAIN grid in the upper Kanda River Watershed (bottom).

4. Storage Function Model in Urban Watersheds

4.1. Conventional SF Models

SF models are event-based lumped models that are used for short-term hydrograph prediction, characterized by a relationship between storage and discharge. Kimura's original SF model in 1961 is a method to divide effective rainfall into storage and runoff processes by separating the watershed into runoff and infiltration areas [24] and is described as the "total storage function method" in the 1985 edition of the Hydroscience Formulary [45]. In general, the relationship between storage and discharge in a watershed during a flood can be described by a two-valued function, where the relationship between storage and discharge differs between the rising and receding parts of the hydrograph, respectively, and often forms a loop [79,80]. Kimura's SF model with two parameters introduced a delay time to eliminate this loop. In 1967, Prasad [81] presented a three-parameter SF model that directly expressed the bivalence between storage and discharge by adding a term for the loop effect. In 1982, Hoshi and Yamaoka [82] developed a four-parameter SF model by adding another parameter to accurately express the bivalence of the loop effect.

Flood runoff models, including conventional SF models, generally target the direct runoff component of the total runoff, except for tank models that incorporate rainwater loss and storage effects. Thus, general flood runoff models require effective rainfall as their input for the prediction of direct runoff. Hence, at first, the separation of baseflow from the total discharge is needed to calculate the effective rainfall when applied to the actual runoff analysis. There is no definitive method for separating direct runoff, and the calculated direct runoff will be different because of the subjective nature of the separation process [83]. As a result, the identification of model parameters may change significantly, which may cause the stability and reliability of the runoff calculations to decrease. Baba et al. [22] and Hoshi et al. [84] developed an SF model that directly uses the observed rainfall and runoff to solve the problems associated with the estimation of direct runoff. The proposal of a model concept that does not require the separation of baseflow is groundbreaking [69].

The four conventional SF models are the Linear, Kimura [18], Prasad [81], and Hoshi [84] models, and their storage equations are shown in Table 1, where s is the storage (mm), Q is the observed river discharge (mm/min), t is the time (min), k_1, k_2 are model parameters that express the constant of proportionality for Q and the time variation of Q, respectively, and p_1, p_2 are model parameters that express the exponent related to Q itself and the time variation of Q, respectively. The associated continuity equations, in which q_l and q_R include parameters k_3, z, and α, respectively, are also listed, and they are explained in the next Section 4.2. Among these models, Hoshi's model was found to be superior owing to an additional parameter, p_2, which may be quantified by numerical experiments and can well define the flow characteristics based on kinematic wave theory [85]. Some simplifications in Hoshi's storage model can lead to Prasad's storage model [81]. If we set $p_2 = 1$ in Hoshi's model, we obtain Prasad's storage model. Similarly, if we set $k_2 = 0$ in Prasad's model, the model can be transformed into Kimura's model [18]. The simplest linear model can be obtained by keeping $p_1 = 1$ in Kimura's model. Kimura's SF model with one storage tank, which is widely used as a special case of Kimura's original model with a delay time of zero [85].

Table 1. Storage function models with their associated continuity equations (PAR represents a parameter).

No.	Models	Storage Equation	Continuity Equation
1	Linear (3-PAR)	$s = k_1 Q$	$\frac{ds}{dt} = R + I - E - O - Q - q_l$
2	Kimura (4-PAR)	$s = k_1 (Q)^{p_1}$	$\frac{ds}{dt} = R + I - E - O - Q - q_l$
3	Prasad (5-PAR)	$s = k_1 (Q)^{p_1} + k_2 \frac{dQ}{dt}$	$\frac{ds}{dt} = R + I - E - O - Q - q_l$
4	Hoshi (6-PAR)	$s = k_1 (Q)^{p_1} + k_2 \frac{d}{dt} (Q)^{p_2}$	$\frac{ds}{dt} = R + I - E - O - Q - q_l$
5	USF (7-PAR)	$s = k_1 (Q + q_R)^{p_1} + k_2 \frac{d}{dt} (Q + q_R)^{p_2}$	$\frac{ds}{dt} = R + I - E - O - (Q + q_R) - q_l$

4.2. Urban Storage Function (USF) Model

The runoff characteristics of urban watersheds are significantly different from those of natural and mountainous watersheds because of the high proportion of impervious areas and the increasing prevalence of sewerage systems. Therefore, it is difficult to accurately reproduce the hydrographs of small-to-medium-sized urban rivers with short flood concentration times by using conventional SF models. Takasaki et al. [86] proposed an SF model including a two-valued function that considered the runoff characteristics of combined sewer systems. Later, Takasaki et al. [23] proposed and constructed an urban storage function (USF) model that does not require the calculation of effective rainfall or the separation of baseflow but directly uses observed rainfall and runoff.

Figure 2 shows the schematic diagram of all the possible inflow and outflow components of an urban watershed with a combined sewer system. Combined sewers are common in small-to-medium-sized river watersheds in old cities, such as the Tokyo Metropolitan District, where sewers have been in place since the 1900s [87]. The inflow components in Figure 2 are represented by rainfall, R (mm/min), and urban-specific and groundwater inflows from other watersheds, I (mm/min). Urban-specific inflows include leakage from water distribution pipes, irrigation flows, etc. The outflow components consist of river discharge, Q (mm/min), evapotranspiration E (mm/min), storm drainage from the watershed through the combined sewer system q_R (mm/min), water extraction from the watershed for, e.g., water supply, agricultural needs, etc., O (mm/min), and groundwater-related loss q_l (mm/min). In addition, domestic sewage q_w is depicted in Figure 2 even though it does not contribute to watershed storage, s (mm).

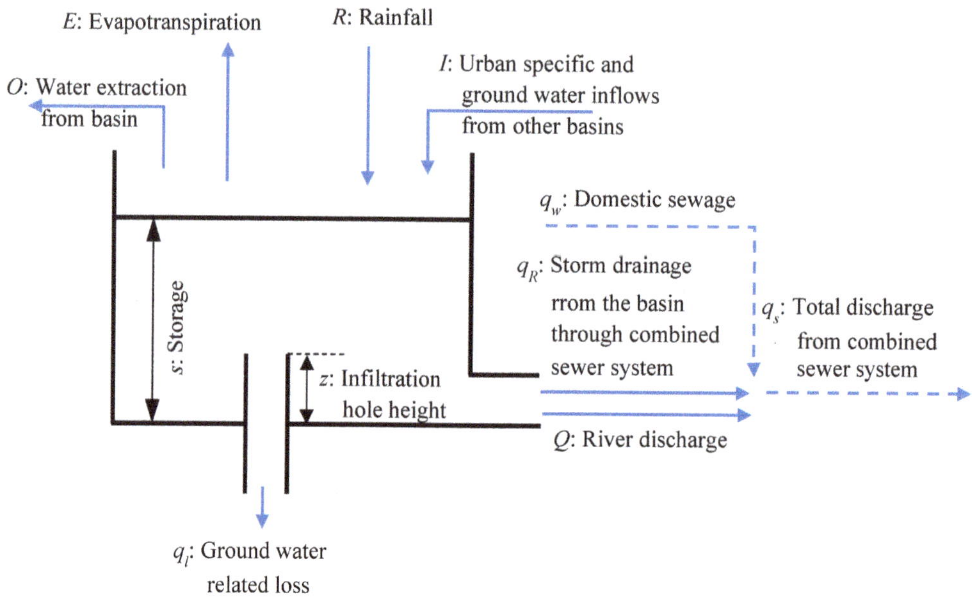

Figure 2. Schematic diagram of all inflow and outflow components of an urban watershed with a combined sewer system.

The USF model is an empirical representation of Hoshi's SF model (shown in Table 1), in which the river discharge, Q, is replaced by the discharge, including storm drainage, $Q + q_R$. The associated continuity equations, which do not require the separation of baseflow for both the conventional SF models and the USF model, are listed in Table 1. Here, the groundwater-related loss (q_l) was defined by considering the infiltration hole height (z) by multiplying k_3, which are the new parameters in the continuity equations in Table 1. Furthermore, the q_R of the USF is defined by using an additional parameter, α, that expresses the slope of the linear relationship between the total discharge, $Q + q_R$, and the drainage, q_R (refer to [23,88] for details).

Thus, seven unknown parameters (k_1, k_2, k_3, p_1, p_2, z, and α) are identified in the USF model. As an optimization method for parameter identification in lumped conceptual models, global search methods using evolutionary computation methods, such as genetic algorithms, have diversified in recent years [89]. Among them, the effectiveness of the shuffled complex evolution University of Arizona (SCE-UA) method developed by Duan et al. [90] has attracted considerable attention [91,92].

4.3. Application of the USF Model in an Urban Watershed

Takasaki et al. [23] applied the USF model to the upper Kanda River watershed (Figure 1), a typical small-to-medium-sized urban watershed in Tokyo that is entirely urbanized and where urban flooding occurs frequently. The small concentration time implies that river runoff occurs within a short period of time immediately after rainfall, and the use of hydrological data with very short time intervals for rainfall-runoff analysis is required. Therefore, FCIS rainfall and water level data at 1-minute intervals (described in Section 3) were used. They used the SCE-UA method for parameter identification and verified the effectiveness of the USF model, including the much-improved reproducibility of the hydrograph [23].

Padiyedath et al. [88] applied the five SF models shown in Table 1 to the upper Kanda River watershed. They also used the SCE-UA method to identify parameters to verify the performance of the SF models. As for the results, the USF model showed not only the highest hydrograph reproducibility among the five SF models, which is not surprising when considering that the hydrograph reproducibility increases with the number of model parameters, but also the most effective SF model from an Akaike information criterion (AIC) [93] perspective. In addition, the uncertainty of the seven parameters of the USF model was evaluated in detail using a residual-based bootstrap approach [94]. Furthermore, Padiyedath et al. [95] expanded the USF model by considering the spatial distribution of rainfall over watersheds and developed a generalized SF model able to directly estimate the water level [96].

Lumped models, such as the USF model with a relatively small number of model parameters, are suitable for real-time flood runoff forecasting by changing the model parameters sequentially. Parameter identification can be almost automated using global search methods like SCE-UA, and hydrograph reproducibility at the target point is generally higher than that of a distributed model. Takasaki et al. [97,98] applied both the particle filter method and the Kalman filter method to real-time flood runoff forecasting using the USF model in the upper Kanda River watershed and obtained accurate forecasting results.

4.4. Runoff Prediction by USF Using XRAIN Data in an Urban Watershed

For localized torrential rainfall events in the upper Kanda River watershed (Figure 1), the USF model has also been used to evaluate hydrograph reproducibility using XRAIN rainfall data and the FCIS rain gauge data [99]. At first, 1-minute XRAIN mesh rainfall data were compared to 1-minute FCIS rain gauge data of the same mesh to check the precision of XRAIN [100,101]. The results showed that 1-minute XRAIN values corresponded well with the 1-minute ground observation rainfall values, but the former were observed 1 to 3 minutes earlier than the ground observation, which was considered mainly due to the falling time of the raindrops. Spatiotemporal correlation analysis between XRAIN and FCIS rain gauge data has also been conducted [102,103]. As for the results, for rainfall during typhoons, where the rainfall area moved clearly along with the typhoon, it was shown that radar rainfall was highly correlated with ground rainfall with a 2-minute time delay. For localized torrential rainfall, it has been shown that the spatiotemporal correlation between XRAIN and ground rainfall is much more complicated than in a typhoon case, and the peak correlation does not always occur in the same mesh as the rain gauge but in the surrounding meshes.

Regarding runoff predictability, the best performance was found for FCIS rainfall observations, whereas the performance was lower for XRAIN [99]. This study also showed that it was possible to improve runoff prediction using XRAIN by considering spatiotemporal correlation characteristics. It should be emphasized that there is a spatial uncertainty in XRAIN rainfall, as it is measured at some elevation in the atmosphere and not at ground level. This uncertainty can significantly affect runoff prediction using XRAIN, particularly in the case of localized torrential rainfall occurring near the watershed boundary in small-to-medium-sized watersheds [99,103].

4.5. AI-Based Flood Runoff Modeling in Urban Watersheds

Machine learning using AI technology has developed remarkably in recent years. In order to utilize machine learning for flood forecasting, many AI-based flood runoff models have been constructed using, e.g., multilayered artificial neural networks (ANN) or deep learning (DL) [69]. Many of these models have been constructed for large river watersheds [104,105] using hourly observation data as input, and some DL applications for small-to-medium-sized mountain river watersheds have used 10-minute data to construct the models [106]. On the other hand, in small-to-medium-sized urban rivers, such as in Tokyo, water level and rainfall observations are carried out at 1-minute intervals, as described in Section 3. It has been concluded that the combination of temporal and volume resolutions in rainfall observations is critical for successful AI modeling. In particular, the hyetographs of 1-minute observations are highly intermittent and differ from the 10-minute and hourly data in large river watersheds.

An attempt was made to emulate the USF model using ANN and DL. Fujizuka et al. [107] tested the performance of a DL model using a virtual hyetograph and a virtual hydrograph and examined the impact of the amount of training data. The DL model was compared with a conventional ANN model on virtual datasets [108]. For both the ANN and DL models, if the number of nodes in the input layer is sufficient, the Nash–Sutcliffe coefficients of the hydrograph of both models are greater than 0.98; therefore, even the conventional ANN model can accurately emulate the USF model. However, when the total number of nodes in the intermediate layer was the same, the accuracy of the DL model was higher than that of the ANN model. Fujizuka et al. [109] also performed a comparison using actual observations and concluded that the DL model was superior when considering the number of parameters in relation to the number of observations.

Recently, hybrid forecasting models that combine an ANN forecasting model with a flood runoff model have been proposed [110]. In conventional ANN models, water level and flow rate are used as the main input data for training and forecasting. In contrast, the hybrid forecasting model incorporates flow rate forecasts from the runoff model into the input layer of the deep ANN model. This is expected to improve the forecasting of unprecedented runoff peaks, which remains a challenge for conventional statistical models.

5. Distributed Physical Models in Urban Watersheds

5.1. Background

In recent years, most of the distributed flood runoff models developed in Japan for urban watersheds have focused on surface flooding and inundation analysis, and few have focused on runoff analysis and the generation of runoff volume directly from urban impervious areas. Since 1980, with significant improvements in numerical techniques and hydrological models for slope flow, surface flow hydrological models have become available for predicting inundation, and inundation hazard maps have been prepared as flood damage reduction countermeasures. Most inundation analysis models are based on shallow water equations as applied, e.g., by Iwasa et al. [111]. Toda et al. [112] used a kinematic wave model for runoff from mountainous areas near urban areas and combined it with an inundation model for urban areas that also consider drainage by sewage systems. Subsequently, a new integrated lowland inundation model was developed and used for flood inundation forecasting, considering slope runoff processes, sewer flows, surface inundation, and runoff from manholes. This model configuration has been established as a standard model for flood damage prediction by flood inundation analysis in urban watersheds.

Furthermore, hydraulic models, such as InfoWorks, MOUSE, and the stormwater management model (SWMM), which are mainly intended for the design and flow analysis of sewer pipelines, have been integrated with surface inundation models and are used to analyze flooding due to poor stormwater pipeline drainage, etc. In Japan, although these models are used in practical applications, they are not widely used for flood runoff analysis in urban watersheds because, as mentioned above, the development of flooding analysis models that take sewer pipeline flow into account has been conducted independently. As an example of an integrated sewerage and inundation model, Shibuo et al. [113] developed a comprehensive analysis model for the Tsurumi River watershed using the water energy-based distributed hydrological model (WEB-DHM) to calculate runoff from the upper catchment and InfoWorks to analyze the sewers and combined them into an integrated river channel and sewage model. In addition, many inundation analyses of urban watersheds have been conducted overseas using the Hydrologic Engineering Center's river analysis system (HEC-RAS), an open source 2D hydraulic model integrated into GIS [114,115]; this model is not widely used in Japan because there is a need to incorporate aspects such as flow in sewer pipelines and/or flooding processes in subways into the modeling. Detailed explanations regarding these aspects will be provided in the next section.

5.2. Representation of Urban Structures in Distributed Physical Models

Urban flooding processes can be described by a dual drainage model that combines a 1D hydrodynamic model of the stormwater drainage network model with a 2D hydrodynamic model of overland flow and flooding [116]. In particular, different representations of urban structures have significant effects on surface flood processes [117]. Table 2 summarizes the computational grid shapes used in different studies. The TSR model proposed by the authors modeled direct runoff generation and flood flows from the perspective of urban formation, whereas the other four models were developed to analyze urban flood inundation.

Table 2. Characteristics of a distributed urban flood runoff models.

Geometry	Image	Characteristics (a) Objective; (b) Features; (c) Land use; (d) Handling of Buildings; (e) Surface Flow Equations
Structured Grid [118–120]	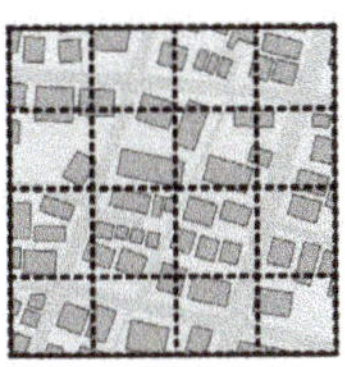	(a) Flood analysis. (b) Model data can be created and calculated efficiently. (c) Usually expressed as a percentage of impermeable area. If the resolution is sufficiently high below 10 m, land use information may be specified in the calculation grid. (d) The impact of the building-blockage effect is considered by building coverage ratio and conveyance reduction factors. If the resolution is fine enough, the height of the building may be set in the grid. (e) 2D diffusive wave model.
Curvilinear Grid [121,122]		(a) Flood analysis. (b) Capable of creating grids suitable for urban structures along roads. Ideally, the road network in the targeted catchment should be like a grid. If the road network is complex, model data preparation becomes more difficult. (c) Usually expressed as a percentage of impermeable area. (d) Resistance is calculated separately for bottom friction, resistance due to plane vortices behind the house, and hydrodynamic forces on the buildings. (e) 2D diffusive wave model.
Unstructured Grid [123,124]	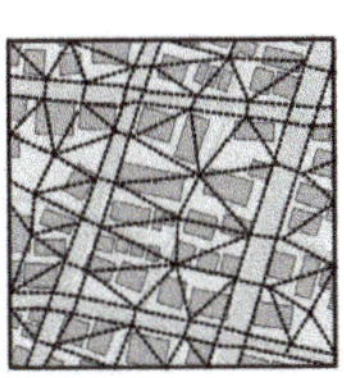	(a) Flood analysis. (b) Represents complex terrain and boundaries on fewer grids. Manual processing, requires GIS data and mesh generation software. (c) Usually expressed as a percentage of impermeable area. If the resolution is fine (less than 20 m), a Digital Surface Model (DSM) may be set for the elevation of the residential block. (d) Buildings of the same scale as the computational grid are subjected to a boundary-fitting process, while smaller buildings are grouped. (e) 2D full shallow-water equations.

Table 2. *Cont.*

Geometry	Image	Characteristics (a) Objective; (b) Features; (c) Land use; (d) Handling of Buildings; (e) Surface Flow Equations
Road Network [125,126]	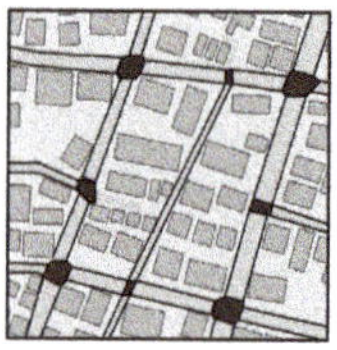	(a) Flood analysis. (b) Capability to represent complex road networks. The ground surface consists of roads and Residential grids. Relatively easy in a GIS environment. (c) Initially developed as a flood inundation model from rivers; rainfall-runoff processes and infiltration processes are not considered. Recently developed models set average building area percentages. (d) Roughness coefficients of large values are set against the flow between the road and the residential block. (e) 1D Saint-Venant equations.
Urban Landscape Delineation [6]		(a) Flood runoff and inundation analysis. (b) Unlike other grids, polygon data is created for roads, buildings, permeable and impermeable ground, etc., classified according to permeability characteristics, with roads, for example, further divided into segments. It is possible to give detailed attributes to individually subdivided polygons (e.g., building structure, use, number of floors, presence of storage facilities, etc.). (c) Land use and grid geometry are consistent. (d) Parameters are set to take into account the building flooding process. (e) 1D Saint-Venant equations.

5.3. Structured Grid

A structured grid is most commonly used for surface models because of the simplicity of grid formation when analyzing ground surface flows as two-dimensional unsteady flows. The surface flow (inundation analysis) model using a structured grid was developed based on a relatively large computational grid for wide-area inundation. At the beginning of the model development, the effects of structures, road networks, etc., were considered by equivalent roughness coefficients based on land use or synthetic roughness coefficients as a function of building occupancy and inundation depth. This was carried out because it was difficult to take into account the influence of roads and buildings, which are likely to have a significant impact on flood flows in urban areas, in the structured grid. Later, building volume (nonflooded space) and flow disturbance by buildings were considered [118].

Currently, this is the standard method used to prepare municipal flood maps in Japan, and the National Institute for Land and Infrastructure Management (NILIM) of the MLIT provides the analysis program NILIM2.0 [119]. The accuracy of inundation analysis can be improved by using high-resolution elevation data. In recent years, with advances in computer performance, analyses using 5 and 1 m mesh elevation data have been made. Inokawa and Kobayashi carried out an inundation analysis on the surface and in an underground station using a 1 m mesh and concluded that it is important to consider slight changes in the inundation level at the station [120].

5.4. Curvilinear Grid

Curvilinear grids are widely used in the numerical analysis of open channel flows, including bends [121]. When this grid model is applied to an urban area, the attributes of roads and buildings can be classified for each grid if the co-ordinate axes are appropriately set. When compared to a structured grid, grid construction requires much more effort. Since there are only two axes, the direction of the road is restricted [121].

Fukuoka et al. focused on structures, such as houses and road networks, in urban areas and developed a model using a curvilinear grid for the road network, achieving the highly accurate reproduction of water levels and flow velocities on roads and between city blocks of houses [122]. However, in recent studies, this co-ordinate system has rarely been used.

5.5. Unstructured Grid

Unstructured grids can be created according to the urban structure, for example, by applying detailed grids in regions with complex boundaries and geometries and coarsening the grids in the regions that are not computationally important. However, as with a curvilinear grid, the time and effort required for grid formation are significant for unstructured grids [121].

Akiyama et al. developed a two-dimensional numerical model of flood flows based on the finite volume method (FVM) with an unstructured grid for the floodplain profile to ensure the proper representation of the road network and houses [123,124]. The flood flow in the river and the runoff on the ground were simulated simultaneously using a 2D free surface flow model based on the flux-difference splitting scheme and an unstructured FVM. For the sewer network, a dynamic network model for free-surface pressure flow based on the flux-difference splitting scheme combined with a stormwater drainage network was used. By performing an inundation analysis on Iizuka City and verifying its accuracy against observations of the waterlogged areas, it was shown that the model could estimate waterlogged areas with sufficient accuracy and estimate detailed inundation properties, such as overflow, internal inundation locations, and flood propagation paths.

5.6. Road Network

Inoue et al. proposed representing roads and buildings as different computational grids and developed an analysis method using a road network model that can represent flooding in real urban areas more faithfully than the method using a structured grid [125]. Road junctions were used as nodes, and the road segments between junctions as links; residential areas were specified as areas surrounded by the network; junctions and residential areas were defined as the analysis grid; and water depths were obtained from the continuity equation. They showed that the road network model could directly parameterize the roughness coefficients at the surface by considering the block walls.

Sekine et al. [126] proposed another flood analysis model using a road network, focusing on flow concentrations on roads and sewers in densely built-up areas. The model is characterized by its ability to faithfully model rainwater runoff from buildings into sewers. The average house in a city block, based on its roof area, was treated as a unit house in the model, and rainwater on the unit house was assumed to flow into the sewers adjacent to the city block using the kinematic wave approximation. Sekine et al. focused on three cities neighboring the 23 wards of Tokyo that were damaged by Typhoon No. 19 in 2019 and performed predictive calculations of inundation in the event of hypothetical maximum rainfall, describing the inundation process in depressed areas and underpasses with significant inundation [127].

5.7. Urban Landscape GIS Delineation

The authors have developed the Tokyo storm runoff (TSR) model that can simulate urban storm runoff and flood inundation using a vector-based watershed description [6]. The TSR model employs so-called "urban landscape GIS delineation" that realistically describes complicated urban land use features in detail. Urban landscape GIS delineation enables an accurate representation of the infiltration and perviousness of each property, making it possible to calculate the impervious area of a target watershed with very high accuracy. In addition, for pervious land, characteristics such as infiltration capacity can be set specifically for each land use type [40].

In Japan, 10 m resolution grid-based data classified by land use type are generally used in distributed runoff modeling in urban watersheds. Although there are 17 land use categories in the classification, including industrial land, commercial/business land, and general low-rise residential land, the original purpose was to obtain the basic data necessary for residential land-related policies in urban areas, and these land use categories were not compiled with differences in infiltration rates in mind. Since estimating the impervious area ratio (IAR) for each land use category is an important component of accurate runoff analysis, Koga et al. [40] overlaid this landscape GIS and grid on a typical urban watershed in Tokyo and calculated the IAR for each grid. The results revealed that the reference value of the IAR for the land use classification had an error of about 7%.

Since the urban landscape GIS considers individual buildings, roads, parking lots, and other factors, Araki et al. [128] added stormwater runoff control facilities for individual buildings and specific permeable paved roads in the upper Kanda River basin as attributes of their maintenance status and examined their effectiveness in controlling flood runoff. The simulation results show that the developed model allows for the quantitative estimation of the efficiency of infiltration facilities.

The construction of urban landscape GIS data currently requires time-consuming manual work using GIS software (ArcGIS 10) [129]. Therefore, research has been conducted on automating urban landscape GIS delineation [130,131]. In addition, current GIS data only represent the planar structure of ground surface objects, and thus cannot represent the three-dimensional structure of subways, underground road tunnels, and the multilevel intersections of roads and bridges, as well as buildings, especially high-rise buildings. In order to more faithfully reproduce rainfall runoff paths physically, it will be necessary to construct three-dimensional GIS data that also take the height direction into consideration. The construction of a four-dimensional GIS that includes not only three spatial dimensions but also time has been pointed out as a challenging issue.

5.8. TSR Model

5.8.1. Outline of Tokyo Storm Runoff (TSR) Model

The Tokyo storm runoff (TSR) model is a distributed physical model that utilizes urban landscape GIS data (for details, see Amaguchi et al. [6]). Figure 3 shows the rainfall-runoff process for the TSR model. Rainfall in the watershed is applied to each land use element, and the rainwater that cannot infiltrate is discharged to nearby roads and further to manholes and sewer pipes based on the infiltration and imperviousness characteristics set individually. In a manhole/sewer pipe element, the water level in the manhole section is first calculated based on the inflow/outflow to/from the road and inflow/outflow to/from the connecting pipe. The flow rate is then calculated based on the water level in the manhole section and the cross-sectional characteristics of the pipe. In this calculation process, water in the manhole overflows onto the road when the water level in the manhole exceeds the ground level of the road. The overflowing water flows down the road and eventually returns into a manhole or sewer pipe with sufficient flow capacity. Thus, water entering the sewer pipes is combined with water in the sewer network and eventually flows out into the river channel. In the inundation analysis, the amount of runoff from land use elements and the amount of overflow through manholes in the sewer system is used to calculate the water flows into residential areas when the road water level is above the surrounding ground level. The water in the residential areas flows back into the road when the water level on the road decreases.

Stormwater runoff processes in urban watersheds are highly dependent on local boundary conditions such as buildings, roads, rivers, and surface flows to sewers. Based on these boundary conditions, the TSR model calculates the direct runoff from roads, buildings, and other impervious areas that constitute a city and simulates the transport to the ground surface, sewerage systems, and rivers. When compared to the other four representations of urban structure (grid, curvilinear, unstructured, and road networks), and the associated models, the TSR model is not suitable for analyzing flooding flows with levee failures from large rivers, but it is intended for inundation phenomena, such as pluvial flooding in urban areas.

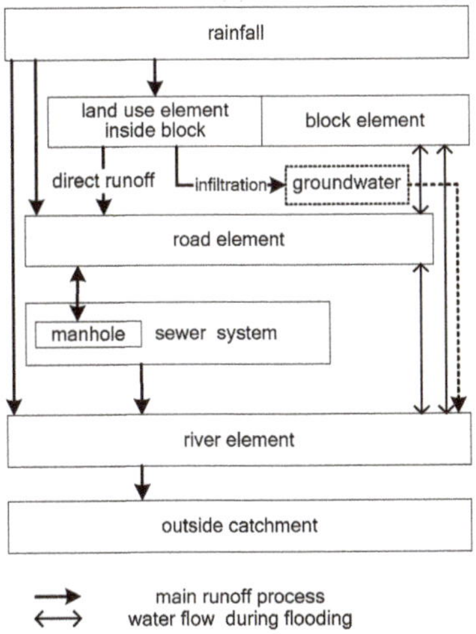

Figure 3. Schematic of the rainfall-runoff process.

5.8.2. Model Application to Urban Watershed

Amaguchi et al. [132,133] used the building element of urban landscape GIS data to set individual values for the site elevation and floor elevation of buildings and made some modifications to the flooding analysis in the TSR model. This made it possible to simulate inundation conditions both inside and outside buildings. The model was set up and evaluated for the upper Kanda catchment in the Tokyo Metropolis, Japan. In the model, the parameters for the foundation height and floor height of each building, as well as the opening ratio below and above the 1st floor, were set for individual buildings based on their structure and purpose (Figure 4). The model was applied to the major flood event of 4 September 2005. There was no flood damage in the target watershed due to the effect of the underground control basins, so the simulation was performed assuming that the control basins were not functioning. The validity of the model was confirmed by comparing the model with river water levels observed at upstream locations not affected by the underground control basins. The validity was confirmed through its ability to overall accurately reproduce the observed river water levels [132]. This model was applied to simulate major flood events and was able to reproduce the differences in the inundation depth of individual buildings, reflecting the additional input parameters. The use of the model to evaluate inundation conditions for specific building properties was also demonstrated. The results show that the suggested approach, which is based on a detailed reproduction of all the relevant elements in an urban watershed, can simulate the inundation of individual buildings (Figure 5).

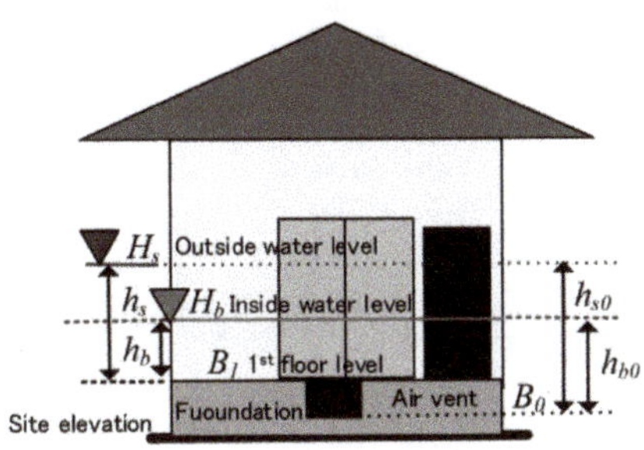

Figure 4. Parameters used for building inundation.

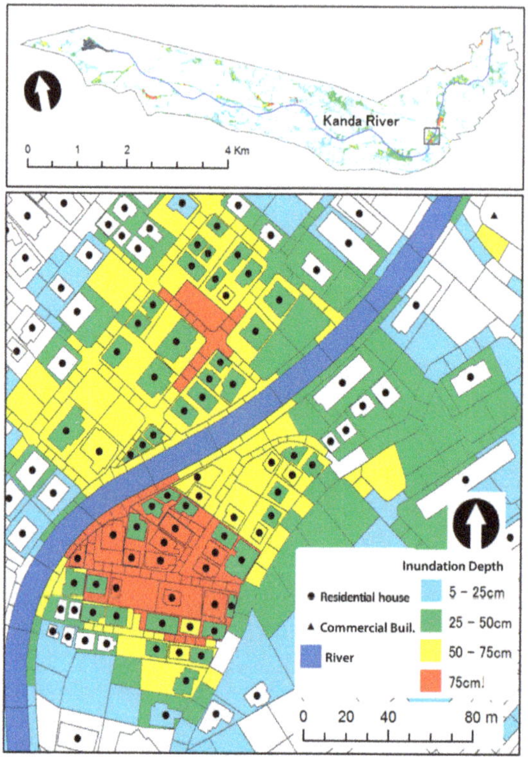

Figure 5. Example of inundation analysis results in the upper Kanda River watershed (The arrows indicate north direction).

6. Concluding Remarks and Future Prospects

In this review, approaches to flood runoff analysis and modeling with a focus on urban rivers in Japan have been reviewed. Relevant essential articles quoted in this paper concerning urban flood modeling in Japan have been summarized chronologically in Table A1 in Appendix A.

The USF model and other lumped conceptual models predict runoff at a given point in a river using rainfall data in the target urban watershed as input. The major advantages of these methods are that the number of parameters is small, the execution time is short, and the reproducibility of the observed hydrograph is generally high.

Concerning distributed modeling, in the TSR model with urban landscape GIS delineation, all land surfaces are represented by polygons divided according to their infiltration characteristics. This concept almost eliminates the problem of the excessive fine graining required in grid-based models to fully represent heterogeneity. However, current GIS data only represent the planar structure of ground surface objects and, thus, cannot represent the three-dimensional structures of subways, underground road tunnels, the multilevel intersections of roads and bridges, and (high-rise) buildings. In order to more faithfully reproduce rainfall runoff paths physically, it will be necessary to construct three-dimensional GIS data that also consider the height direction.

Furthermore, in flood runoff analysis using distributed physical models, it is important to improve observation systems for model calibration and the real-time prediction of flooding and inundation. These observations include spatial real-time rainfall radar information, the water levels in sewers, and the inundation depth during flooding. For this purpose, technology to quickly collect and utilize accurate information is needed.

Additionally, runoff and inundation forecasts need to be transmitted in an easy-to-use format and be presented in an easy-to-understand manner to different recipients. Recently, the use of drones to observe early flood damage has been proposed to help local residents decide about evacuation and efficiently promote self-help. Finally, to enable highly accurate real-time flood forecasting with sufficient lead time using any type of runoff model, accurate and reliable high-resolution rainfall forecasts are required at least one hour ahead of an event in the case of urban watersheds.

Especially in lower urban areas, seamless, coupled models will be developed that incorporate not only runoff, channel, and sewerage models but also tidal models, which provide boundary conditions for flooding and inundation, and coastal hydraulic models to simulate storm surges, tidal waves, and tsunamis. In addition, this seamless, coupled modeling will incorporate atmospheric observations and simulations, such as radar rainfall information and land surface schemes. These seamless models, or "Digital Twins", will move beyond the scope of a single urban watershed to encompass a wide range of watersheds, including rural and mountainous watersheds (ultimately, the entire globe).

Finally, it is crucial to recognize that flood runoff models play a significant role in mitigating the impact of destructive floods through the implementation of appropriate protection measures, including flood control planning and policy. As there are very different kinds of urban floods worldwide, they require different protection measures [134]. The implementation of integrated flood risk management (IFRM) is also adopted in both developed and developing countries, yet some countries have encountered barriers to IFRM adaptation [135]. In Japan, a new policy called 'River Basin Disaster Resilience and Sustainability by All' was recently introduced in response to climate change risks and social changes [136]. It takes comprehensive measures, mainly consisting of flood prevention, exposure reduction, and appropriate evacuation, response, and recovery, aiming to strengthen disaster resilience and achieve sustainability through concerted efforts among all stakeholders [136]. Under the present circumstances, various rainfall-runoff models are becoming increasingly important tools, especially in urban watersheds.

Author Contributions: Investigation, H.A. and A.K.; Writing—original draft, A.K.; Data curation, H.A., J.O., A.K. and H.T.; Methodology, H.A. and J.O.; Funding acquisition, H.A. and J.O.; Writing—review and editing, H.A., J.O. and A.K.; Supervision, A.K. and H.A. All authors have read and agreed to the published version of the manuscript.

Funding: This research was funded by the Japan Society for the Promotion of Science (BR200501) and the Scandinavia-Japan Sasakawa Foundation (GA20-SWE-0065).

Data Availability Statement: All data are provided as tables and figures.

Acknowledgments: This work was supported by Tokyo Metropolitan Government Advanced Research Grant Number (R4-2).

Conflicts of Interest: The authors declare no conflict of interest.

Appendix A

Table A1. Chronological list of essential articles relevant to urban flood modeling in Japan. (Parenthetical Ref. No. is not developed in Japan).

Year	Ref. No.	Model/Method Name	Remarks
1954	[58]	Kinematic Wave Theory	Hillslope runoff analysis.
1955	[59]	Kinematic Wave Theory	Characteristic curve method is used for hillslope runoff analysis.
1955	([60])	Kinematic Wave Theory	River flow tracking analysis.
1961	[18]	Storage Function Model	The original storage function model was developed.
1967	[1]	-	The first paper in Japan that quantitatively describe urban flood runoff mechanisms.
1967	([81])	Storage Function Model	The Prasad storage function model was developed.

Table A1. *Cont.*

Year	Ref. No.	Model/Method Name	Remarks
1972	[17]	Tank Model	The original tank model was developed.
1972	[55]	Modified RRL Method	The model was applied to urban watersheds.
1976	[51]	Kinematic Wave Theory	Concentration time of flood by urbanization was studied.
1977	[54]	Quasi-Linear Storage Model	The model was applied to urban watersheds.
1979	([30])	TOPMODEL	A watershed hydrologic model that combines the advantages of a concentrated constant model with the effects of runoff contribution rate changes and the dispersion of stormwater due to channel network runoff.
1980	[79]	Storage Function Model	Explanatory paper of the model.
1980	[111]	2D Diffusive Model (Structured Grid)	Pioneering model for inundation calculations.
1982	[2]	-	A study of flood runoff changes in hilly basins by urbanization.
1982	[57]	Kinematic Wave Theory	The effects of urbanization on flood runoff was analyzed.
1982	[50]	Rational Method	Application of Rational Method to urbanized channels.
1982	[82]	Storage Function Model	Relationship with kinematic wave model was studies
1985	[3]	-	The effect of field moisture distribution on runoff was analyzed.
1985	[4]	-	A review of runoff changes due to urbanization is made.
1985	[85]	Storage Function Model	Comparison of various SF models
1991	[63]	Kinematic Wave Model	Application for an urban watershed by dividing it into rectangular sub-watersheds.
1992	([31])	Xinanjiang Model	Rainfall-runoff, distributed, basin model.
1995	([33])	SHE/SHESED	Basin-scale water flow and sediment transport modeling system.
1995	([34])	Institute of Hydrology Distributed Model	The model uses established flow equations for surface and subsurface.
1995	([35])	HBV Model	Lumped (or semi-distributed) bucket-type (conceptual) catchment model.
1995	[64]	Diffusion Wave Model Dynamic Wave Model	Comparison with kinematic wave model for nonstationary and unsteady river channel flow.
1995	[91]	Tank Model	Tank model parameter estimation using global optimization methods,
1996	[65]	Doken Model	Distributed model for real-time flood forecasting.
1997	[67]	Kinematic Wave Model	Model parameters were obtained using GIS information.
1998	[52]	Rational Model	Application for small urban rivers.
1998	[122]	2D Diffusive Wave Model (Curvilinear Grid)	Pioneer research on flood flow considering the influence of buildings and road structures.
1999	[22]	Storage Function Model	Proposal of SF model considering loss mechanisms that directly uses the observed rainfall and runoff.
1999	[25]	Distributed Runoff Model	Evaluation of the permeable pavement and the house infiltration system.
1999	[125]	2D Diffusive Wave Model (Road Network)	Proposal of road network model in an urban area.
2000	[84]	Storage Function Model (Hoshi's Model)	Proposal of four-parameter SF model that directly uses the observed rainfall and runoff.
2000	[112]	2D Diffusive Model (Structured Grid)	The model of [112] was extended to include inflows from mountain watersheds and sewer overflows.
2001	[26]	A Distributed Hydrological Model-WEP Model	WEP model that evaluates the effects of stormwater detention ponds and infiltration trenches in urban watersheds.
2002	[121]	2D Diffusive Model (Structured/Curvilinear /Unstructured Grid)	Proposal of a flood runoff analysis model considering urban area characteristics.
2004	[119]	NILIM2.0 2D Diffusive Model	A standard method used to prepare municipal flood maps in Japan.
2007	[39]	TSR Model	TSR model in the early development stage. The model is a physically based distributed model, which was applied for an urban catchment using polygon feature GIS data.

Table A1. *Cont.*

Year	Ref. No.	Model/Method Name	Remarks
2008	[86]	Storage Function Model	Proposal of SF model considering outflow through combined sewer system.
2008	[123]	Flood Disaster Deduction System (Dual Drainage, Unstructured Grid)	Consideration of sewer flows in surface flooding models that use 2D full shallow-water equations.
2009	[23]	Urban storage function (USF) Model	Proposal of USF Model that directly uses the observed rainfall and runoff.
2010	([36])	HYPE Model	Dynamic and semi-distributed hydrological model integrating fluxes of water, nutrients, and other substances.
2010	[124]	Flood Disaster Reduction System (Dual Drainage, Unstructured Grid)	The numerical simulation for inundation flows with flood control system was carried out.
2011	[118]	2D Diffusive Model (Structured Grid)	Formulating the impact of buildings on flood flows.
2011	[126]	1D Saint-Venant Equations (Dual Drainage, Road Network)	Numerical analysis of inundation in the region of downtown Tokyo with residence area.
2012	[6]	Dual drainage model TSR model (Urban Landscape GIS)	TSR model was proposed that is a distributed urban storm runoff event model with a vector-based catchment.
2012	[28]	RRI model	RRI model was proposed that is a 2D model capable of simultaneously analyzing rainfall-runoff and flood inundation.
2012	[38]	Urban Landscape GIS Delineation	Permeable area ratio of 10m-mesh land use classification was studied.
2012	[53]	Synthesized Rational Model	Theoretical derivation of the synthesized rational formula.
2012	[97]	USF Model	Real-time runoff forecasting of the model using particle filter.
2013	[98]	USF Model	Real-time runoff forecasting of the model using the Kalman filter.
2013	[132]	TSR model (Urban Landscape GIS)	TSR model considering the flooding process of buildings.
2015	[92]	Distributed Model SCE-UA Method	Parameter identification of a distributed runoff model using SCE-UA method.
2016	[40]	Urban Landscape GIS Delineation	Impervious area ratios of grid-based land use classifications were assessed.
2016	[100]	XRAIN Data	Precision evaluation of XRAIN data in a small urban watershed.
2016	[104]	Deep Learning (DL) Method	Development of the real-time river stage prediction method using DL approach.
2017	[62]	Various Numerical Methods	Review paper for surface water flow numerical analysis.
2017	[83]	Baseflow Estimation Method	Baseflow estimation using recursive digital filters.
2017	[102]	XRAIN Data	Spatiotemporal characteristic analysis of XRAIN data.
2017	[103]	XRAIN Data	Spatiotemporal correlation analysis of XRAIN data focused on torrential rainfall events at each ground observation point.
2017	[106]	Machine Learning (ML) Method	Real-time forecasting of water levels in sewer system by the method.
2017	[110]	Deep Neural Network (DNN) Model Distributed Model	Proposal of real-time river stage prediction model by combining DNN and a distributed model.
2018	([43])	SWMM	Drainage area characterization for evaluating green infrastructure.
2018	([66])	IFAS (Integrated Flood Analysis System)	Flood forecasting in watersheds with inadequate water level observations.
2018	[88]	Storage Function Model	An effective storage function model for an urban watershed was investigated.
2018	[99]	USF Model XRAIN Data	Reproducibility of hydrograph by USF model using different rainfall datasets was studied.
2018	[101]	XRAIN Data	Precision of 1-minute XRAIN data in a small urban watershed was studied.
2018	[105]	DL Method	Application of DL method to long-term prediction of dam inflow.
2018	[113]	Integrated Model with InfoWorks and WEB-DHM	A river channel and sewer model were built by integrating InfoWorks into WEB-DHM.
2019	[21]	HYPE Model	HYPE model using EEA Urban Atlas.

Table A1. *Cont.*

Year	Ref. No.	Model/Method Name	Remarks
2019	[94]	USF Model	Parameter uncertainty of USF model was evaluated.
2019	[95]	USF Model	Proposal of a generalized USF model considering spatial rainfall distribution.
2019	[107]	USF Model DL Method	Possibility of emulating USF model by DL method was studied for benchmark virtual hyeto and hydrograph.
2019	[108]	Neural Network Method DL Method	Comparison of emulation ability for the USF model by the two methods.
2019	([114])	HE-C-RAS	HEC-RAS was used to generate flood inundation and develop a risk map under the different rainfall scenarios.
2020	[27]	TSR model	TSR model, considering green infrastructure.
2020	[96]	Generalized Storage Function Model	Proposal of a generalized storage function model for the water level estimation.
2020	[109]	DL Method	Prediction evaluation of hydrograph by the method using an observed urban dataset.
2021	([115])	HEC-RAS	Building representation method by HEC-RAS 2-D and evaluation of the influence of fluid forces on buildings.
2023	[77]	Time Series Method	Water level forecasting based on time series analysis in an urban watershed.

References

1. Kinoshita, T. Runoff changes due to urbanization. *Civ. Eng. Technol. Mat.* **1967**, *9*, 11–15. (In Japanese)
2. Takahashi, Y.; Izumi, K.; Ando, Y.; Kanao, K. Change of flood runoff in hilly basins by urbanization. *Proc. Hydraul. Conf. Jpn. Soc. Civ. Eng.* **1982**, *26*, 261–266. (In Japanese)
3. Yamada, T.; Ishii, F.; Yamazaki, K.; Iwatani, K. On relation between distribution of field moisture capacity and runoff characteristics in small basins. *Proc. Hydraul. Conf. Jpn. Soc. Civ. Eng.* **1985**, *29*, 25–30. (In Japanese)
4. Kadoya, M. A review of the study on runoff changes due to urbanization. *J. Jpn. Soc. Civ. Eng.* **1985**, *363*, 23–34. (In Japanese) [CrossRef] [PubMed]
5. Takahashi, Y. *City and Water*; Iwanami Shinsho: Tokyo, Japan, 1988; pp. 1–34. (In Japanese)
6. Amaguchi, H.; Kawamura, A.; Olsson, J.; Takasaki, T. Development and testing of a distributed urban storm runoff event model with a vector-based catchment. *J. Hydrol.* **2012**, *420–421*, 205–215. [CrossRef]
7. Tamai, N. *River Engineering*; Ohmsha: Tokyo, Japan, 1999; pp. 55–95. (In Japanese)
8. Ishihara, S.; Takasaki, T.; Kawamura, A.; Amaguchi, H. Background of new development plan and future development methods of urban rivers in Tokyo metropolis. *Adv. River Eng.* **2014**, *20*, 437–442. (In Japanese)
9. Kanae, S. Characteristics of localized torrential rainfall (so-called guerrilla rainfall). *J. Hydrol. Syst.* **2009**, *73*, 11–16. (In Japanese)
10. Nakakita, E.; Yamaguchi, K.; Yamabe, H. A study on detecting origin of localized torrential rainfall using volume scanning radar. *DPRI Annu.* **2009**, *52*, 547–562. (In Japanese)
11. Ushiyama, M. A study of relationship between "Guerilla Heavy Rainfall" and disaster. *J. Jpn. Soc. Civ. Eng. Ser. B1 (Hydraul. Eng.)* **2011**, *67*, I_505–I_510. (In Japanese) [CrossRef]
12. Fujibe, F.; Togawa, H.; Sakata, M. Long-term change and spatial anomaly of warm season afternoon precipitation in Tokyo. *SOLA* **2009**, *5*, 17–20. [CrossRef]
13. Seino, N.; Aoyagi, T.; Tsuguti, H. Numerical Simulation of Urban Impact on Precipitation in Tokyo: How does Urban Temperature Rise Affect Precipitation? *Urban Clim.* **2018**, *23*, 8–35. [CrossRef]
14. Kawamura, A. Urban flooding and its countermeasures. In *Urban Technology*, Revised ed.; Graduate School of Urban Infrastructure and Environment, Tokyo Metropolitan University, Ed.; Gihodo Pub.: Tokyo, Japan, 2016; pp. 235–249. (In Japanese)
15. Singh, V.P. Watershed modeling. In *Computer Models of Watershed Hydrology*; Singh, V.P., Ed.; Water Res. Pub.: Denver, CO, USA, 1995; pp. 1–22.
16. JSCE Committee on Hydroscience and Hydraulic Engineering. *Hydroscience Formulary*; JSCE: Tokyo, Japan, 1999; pp. 35–47. (In Japanese)
17. Sugawara, M. *Rainfall-Runoff Analysis*; Kyoritsu Pub: Tokyo, Japan, 1972; p. 257. (In Japanese)
18. Kimura, T. Storage Function Model. *Civ. Eng. J.* **1961**, *3*, 36–43. (In Japanese)
19. Kawamura, A. Inverse problems in hydrology. In *Introduction to Inverse Problems in Civil Engineering*; Jpn. Soc. Civil Eng.; Maruzen: Tokyo, Japan, 2000; pp. 24–30. (In Japanese)
20. Hino, M. What have hydrologists done and what should we do? What should we do from now on? *J. Jpn. Soc. Hydrol. Water Res.* **1996**, *9*, 3–16. (In Japanese)
21. Tanouchi, H.; Olsson, J.; Lindstrom, G.; Kawamura, A.; Amaguchi, H. Improving urban runoff contributions in multi-basin hydrological simulation by the HYPE model using EEA Urban Atlas. *Hydrology* **2019**, *6*, 28. [CrossRef]

22. Baba, H.; Hoshi, K.; Hashimoto, N. Synthetic storage routing model coped with loss mechanisms. *Proc. Hydraul. Conf. Jpn. Soc. Civ. Eng.* **1999**, *43*, 1085–1090. (In Japanese) [CrossRef]
23. Takasaki, T.; Kawamura, A.; Amaguchi, H.; Araki, K. Storage function model considering urban runoff process. *J. Jpn. Soc. Civ. Eng. B* **2009**, *65*, 217–230. (In Japanese) [CrossRef]
24. Shiiba, M.; Tachikawa, Y.; Ichikawa, Y. *Hydrology and Hydraulic Engineering Planning Studies*; Kyoto University Press: Kyoto, Japan, 2013; pp. 351–371. (in Japanese)
25. Hanaki, T.; Suharyanto, A.; Sugio, S. Investigation of flood control by rainwater infiltration facilities in urbanized watershed. *Proc. Hydraul. Conf. Jpn. Soc. Civ. Eng.* **1999**, *48*, 37–42. (In Japanese) [CrossRef]
26. Jia, Y.; Ni, G.; Kinouchi, T.; Yoshitani, J.; Kawahara, Y.; Suetsugi, T. Study on effects of storm-water detention facilities in an urbanized watershed using distributed model. *Jpn. Soc. Civ. Eng. Hydraul. Conf.* **2001**, *45*, 109–114. (In Japanese) [CrossRef]
27. Amaguchi, H.; Kawamura, A. A proposal of urban runoff model considering green infrastructures and its application. *J. Jpn. Soc. Civ. Eng. Ser. G (Environ. Res.)* **2020**, *76*, I_319–I_325. (In Japanese) [CrossRef]
28. Sayama, T.; Ozawa, G.; Kawakami, T.; Nabesaka, S.; Fukami, K. Rainfall-runoff-inundation analysis of the 2010 Pakistan flood in the Kabul River basin. *Hydrol. Sci. J.* **2012**, *57*, 298–312. [CrossRef]
29. Public Works Research Institute. RRI Model. Available online: http://www.icharm.pwri.go.jp/research/rri/rri_top.html (accessed on 18 November 2022).
30. Beven, K.J.; Kirkby, M.J. A physically based variable contributing area model of basin hydrology. *Hydrol. Sci. Bull.* **1979**, *24*, 43–69. [CrossRef]
31. Zhoa, R.J. The Xinanjiang model applied in China. *J. Hydrol.* **1992**, *135*, 371–381.
32. Ren, L.; Yuan, F. The Xinanjiang model on a digital basin platform. In *Watershed Models*; Singh, V.P., Frevert., D.K., Eds.; CRC Press: Boca Raton, FL, USA, 2006; pp. 179–208.
33. Bathurst, J.C.; Wicks, J.M.; O'Connell, P.E. The SHE/SHESED basin scale water flow and sediment transport modelling system. In *Computer Models of Watershed Hydrology*; Singh, V.P., Ed.; Water Res. Pub.: Denver, CO, USA, 1995; pp. 563–594.
34. Calver, A.; Wood, W.L. The Institute of Hydrology Distributed Model. In *Computer Models of Watershed Hydrology*; Singh, V.P., Ed.; Water Res. Pub.: Denver, CO, USA, 1995; pp. 595–626.
35. Bergström, S. The HBV Model. In *Computer Models of Watershed Hydrology*; Singh, V.P., Ed.; Water Res. Pub.: Denver, CO, USA, 1995; pp. 443–476.
36. Lindström, G.; Pers, C.P.; Rosberg, R.; Strömqvist, J.; Arheimer, B. Development and test of the HYPE (Hydrological Predictions for the Environment) model—A water quality model for different spatial scales. *Hydrol. Res.* **2010**, *41*, 295–319. [CrossRef]
37. SMHI, HYPE. Available online: https://www.smhi.se/en/research/research-departments/hydrology/hype-1.7994 (accessed on 18 November 2022).
38. Koga, T.; Kawamura, A.; Amaguchi, H. Study on permeable area ratio of 10m-mesh land use classification using advanced GIS delineation in the upper Kanda River basin. *J. Jpn. Soc. Civ. Eng. Ser. B1 (Hydraul. Eng.)* **2012**, *68*, I_505–I_510. (In Japanese)
39. Amaguchi, H.; Kawamura, A.; Takasaki, T. Physically based distributed flood runoff model for an urban catchment using polygon feature GIS data. *J. Jpn. Soc. Civ. Eng. B* **2007**, *63*, 206–223. (In Japanese) [CrossRef]
40. Koga, T.; Kawamura, A.; Amaguchi, H.; Tanouchi, H. Assessing impervious area ratios of grid-based land-use classifications on an example of an urban watershed. *Hydrol. Sci. J.* **2016**, *61*, 1728–1739. [CrossRef]
41. Sample, D.J.; Heaney, J.P.; Wright, L.; Koustas, R. Geographic information systems, decision support systems, and urban storm-water management. *J. Water Plan. Man.* **2001**, *127*, 155–161. [CrossRef]
42. Rodriguez, F.; Andrieu, H.; Creutin, J.D. Surface runoff in urban catchments: Morphological identification of unit hydrographs from urban databanks. *J. Hydrol.* **2003**, *283*, 146–168. [CrossRef]
43. Lee, J.G.; Nietch, C.T.; Panguluri, S. Drainage area characterization for evaluating green infrastructure using the storm water management model. *Hydrol. Earth Syst. Sci.* **2018**, *22*, 2615–2635. [CrossRef]
44. Japan Society of Hydrology and Water Resources. *Handbook of Hydrology and Water Resources*; Asakura: Tokyo, Japan, 1997; p. 636. (In Japanese)
45. JSCE Committee on Hydroscience and Hydraulic Engineering. *Hydroscience Formulary, 1985 ed.*; JSCE: Tokyo, Japan, 1985; pp. 151–172, 713. (In Japanese)
46. JSCE Committee on Hydroscience and Hydraulic Engineering. *Hydroscience Formulary, 1971 ed.*; JSCE: Tokyo, Japan, 1971; p. 616. (In Japanese)
47. JSCE Committee on Hydroscience and Hydraulic Engineering. *Hydroscience Formulary, 1963 ed.*; JSCE: Tokyo, Japan, 1963; p. 603. (In Japanese)
48. Morita, S. *Urban Flood Risk Analysis*; Forum 8: Tokyo, Japan, 2014; pp. 3–75. (In Japanese)
49. Matsubayashi, U. Runoff mechanism and characteristics in urban areas. In *Handbook of Hydrology and Water Resources*; Jpn Soc. Hydrol. Water Res.; Asakura: Tokyo, Japan, 1997; pp. 146–148. (In Japanese)
50. Ishikawa, K.; Sato, K.; Izumi, K. Applicability of rational method to the urbanized channels in Tokyo. *Proc. Hydraul. Conf. Jpn. Soc. Civ. Eng.* **1982**, *26*, 267–272. (In Japanese)
51. Kadoya, M.; Fukushima, A. Concentration time of flood in small or medium river basin. *DPRI Annu.* **1976**, *19B*, 143–152. (In Japanese)

52. Tanioka, Y.; Fukuoka, S.; Taniguchi, M.; Koyama, Y. Characteristics of floods in small urban rivers. *J. Jpn. Soc. Civ. Eng.* **1998**, *586*, 1–12. (In Japanese) [CrossRef] [PubMed]
53. Watanabe, A.; Sasada, T.; Watanabe, N.; Yamada, T. Theoretical derivation of synthesized rational formula. *J. Jpn. Soc. Civ. Eng. Ser. B1 (Hydraul. Eng.)* **2012**, *68*, I_499–I_504. (In Japanese)
54. Hashimoto, T.; Hasegawa, T. A runoff model to evaluate land use change. *Civ. Eng. J.* **1977**, *19*, 221–226. (In Japanese)
55. Yamaguchi, T.; Matsubara, S.; Yamamori, T. Second report of rainfall-runoff study in an urban area -Runoff estimation by modified R.R.L. method. *Civ. Eng. J.* **1972**, *14*, 34–39. (In Japanese)
56. Matsubayashi, U. Runoff model for urban watersheds. In *Handbook of Hydrology and Water Resources*; Jpn. Soc. Hydrol. Water Res.; Asakura: Tokyo, Japan, 1997; pp. 148–151. (In Japanese)
57. Sukegawa, N.; Kitagawa, Y. A rainfall-runoff simulation model for assessing the effect of urbanization on flood runoff. *Proc. Jan. Soc. Civ. Eng.* **1982**, *325*, 51–59. (In Japanese) [CrossRef]
58. Iwagaki, Y.; Sueishi, T. On the unsteady flow in open channels with uniform lateral inflow—Hydraulic studies on the run-off phenomena of rain water, 1st report. *J. Jpn. Soc. Civ. Eng.* **1954**, *39*, 575–583. (In Japanese)
59. Sueishi, T. On the run-off analysis by the method of characteristics hydraulic studies on the run-off phenomena of rain water, 2nd report. *J. Jpn. Soc. Civ. Eng.* **1955**, *29*, 74–87. (In Japanese) [CrossRef]
60. Lighthill, M.J.; Whitham, G.B. On kinematic waves: 1. Flood movement in long rivers. *Proc. R. Soc.* **1955**, *229*, 281–316.
61. Singh, V.P.; Jain, S.K. Rainfall-Runoff Modeling. In *Handbook of Applied Hydrology*, 2nd ed.; Singh, V.P., Ed.; McGraw-Hill Education: New York, NY, USA, 2016; Chapter 59; pp. 59-1–59-8.
62. Kimura, M.; Tanaka, T.; Azechi, I.; Nakatani, K.; Yamazaki, D.; Yoshioka, H. Interdisciplinary perspective of surface water flow numerical analysis. *J. Jpn. Soc. Hydrol. Water Res.* **2017**, *30*, 307–334. (In Japanese) [CrossRef]
63. Ando, Y.; Nabeyama, T.; Nishijima, S. Flood runoff analysis of urban watersheds. *J. Urban Res.* **1991**, *41*, 69–89. (In Japanese)
64. Fujimura, K.; Ando, Y.; Maeda, M. A study on stormwater runoff model of urban lowland basin. *Proc. Hydraul. Conf. Jpn. Soc. Civ. Eng.* **1995**, *39*, 61–66. (In Japanese) [CrossRef]
65. Suzuki, T.; Terakawa, Y.; Matsuura, T. Development of physics-based distributed model for operational hydrological forecasting. *Civ. Eng. J.* **1996**, *38*, 26–31. (In Japanese)
66. Shahzad, A.; Gabriel, H.F.; Haider, S.; Mubeen, A.; Siddiqui, M.J. Development of a flood forecasting system using IFAS: A case study of scarcely gauged Jhelum and Chenab river basins. *Arab. J. Geosci.* **2018**, *11*, 383.
67. Chikamori, H.; Oka, T.; Takara, K.; Okubo, G. Fundamental Study on Application of Geographic Information System to Construction of Rainfall-Runoff Model of Urbanized Area. *DPRI Annu.* **1997**, *B 40*, 137–144. (In Japanese)
68. Ministry of Land, Infrastructure, Transport and Tourism. *Investigation Guidelines for River Sand and Gravel Management Technology Standards, Survey Edition*; MLIT: Tokyo, Japan, 2016; Chapter 3.
69. Kawamura, A. Status quo and perspectives of flood runoff analysis for urban watersheds. *J. Jpn. Soc. Hydrol. Water Res.* **2018**, *31*, 451–466. (In Japanese) [CrossRef]
70. Shibuo, Y.; Furumai, H. Advances in Urban Stormwater Management in Japan: A Review. *J. Disaster Res.* **2021**, *16*, 310–320. [CrossRef]
71. Water Management and Land Conservation Bureau. Available online: https://www5.river.go.jp/ (accessed on 27 March 2023).
72. Water Management and Land Conservation Bureau. River Disaster Prevention Information. Available online: https://city.river.go.jp/kawabou/reference/index01.html (accessed on 27 March 2023).
73. Japan Meteorological Agency. Automated Meteorological Data Acquisition System 2020. Available online: https://www.jma.go.jp/jma/en/Activities/amedas/amedas.html (accessed on 10 November 2022).
74. Ministry of Land, Infrastructure, Transport and Tourism. X-Band MP Radar Network. Available online: http://mp-radar.bosai.go.jp/ (accessed on 18 November 2022).
75. Schilling, W. Rainfall data for urban hydrology: What do we need? *Atmos. Res.* **1991**, *27*, 5–21. [CrossRef]
76. Fabry, F.; Bellon, A.; Duncan, M.R.; Austin, G.L. High resolution rainfall measurements by radar for very small basins: The sampling problem reexamined. *J. Hydrol.* **1994**, *161*, 415–428. [CrossRef]
77. Koyama, N.; Sakai, M.; Yamada, T. Study on a Water-Level-Forecast Method Based on a Time Series Analysis of Urban River Basins—A Case Study of Shibuya River Basin in Tokyo. *Water* **2023**, *15*, 161. [CrossRef]
78. Tokyo Metropolitan Government—Flood Control Integrated Information System. Available online: https://www.kasen-suibo.metro.tokyo.lg.jp/im/uryosuii/tsim0102g_en.html (accessed on 10 November 2022).
79. Kadoya, M.; Nagai, A. Runoff analysis methods (Part 10) Storage method—Flood runoff analysis by the storage function model. *J. Jpn. Soc. Agric. Civ. Eng.* **1980**, *48*, 43–50.
80. Holmes, R.R., Jr. Streamflow Ratings. In *Handbook of Applied Hydrology*, 2nd ed.; Singh, V.P., Ed.; McGraw-Hill Education: New York, NY, USA, 2016; Chapter 6; pp. 6-1–6-14.
81. Prasad, R. A nonlinear hydrologic system response model. *J. Hydraul. Div.* **1967**, *93*, 201–222. [CrossRef]
82. Hoshi, K.; Yamaoka, T. A relationship between kinematic wave and storage routing models. *Proc. Hydraul. Conf. Jpn. Soc. Civ. Eng.* **1982**, *26*, 273–278. (In Japanese)
83. Padiyedath, S.G.; Kawamura, A.; Amaguchi, H.; Azhikodan, G. Baseflow estimation for tropical wet and dry climate region using recursive digital filters. *J. Jpn. Soc. Civ. Eng. Ser. G (Environ. Res.)* **2017**, *73*, 9–16. [CrossRef]

84. Hoshi, K.; Baba, H.; Hashimoto, A. Development of a storage function method for tracking floods in watershed and channel systems. *Adv. River Eng.* **2000**, *6*, 297–302. (In Japanese)
85. Takasao, T.; Takara, K.; Kusuhashi, H. Comparative study of storage function models of flood runoff. *Proc. Hydraul. Conf. Jpn. Soc. Civ. Eng.* **1985**, *29*, 245–250. (In Japanese)
86. Takasaki, T.; Kawamura, A.; Amaguchi, H. Development of the storage function model for urban flood considering the outflow through combined sewer system. *J. Jpn. Soc. Hydrol. Water Res.* **2008**, *21*, 228–241. (In Japanese) [CrossRef]
87. Ministry of Land, Infrastructure, Transport and Tourism. History of the Sewerage System. Available online: https://www.mlit.go.jp/crd/city/sewerage/data/basic/rekisi.html (accessed on 31 January 2023). (In Japanese).
88. Padiyedath, S.G.; Kawamura, A.; Takasaki, T.; Amaguchi, H.; Azhikodan, G. An effective storage function model for an urban watershed in terms of hydrograph reproducibility and Akaike information criterion. *J. Hydrol.* **2018**, *563*, 657–668. [CrossRef]
89. Kawamura, A.; Marabtene, T. Evolutionary computing: Genetic algorithms. In *Handbook of Applied Hydrology*, 2nd ed.; Singh, V.P., Ed.; McGraw-Hill Education: New York, NY, USA, 2016; Chapter 13; pp. 13-1–13-4.
90. Duan, Q.; Sorooshian, S.; Gupta, V.K. Optimal use of the SCE-UA global optimization method for calibrating watershed models. *J. Hydrol.* **1994**, *158*, 3–4, 265–284. [CrossRef]
91. Tanakamaru, H. Parameter estimation for the tank model using global optimization. *J. Jpn. Soc. Agric. Civ. Eng.* **1995**, *178*, 503–512. (In Japanese)
92. Matsubara, T.; Tsuchida, K.; Hibiya, M. Study on parameter identification of distributed runoff model using SCE-UA method. *J. Jpn. Soc. Civ. Eng. Ser. B1 (Hydraul. Eng.)* **2015**, *71*, I_265–I_270. (In Japanese)
93. Akaike, H. Likelihood of a model and information criteria. *J. Econom.* **1981**, *16*, 3–14. [CrossRef]
94. Padiyedath, S.G.; Kawamura, A.; Takasaki, T.; Amaguchi, H.; Azhikodan, G. A bootstrap approach for the parameter uncertainty of an urban-specific rainfall runoff model. *J. Hydrol.* **2019**, *579*, 1–18. [CrossRef]
95. Padiyedath, S.G.; Kawamura, A.; Takasaki, T.; Amaguchi, H.; Azhikodan, G. A generalized urban storage function model considering spatial rainfall distribution. *J. Jpn. Soc. Civ. Eng. Ser. B1 (Hydraul. Eng.)* **2019**, *75*, I_223–I_228. [CrossRef] [PubMed]
96. Padiyedath, S.G.; Kawamura, A.; Amaguchi, H.; Takasaki, T.; Azhikodan, G. A generalized storage function model for the water level estimation using rating curve relationship. *Water Res. Man.* **2020**, *34*, 2603–2619.
97. Takasaki, T.; Kawamura, A.; Amaguchi, H.; Ishihara, S. Real-time runoff forecasting characteristics of urban storage function model using particle filter. *J. Jpn. Soc. Civ. Eng. Ser. B1 (Hydraul. Eng.)* **2012**, *68*, I_511–I_516. (In Japanese)
98. Takasaki, T.; Kawamura, A.; Amaguchi, H.; Ishihara, S. Real-time runoff forecasting characteristics of Urban Storage Function Model using Kalman filter. *J. Jpn. Soc. Civ. Eng. Ser. B1 (Hydraul. Eng.)* **2013**, *69*, I_457–I_480. (In Japanese)
99. Yonese, Y.; Kawamura, A.; Tonotsuka, A.; Amaguchi, H. Runoff characteristics using XRAIN in a small urban watershed. *J. Jpn. Soc. Civ. Eng. Ser. B1 (Hydraul. Eng.)* **2018**, *74*, I_97–I_102. (In Japanese)
100. Yonese, Y.; Kawamura, A.; Amaguchi, H.; Tonotsuka, A. Precision evaluation of X-band MP radar rainfall in a small urban watershed by comparison to 1-minute ground observation rainfall data. *J. Jpn. Soc. Civ. Eng. Ser. B1 (Hydraul. Eng.)* **2016**, *72*, I_217–I_222. (In Japanese)
101. Yonese, Y.; Kawamura, A.; Amaguchi, H.; Tonotsuka, A. Study on the precision of 1-minute X-band MP radar rainfall data in a small urban watershed. *Int. J. Sustain. Dev. Plan.* **2018**, *13*, 614–625. [CrossRef]
102. Yonese, Y.; Kawamura, A.; Amaguchi, H.; Tonotsuka, A. Spatiotemporal characteristic analysis of X-band MP radar rainfall in a small urban watershed on the movement of rainfall area. *J. Jpn. Soc. Civ. Eng. Ser. B1 (Hydraul. Eng.)* **2017**, *73*, I_217–I_222. (In Japanese)
103. Tonotsuka, A.; Kawamura, A.; Yonese, Y.; Amaguchi, H. Spatiotemporal correlation characteristics of X-band MP radar rainfall in a small urban watershed focused on torrential rainfall events at each ground observation point. *J. Jpn. Soc. Civ. Eng. Ser. G (Environ. Res.)* **2017**, *73*, I_261–I_268. (In Japanese)
104. Hitokoto, M.; Sakuraba, M.; Sei, Y. Development of the real-time river stage prediction method using deep learning. *J. Jpn. Soc. Civ. Eng. Ser. B1 (Hydraul. Eng.)* **2016**, *72*, I_187–I_192. (In Japanese) [CrossRef] [PubMed]
105. Tamura, K.; Nanou, S.; Miura, S.; Yamawaki, M.; Kaneko, H. Application of deep learning to long-term prediction of dam inflow toward efficiency of the flood control operation. *J. Jpn. Soc. Civ. Eng. Ser. B1 (Hydraul. Eng.)* **2018**, *74*, I_1327–I_1332. (In Japanese) [CrossRef]
106. Hida, Y.; Chiha, H.; Asaoka, Y.; Nagabayashi, H. Real-time forecasting for water levels in sewer by machine learning and inundation simulation. *J. Jpn. Soc. Civ. Eng. Ser. B1 (Hydraul. Eng.)* **2017**, *73*, I_649–I_654. (In Japanese) [CrossRef]
107. Fujizuka, S.; Kawamura, A.; Amaguchi, H.; Takasaki, T. Emulation of urban runoff model by deep learning for benchmark virtual hyeto and hydrograph. *J. Jpn. Soc. Civ. Eng. Ser. G (Environ. Res.)* **2019**, *75*, I_289–I_296. (In Japanese) [CrossRef]
108. Fujizuka, S.; Kawamura, A.; Amaguchi, H.; Takasaki, T. Emulation performance evaluation of urban runoff model by neural network and deep learning. *J. Jpn. Soc. Civ. Eng. Ser. B1 (Hydraul. Eng.)* **2019**, *75*, I_229–I_234. (In Japanese) [CrossRef]
109. Fujizuka, S.; Kawamura, A.; Amaguchi, H.; Takasaki, T. Rainfall runoff benchmark test by deep learning model using urban medium and small river basin dataset. *J. Jpn. Soc. Civ. Eng. Ser. B1 (Hydraul. Eng.)* **2020**, *76*, I_355–I_360. (In Japanese) [CrossRef]
110. Hitokoto, M.; Sakuraba, M. Hybrid deep neural network and distributed rainfall-runoff model for the real-time river stage prediction. *J. Jpn. Soc. Civ. Eng. Ser. B1 (Hydraul. Eng.)* **2017**, *73*, I_22–I_34. (In Japanese) [CrossRef]
111. Iwasa, Y.; Inoue, K.; Mizutori, M. Hydraulic analysis of overland flood flows by means of numerical method. *DPRI Annu.* **1980**, *23*, 305–317. (In Japanese)

112. Toda, K.; Inoue, K.; Murase, S.; Ichikawa, Y.; Yokoo, H. Inundation analysis due to heavy rainfall in urban area. *J. Jpn. Soc. Civ. Eng.* **2000**, *663*, 1–10. (In Japanese) [CrossRef]
113. Shibuo, Y.; Lee, S.; Sanuki, H.; Yoshimura, K.; Tajima, Y.; Sato, S.; Furumai, H. Development of couped river and sewer network model for improved water level prediction in urban rivers. *J. Jpn. Soc. Civ. Eng. Ser. B1 (Hydraul. Eng.)* **2018**, *74*, I_1357–I_1362. (In Japanese)
114. Rangari, V.A.; Umamahesh, N.V.; Bhatt, C.M. Assessment of inundation risk in urban floods using HEC RAS 2D. *Mod. Earth Syst. Environ.* **2019**, *5*, 1839–1851. [CrossRef]
115. Mustafa, A.; Szydłowski, M. Application of different building representation techniques in HEC-RAS 2-D for urban flood modeling using the Toce River experimental case. *PeerJ* **2021**, *2*, e11667. [CrossRef] [PubMed]
116. Rosenzweig, B.R.; Cantis, P.H.; Kim, Y.; Cohn, A.; Grove, K.; Brock, J.; Yesuf, P.; Mistry, P.; Welty, C. The value of urban flood modeling. *Earth's Future* **2021**, *9*, e2020EF001739. [CrossRef]
117. Amaguchi, H.; Tsubaki, R. On the inundation prediction, present state and matters to be resolved. *Adv. River Eng.* **2015**, *21*, 425–430. (In Japanese)
118. Miura, S.; Kawamura, I.; Kimura, I.; Miura, A. Study on inundation flow analysis method in densely populated urban area on alluvial fan. *J. Jpn. Soc. Civ. Eng. Ser. B1 (Hydraul. Eng.)* **2011**, *67*, I_979–I_984. (In Japanese) [CrossRef]
119. Nakamura, T.; Sasaki, Y.; Mizukusa, K. A guide of application for flood analysis models in the urban areas. Technical Note of NILM 2004, No. 202. Available online: https://www.nilim.go.jp/lab/bcg/siryou/tnn/tnn0202pdf/ks0202.pdf (accessed on 10 November 2022).
120. Inokawa, N.; Kobayashi, K. Analysis of flooding in underground space using 1m mesh elevation data and study of evacuation risk. *J. Jpn. Soc. Civ. Eng. Ser. B1 (Hydraul. Eng.)* **2022**, *78*, I_1415–I_1420. (In Japanese) [CrossRef]
121. Kawaike, K.; Inoue, K.; Hayashi, H.; Toda, K. Development of inundation flow model in urban area. *J. Jpn. Soc. Civ. Eng.* **2002**, *698*, 1–10. (In Japanese) [CrossRef]
122. Fukuoka, S.; Kawashima, M.; Yokoyama, H.; Mizuguchi, M. The numerical simulation of flood-induced flows in urban residential area and the study of damage reduction. *J. Jpn. Soc. Civ. Eng.* **1998**, *600*, 23–36. (In Japanese) [CrossRef]
123. Akiyama, J.; Shigeeda, M.; Tanabe, T. Dynamic network model for free-surface-pressurized flows and its applicability to sewer network in the urban area. *Adv. River Eng.* **2008**, *14*, 425–430. Available online: http://library.jsce.or.jp/jsce/open/00906/2008/14-0241.pdf (accessed on 25 July 2023). (In Japanese)
124. Akiyama, J.; Shigeeda, M.; Kozono, Y. Numerical model for rainfall-runoff and flood inundation processes and its application for Iizuka city. *Proc. Hydraul. Conf. Jpn. Soc. Civ. Eng.* **2010**, *54*, 919–924. Available online: http://library.jsce.or.jp/jsce/open/00028/2010/54-0154.pdf (accessed on 25 July 2023). (In Japanese)
125. Inoue, K.; Kawaike, K.; Hayashi, H. Inundation flow modeling in urban area. *Proc. Hydraul. Conf. Jpn. Soc. Civ. Eng.* **1999**, *43*, 533–538. (In Japanese) [CrossRef]
126. Sekine, M. Numerical analysis of inundation in the region of downtown Tokyo with residence area. *J. Jpn. Soc. Civ. Eng. Ser. B1 (Hydraul. Eng.)* **2011**, *67*, 70–85. (In Japanese) [CrossRef]
127. Sekine, M.; Kamata, T.; Hosono, Y.; Shibuya, Y. Inundation process in three cities of Musashino, Mitaka, and Chofu for the largest possible torrential rainfall. *J. Jpn. Soc. Civ. Eng. Ser. B1 (Hydraul. Eng.)* **2022**, *78*, I_427–I_432. (In Japanese) [CrossRef]
128. Araki, T.; Amaguchi, H.; Kawamura, A.; Takasaki, T. Development of a groundwater recharge model for an urban catchment using urban landscape GIS delineation and its simulation for actual catchment. *J. Jpn. Soc. Civ. Eng. Ser. B1 (Hydraul. Eng.)* **2012**, *68*, 109–124. (In Japanese)
129. Tanouchi, H.; Amaguchi, H.; Kawamura, A.; Nakagawa, N. Development of an automated construction algorithm of advanced delineation GIS data using 1:2500 topological map. *J. Jpn. Soc. Civ. Eng. Ser. B1 (Hydraul. Eng.)* **2013**, *69*, I_523–I_528. (In Japanese)
130. Tanouchi, H.; Amaguchi, H.; Kawamura, A.; Nakagawa, N.; Koga, T. Study on an automated construction method of minute road segments aiming at urban storm runoff analysis. *Theory Appl. GIS* **2014**, *22*, 25–34. (In Japanese) [CrossRef]
131. Tanouchi, H.; Amaguchi, H.; Kawamura, A.; Koga, T.; Hagiwara, Y. Automated construction method of polygonal blocks for an urban watershed. *J. Jpn. Soc. Hydrol. Water Res.* **2015**, *26*, 298–303. (In Japanese) [CrossRef]
132. Amaguchi, H.; Nagasaka, T.; Kawamura, A.; Takasaki, T.; Nakagawa, N. A proposal of storm runoff model considering process of building inundation for an urban catchment. *Adv. River Eng.* **2013**, *19*, 211–216. (In Japanese)
133. Amaguchi, H.; Kawamura, A. Development of a storm runoff model considering process of individual building inundation. *Int. J. Saf. Secur. Eng.* **2016**, *6*, 570–581.
134. Bonacci, O. Floods: New Concepts merge—Old problems remain. In Proceedings of the 15th International Symposium on Water Management and Hydraulic Engineering, Primošten, Croatia, 6–8 September 2017.
135. Mercado, J.M.R.; Kawamura, A.; Medina, R. An expanded interpretive structural modeling analysis of the barriers to integrated flood risk management adaptation in Metro Manila. *Water* **2023**, *15*, 1029. [CrossRef]
136. Koike, T. Evolution of Japan's flood control planning and policy in response to climate change risks and social changes. *Water Policy* **2021**, *23*, 77–84. [CrossRef]

water

MDPI

Review

Real-Time Urban Flood Forecasting Systems for Southeast Asia—A Review of Present Modelling and Its Future Prospects

Detchphol Chitwatkulsiri * and Hitoshi Miyamoto

Department of Civil Engineering, Shibaura Institute of Technology, 3-7-5 Toyosu, Koto-ku, Tokyo 135-8548, Japan
* Correspondence: na20109@shibaura-it.ac.jp

Abstract: Many urban areas in tropical Southeast Asia, e.g., Bangkok in Thailand, have recently been experiencing unprecedentedly intense flash floods due to climate change. The rapid flood inundation has caused extremely severe damage to urban residents and social infrastructures. In addition, urban Southeast Asia usually has inadequate capacities in drainage systems, complicated land use patterns, and a large vulnerable population in limited urban areas. To reduce the urban flood risk and enhance the resilience of vulnerable urban communities, it has been of essential importance to develop real-time urban flood forecasting systems for flood disaster prevention authorities and the urban public. This paper reviewed the state-of-the-art models of real-time forecasting systems for urban flash floods. The real-time system basically consists of the following subsystems, i.e., rainfall forecasting, drainage system modelling, and inundation area mapping. This paper summarized the recent radar data utilization methods for rainfall forecasting, physical-process-based hydraulic models for flood inundation prediction, and data-driven artificial intelligence (AI) models for the real-time forecasting system. This paper also dealt with available technologies for modelling, e.g., digital surface models (DSMs) for the finer urban terrain of drainage systems. The review indicated that an obstacle to using process-based hydraulic models was the limited computational resources and shorter lead time for real-time forecasting in many urban areas in tropical Southeast Asia. The review further discussed the prospects of data-driven AI models for real-time forecasting systems.

Keywords: urban floods; real-time forecasting; methodology; physical-process-based models; artificial-intelligence-based models; regional implementation

Citation: Chitwatkulsiri, D.; Miyamoto, H. Real-Time Urban Flood Forecasting Systems for Southeast Asia—A Review of Present Modelling and Its Future Prospects. *Water* **2023**, *15*, 178. https://doi.org/10.3390/w15010178

Academic Editors: Akira Kawamura and Kei Nakagawa

Received: 12 November 2022
Revised: 22 December 2022
Accepted: 29 December 2022
Published: 1 January 2023

1. Introduction

Urban areas in Southeast Asia have been expanding and changing more rapidly than constructing and maintaining their storm drainage systems. If adequate flood risk management is not carried out, threats from climate change, growing urbanization, and deteriorated drainage infrastructure could increase flood disasters. Urban flood management has significantly benefited from the hydroinformatics tools such as rainfall predictions and flood modelling. However, data availability is one of the primary limitations of using hydroinformatics tools. In addition, various variables can affect the choice of rainfall prediction methods and flood modelling tools, resulting in a potential limitation for real-time flood forecasting. In most urban areas in Southeast Asia, gathering information and creating weather forecasts could be more feasible than making significant adjustments to drainage network infrastructures. Consequently, having a real-time flood forecast available with sound weather forecasts would be a great way to deal with the uncertainties associated with the increasing frequency of flood disasters under current climate change.

Procedures for simulating urban flood dynamics that can mitigate flood risks have been well established for urban areas in the Global North [1–3]. However, for urban areas in the Global South, including Southeast Asia, there are still many obstacles and gaps in our understanding of urban floods [4–9], although there has been growing expertise in urban flood modelling and simulation. The Global South differs significantly from the

Global North, where the modelling technique has been well established, including drainage systems, data availability, design, maintenance, and land use patterns [10]. Moreover, there are significant differences in the hydrologic cycle and rainfall/runoff processes between tropical climates in the Global South and temperate temperatures typical of the Global North [11–15]. Therefore, it is essential to investigate how well current modelling techniques can adequately capture urban flood situations in the urban areas of the Global South.

Real-time forecasting of hydro-meteorological systems can improve urban resilience in two crucial viewpoints, i.e., the issues of real-time control and alert for urban floods [16–20]. Urban drainage systems' real-time management has been studied during the past few decades, focusing mainly on combined sewer systems to reduce overflows in wet weather [21]. The real-time control regulates the discharge, filling, and emptying rate and the available capacity of the conveyance and storage components in the system based on previously defined rules for meteorological, hydrologic, and hydraulic conditions [22]. Therefore, accurate forecasting of meteorological, hydrologic, and hydraulic variables could be a crucial part of the real-time forecasting system, initiating an automated response for flow diversion or storage strategies.

As for urban flood simulations, there is a wide range of modelling techniques. In general, physical-process-based models can simulate accurate flood forecasts, so they have been giving us useful methodologies for urban flood applications [23]. However, compared to fluvial flooding with a typical open dendritic structure, urban pluvial flooding necessitates distinct modelling of small-scale structures in urban environments, e.g., streets and buildings. In addition, urban drainage systems have intricate matrix architecture and a wide variety of flood control devices, i.e., pumps, weirs, gates, and retarding storages. Therefore, a finer spatial resolution is required to represent urban terrain characteristics with social infrastructure. Moreover, in urban areas of Southeast Asia, a few drainage networks have been under construction, and their infrastructure data archives and maps have barely been updated. Under these circumstances, simulating urban pluvial flooding needs a longer computational time. On the other hand, a shorter computation time for real-time urban flood forecasting systems is be essential because it gives the system controls and flood alerts a longer lead time. In this sense, empirical models, including data-driven artificial intelligence (AI), have an advantage over traditional physical-process-based models [24].

This study tried to provide readers with a state-of-the-art review of current development for real-time urban flood forecasting systems. It includes the above-mentioned hydroinformatics tools, i.e., rainfall forecasting and pluvial flood modelling, as shown in Figure 1. Currently available approaches used for rainfall forecasting and pluvial flood modelling were listed and discussed regarding their advantages and limitations for urban areas in Southeast Asia. Then, their standard architecture was featured for constituting real-time urban flood forecasting systems. Moreover, this paper also tried to discuss further recent research trends of data-driven AI approaches for real-time urban flood forecasting systems.

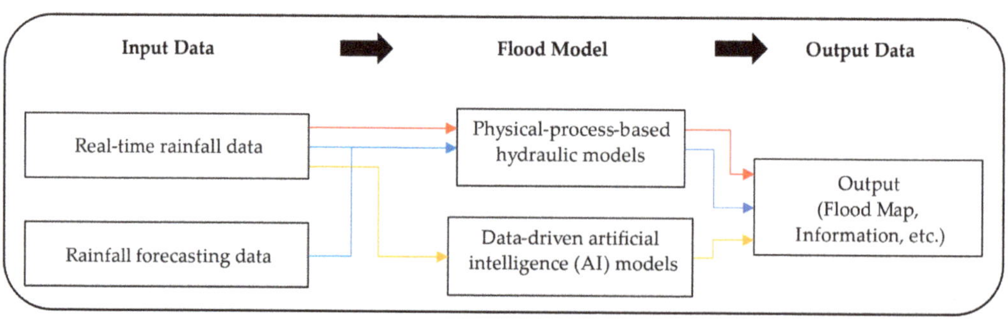

Figure 1. Conceptual diagram of the current real-time urban flood forecasting systems.

2. Rainfall Information for Real-Time Urban Flood Forecasting Systems

2.1. Rainfall Data Sources

Precipitation is the primary input for real-time urban flood forecasting systems. However, it is challenging to quantify it precisely due to its discontinuous behavior, spatiotemporal unpredictability, and susceptibility to climatic variables [25]. Rain gauges are the most popular in situ tool when measuring precipitation intensity and duration using telemetry. Rain gauges' continual recording of rainfall is crucial from a hydrological perspective. However, the sparse distribution of rain gauges, particularly in areas with high spatial variability such as mountain ranges, results in poor areal rainfall estimates.

Weather radars are remote sensing devices frequently used in hydrological fields to estimate areal precipitation with high spatial and temporal resolution. The spatial and temporal characteristics of a particular storm event—particularly its velocity and temporal changeability—and the catchment extent reveal the minimal resolutions of rainfall data for the given storm cell. When radar reflectivity data predict precipitation, it can instantly deliver the precipitation information with excellent temporal resolution. The highest temporal resolution for short-range local radars, employed particularly in cities, is one minute with a streamlined scanning technique constrained to the lowest elevation angles. The considerable variability of temporal precipitation, particularly for brief rainfall episodes, is frequently ignored when calculating precipitation intensity from weather radar data using a standard scan approach (typically with a temporal resolution of roughly 5 min).

As for spatial resolution, weather radars scan the atmosphere over a volume with a projected area of about 1 km^2 for a typical C-band radar. On the other hand, rain gauges measure rainwater at ground level over a circular region with a diameter of 20 cm. The spatial scales of the two devices are very different from each other. Consequently, it is not easy to compare the results of a rain gauge with those of weather radar [26]. Numerous studies have combined radar precipitation estimates with rain gauge data, either for quality control of weather radar data or to produce high-resolution merged products with greater accuracy than either rain gauge data or weather radar data alone [27–30]. However, the geographical and temporal inconsistency between radar estimations and rain gauge measurements continues to pose challenges to accurately determining real rainfall [31].

2.2. Nowcasting Techniques in Radar Rainfalls

Flash floods are brought on by intense convective storms and severe precipitation, resulting in considerable economic losses and, in some circumstances, fatalities. Furthermore, heavy rains are extremely challenging to predict since they change quickly and influence a small area. Although predictions, in general, are often successful in predicting favorable atmospheric conditions for severe storms, we are currently unable to accurately predict their precise positions and occurrence time, as well as the different characteristics of specific storms, e.g., heavy rains, downbursts, and lightning. As the primary factor could be due to their rapid development, storms are only detected by weather radar a few tens of minutes before they have a catastrophic appearance. Convection spots could be detected earlier when using satellite data. However, it would still be challenging to predict whether the convection produces a storm with a dangerous impact.

Nowcasting techniques for radar rainfall are categorized into a few groups. Extrapolation techniques are the first group, which calculates motion vectors and their extrapolated ones along the Lagrangian trajectories based on the current atmospheric conditions. They use only radar data without numerical weather prediction (NWP) models. The primary drawback of extrapolation techniques is that they cannot be extrapolated from the existing atmospheric conditions if any further atmospheric development occurs. The second group refers to blending techniques, which combine the extrapolation techniques with the NWP model results. The third group is artificial-intelligence-based models. They do not explicitly employ extrapolation along Lagrangian trajectories, in contrast to extrapolation approaches. Examples of nowcasting techniques were thoroughly examined in Sydney and Beijing [28,29].

2.3. Forecasting Techniques in Radar Rainfalls

Forecasting techniques for radar rainfalls are classified into several groups, i.e., motion field predictions, deterministic, ensemble, and probabilistic forecasting methods, and conceptual methods. The motion field required for forecasting is predicted from the series of radar reflectivity data. Then, weather radar data forecasts the deterministic, probabilistic, and ensemble predictions of precipitation [32]. Different forecasting techniques have been developed with variations in the motion field calculations and Lagrangian trajectory utilization. The evolution of areal echoes in the radar reflectivity yields the motion fields. The optical flow method (OFM) is most frequently used to calculate the motion fields [33]. In addition, there are prediction techniques, e.g., the Storm Cell Identification and Tracking (SCIT) algorithms [34] and the Thunderstorm Identification Tracking Analysis and Nowcasting (TITAN), to monitor the development of individual storms [35]. A finer estimate is required for motion fields to improve the efficiency of algorithms. Currently, the forward-in-time technique is preferably used when forecasting individual storm movements. Most motion field predictions employ a regular grid point to calculate the motion fields and forecast the rainfall.

Quantitative precipitation forecasts attempt to predict the precipitation intensity in a target location at a time or the accumulated precipitation in a specific time interval. The abovementioned extrapolation techniques compute the location and time deterministically by transferring the existing precipitation field along the Lagrangian trajectories. They assume precipitation does not change along the storm courses [36–42]. This assumption might be acceptable for forecasts for stratiform precipitation with lead times of several hours. However, rainfall often changes significantly along the trajectories for convective precipitation. Moreover, the forecasting improvements are debatable if they have reduced accurate predictions for longer lead times by filtering smaller storms out from forecasting due to a radar's coarse spatial resolution. In general, it is impossible to quantify the precipitation fields ranging from large objects to small features because their spatial dimensions vary widely depending on meteorological situations and the resolution and quality of radar data.

Ensemble and probabilistic forecasts are closely related because the ensemble members can yield a probabilistic forecast. Ensemble forecasts can often assess potential economic losses due to forecasting uncertainties. At the same time, probabilistic forecasts can provide the precipitation occurrence likelihood with an arbitrary threshold being exceeded. A basic technique of probabilistic forecasts considers forecasted values in the vicinity of a point as a possible future for computation [43]. Precipitation forecasts can also be performed using conceptual models describing precipitation developments. The automated forecasting system of convective precipitation GANDOLF [44] is a well-known example of this approach, attempting to simulate storm developments. GANDOLF includes a process of convective cloud life cycles. It identifies and forecasts convective cells using satellite and radar data, and forecasting values with NWP models. Other examples of conceptual model applications can be found in [45,46], which has been used in AutoNowCaster [47,48].

2.4. Forecasting Application and Its Future Perspective in Southeast Asia

Radar rainfall measurements have already been of great importance in improving basic knowledge of flash flood occurrence, providing critical information for flood risk management for social infrastructure and the public. However, pluvial flood risk management for urban cities needs finer spatial and temporal resolutions than fluvial flood risk management, e.g., up to 1 km in spatial resolution with time series of 5 min in temporal resolution. On the other hand, existing rainfall forecasting research has provided credible estimates for a lead time of up to 2 h with a 10 min interval. Nowcasting has provided a reasonable forecast for 10 to 20 min in convective precipitation. Therefore, radar rainfall data for urban flash flood modelling should be expanded through tactics integrating expert knowledge with hydrological models. As for real-time forecasting applications, the utilization of weather radar data has a quality constraint. The rainfall forecasting techniques should process radar data using statistical and/or artificial intelligence approaches, reducing the high PC time consumption of sophisticated mathematical models used for usual weather forecasts.

Most cities in Southeast Asia have a radar rainfall measurement system. It has a different main purpose for each city using the radar system. For example, in Bangkok, the Department of Drainage and Sewerage (DDS) of the Bangkok Metropolitan Administration (BMA) has a radar station operating to monitor and track storms [49]. The rainfall forecasting system provided by the Thai Meteorological Department uses the weather research and forecasting (WRF) model to forecast hourly and one-day-ahead weather predictions [50]. In Singapore, the Meteorological Service Singapore (MSS), Singapore's national authority on the weather and climate, operates a radar rainfall measurement system with 2 h and one-day weather forecasting [51]. Other countries also provide radar rainfall measurements and weather forecasting data, e.g., Malaysia [52]. Consequently, most Southeast Asian countries, especially in urban cities, operate a radar rainfall measurement system and provide weather and rainfall forecasting data. However, the radar rainfall forecasting data are still limited to the one-sided purpose of weather monitoring and tracking. Therefore, their applications to real-time urban flood forecasting systems, based on physical-process-based/data-driven artificial-intelligence-based models discussed in this paper, have not yet been fully developed and will be a future challenge for urban cities in Southeast Asia.

3. Physical-Process-Based Hydraulic Models for Real-Time Urban Flood Forecasting Systems

3.1. Overview

Many urban storm management models, such as SWMM (Storm Water Management Model), InfoWorks ICM, and MIKE MOUSE, have been developed and applied to urban areas over the last several decades [53]. Because of its ability to completely simulate the urban drainage process, the open-source SWMM model is widely used for urban pluvial flood modelling and urban drainage planning. Commercial software has also been used, such as InfoSWMM and MIKE MOUSE [54]. These urban storm management models simulate the rainfall-runoff process, pipe flows in urban drainage networks, and overflow phenomena over pipe nodes [55]. Computational urban drainage models are beneficial to direct efforts to reduce disaster risk. They allow for the reproduction of historical events for which validation data are available, as well as the simulation of expected changes in climate and land usage while considering possibilities for varying return periods. There are several reviews of urban flood modeling [10,56,57] that include access to input data (such as sewage systems, building locations, topography from GIS databases, and precipitation data), data processing (such as building treatment methods), coupling techniques with overland flow models (such as sewer pipe systems, groundwater models, and green infrastructure), access to validation data for historical occurrences (such as in situ measurements, crowdsourcing, and flow characteristics), and more [58]. Recently, research has emphasized improving the accuracy of urban flood models, and urban flood models' computing time is being reduced. Several techniques are used for improving model accuracy, including topographic data improvements, 1D–2D model connections, and validation data collection.

3.1.1. Model Improvement—Topographic Data Treatment

Several studies have focused on the use of a 2D shallow water model. Mustafa and Szydłowski [59] evaluated building resistance models for predicting water level time series in a diminished experiment with a possible network of buildings in a flood. They confirmed the superiority of the building resistance for simulating flow fields in an urban area. Bermúdez et al. [60] conducted a similar comparison on damage estimation at the micro-scale building level. Arrighi et al. [61] compared flood extents, flow depths, and building damage estimations for a small urban area in Italy using four DEMs from a LiDAR survey. Local differences between the DEMs and the field measurements were generally small, ranging from 0.26 to 2.5 m. Given the great importance of topographic data and its uncertainty analyses, these recent studies have emphasized the importance of combining multiple data sources.

Regarding the topographic data resampling approach, used to shorten the calculation time, the findings typically showed blurry representations of small-scale urban topography features, such as narrow paths between buildings. To emulate the flow patterns of a high-resolution model, Ramsauer et al. [62] solved this problem by implementing virtual surface linkages between structures. The virtual surface linkages greatly enhanced the model's performance when used in a case study including synthetic rainfall. The size, density, and arrangement of the structures in each floodplain continued to determine the improvement.

One study from De Almeida et al. [63] showed that changes in street elevation on the decimetre scale might produce striking variation in anticipated flow statistics. In a UK urban flood, they employed extremely fine-resolution topographical data with a 10 cm precision. Due to the excessively high computing costs and rarity of such small-scale topographic data for practical engineering research, this hampers urban flood modelling. On the other hand, little topographic alterations in carefully selected areas were enough to guide flows to low-impact zones, such as parks. This might create opportunities for flood mitigation; however, as suggested by Yalcin et al. [64], who explored results that differed from those of [63], such conclusions are heavily dependent on the analyzed events and the morphology of the floodplain.

As for 3D modelling approaches, Rong et al. [63] demonstrate the power of 3D computational fluid dynamics compared to traditional 2Ds. When the 3D flow model and a regular DEM were merged, especially in the presence of small streets and intricate terrain features, digital aerial photogrammetry was able to recreate realistic flow patterns in a coastal flood caused by storm surges. Furthermore, using a building information model (BIM) in conjunction with 3D computational fluid dynamics allows the flood simulation to be extended inside buildings.

Using high-resolution GIS terrain data representing finer urban surfaces has enhanced the usefulness of 2D flood inundation models [65]. The high-resolution raster data in GIS (Geographic Information System) successfully replicate micro-scale urban flood characteristics with building and roadway topographic aspects. Therefore, topographic data have been recognized as one of the key contributors to flood simulation accuracy among the different sources of uncertainty in model structure and model parameters [66–68]. A flood inundation region is often defined using topographic data given by a DEM (digital elevation model). The properties of DEM have significant impacts on model performance [68]. Advances in high-resolution data from LiDAR (Light Detection And Ranging) technology have expanded the widespread use of DEM products with various resolutions and vertical precision for numerical flooding modelling [69]. For example, Annis et al. [69] created a new DEM product used as a geospatial tool to define a floodplain extent.

Furthermore, quantitative comparisons revealed that DEMs obtained from data collected using UAVs (unmanned aerial vehicles) could be a suitable alternative to LiDAR-derived products for small-scale flood mapping [70]. High-resolution DEM data are thought to be more accurate in defining the underlying surface and recreating small-scale flow channels, resulting in lower error and uncertainty in simulation results [66,71,72]. Previous research that investigated the variation in flood simulation performance between DEM data products and sources found that the quality and accuracy of the DEM were more important than the DEM resolution [73,74]. Furthermore, fine-resolution models are susceptible to vertical inaccuracies in DEM data, although this sensitivity decreases with resolution [73]. Given the financial problems of manufacturing high-resolution DEMs with a broad spatial extent, particularly in developing countries, e.g., in Southeast Asia, much research has looked into how to identify the best resolution for a given size and processing cost [75]. Prior research, for example, examined the trade-off connection between DEM resolution, vertical accuracy, and simulation error in a wide floodplain [76]. Previous research proved that a 5 m resolution DEM was adequate to offer satisfactory performance for numerical simulation of urban floods on a modest scale, as shown in Figure 2 [77–80].

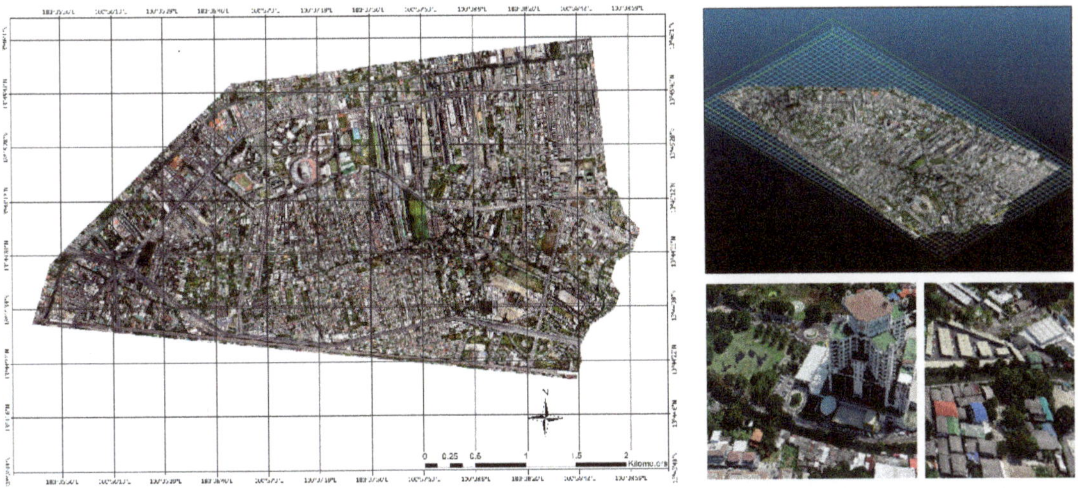

Figure 2. Digital Elevation Model (DEM) of the Ramkhamhaeng polder, Bangkok, Thailand [80].

Previous research on the influence of DEM resolution on surface flood modelling has mostly focused on two types of flood events and scales: urban flash flooding and fluvial overbank flood modelling. Several building treatment approaches have been presented for flood modelling in urban environments, including techniques for increasing resistance, blocks, holes, and porosity [81–84]. Schubert et al. [85,86] summarize a complete comparison of the simulation results obtained for each building treatment approach at different DEM resolutions. As the grid resolution coarsens, the sensitivity of the building block approach increases, resulting in a more extensive underestimated range. As a result, the computational cell size and DEM resolution must be lower than the building scale to effectively describe flood behaviour between buildings and roadways [78]. Chen et al. [87] proposed a unique building feature layer that uses an area/width-reduction factor approach to assess the impact of building storage and resistance, which has been proven to produce an acceptable performance at coarse resolutions.

Surface flooding behaviour has an inverse relationship with DEM resolution, i.e., peak flow becomes lower with a coarser resolution for larger storms but higher for smaller storms depending on the DEM resolution in the sub-catchments [88,89]. Even for the same inundation indicators, studies have found a difference in model performance with coarser DEM resolution. According to Muthusamy et al. [75], inundation depth rises with DEM resolution ranging from 1 to 50 m. However, in the field of 0.1–1 m, Ozdemir et al. [90] discovered that inundation depth decreases with DEM resolution.

The numerical models available to simulate surcharging sewer networks and overland surface flows are listed in [10,57]. Hossain et al. [91] discovered that whether or not stormwater infrastructures were included in the modelling for a pluvial flooding event in the United States changed flood volumes by factors of eight to 20, depending on the return period. The spatial explicitness of soil input data was also a factor in the flood simulations (pervious vs. impervious). The scarcity of data on the location and characteristics of stormwater infrastructures remains a significant challenge for such urban flood simulations.

3.1.2. Accuracy Improvement—Model Validation Data

Another major impediment to flood modelling and management model development is a lack of data for model elaboration, calibration, and validation [92]. Significant efforts were made to retrieve valuable data from unusual sources, such as crowdsourced data to address this issue. Macchione et al. [93] combined amateur videos, photographs, traditional topographic surveys, and news reports, among other sources, to reconstruct hydraulic data from an urban flood event in Italy. They proposed a method of combining various conventional and unconventional information sources. Scotti et al. [94] presented another integrated approach for reconstructing temporal and spatial patterns of flood events. They combined satellite images and markers from social media to effectively calibrate and validate the outcomes of their hydraulic model in the case of Hurricane Harvey in Houston, Texas, in 2017. The authors demonstrated that combining multiple data sources could aid in dealing with the high uncertainties associated with each source of information. Molinari et al. [92] used existing surveillance cameras to gain insights into flood-level evolution. Their image-processing method was based on a deep convolutional neural network trained using over 12,000 flooding images. The method was deemed inexpensive, versatile, scalable, and portable to other sites.

3.1.3. Computational Time Improvement

Improvement of the computational performance is one of the most important issues for real-time urban flood modelling [92]. It helps with larger computational domains, higher spatial resolution, and/or longer time horizons. Moreover, because model outputs should be in real-time, it is especially important in the planning and the forecasting and crisis management phases. Several strategies have currently been investigated to accelerate urban flood calculations. A massive parallelization, an adaptative mesh, a porosity model, and a machine learning application were all presented in recent research.

One method for speeding up urban flood calculations is to use massive parallelization techniques. The use of Graphic Processing Units (GPU) is exceptionally efficient because it allows for the benefit of thousands of processors within a single device. Fernández et al. [95] presented the first GPU implementation of a coupled dual drainage model that included the overland and sewer fullwidth, their exchanges, and pollutant transports. In general, when GPU was introduced to real-world modelling scenarios, it could result in a speed-up factor of several hundred compared to a standard CPU computation.

An alternative for accelerating urban flood computation is a dynamic adaptive meshing technique, allowing the mesh's capacity to adjust flows to evolve in space and time. Hu et al. [96] used the Gmsh algorithm to adjust the mesh every ten time steps using the Hessian matrix of the flow depth and velocity [97]. The technique produced accurate results for a case study while shortening the computation time by a factor of two. Porosity models employed sub-grid modelling to accelerate overland flow simulations. They are classified into "single-" or "dual-porosity" based on the porosity parameters considered, as well as their mathematical representation into "differential" or "integral." These models have shown significant advancements in recent years [98]. In addition, many studies have investigated machine learning approaches to accelerate urban flood computations. A recent review of machine-learning-based urban drainage models [99] highlights the widespread use of data-driven models to improve their performance and efficiency. Machine learning models, in general, are trained using detailed hydrodynamic data, primarily water depths and velocities. The trained models can then predict large-scale flow patterns for situations that differ slightly from the configurations used for training. Two machine learning approaches are based on the origin of the data used to train the model. The first type considers the results of physical-process-based numerical calculations performed for various input conditions such as topography and boundary settings. The second category focuses on data collected on-site and from historical events but does not include any hydraulic calculations [100–102].

3.2. *Current Model Applicability and Its Future Perspective*

Real-time urban flood forecasting systems can provide lifesaving information to residents and emergency services and are valuable for mitigating the devastation of flash flood disasters. Understanding high-risk areas allows emergency responders to prioritize evacuations, being best at the onset of an extreme event [61]. Moreover, climate change accelerates the demand for modelling suitable for real-time applications with advances in high-resolution numerical weather predictions against the increasing frequency of heavy rainfall events. As mentioned above, several approaches have already been proposed to improve computational time so that physical-process-based models can simulate real-time flood forecasting [100]. GPU has also been successfully used to shorten the computation time for both 1D and 2D models [103,104]. Alternative approaches for real-time flood forecasting have been proposed to overcome the difficulties of physical-process-based models, such as simplifying the 2D shallow water equations by omitting inertia terms [105], using cellular automata approaches [106], using simplified, non-physical-based methods [3], and applying empirical/data-driven surrogate models [107]. Furthermore, parallel computing, including cloud servers and code parallelization, has further been used [100]. Despite advances in computer capability and computational efficiency of hydrodynamic models, using these methods for real-time urban flood forecasting remains challenging [3].

Another challenge in developing a real-time urban flood forecasting system is the high-resolution data requirements and the associated setup costs for 1D–2D physical-process-based models. Required data, such as detailed sewer drainage system properties, are frequently unavailable in urban areas in Southeast Asia. A measuring campaign is required before the models are prepared for development. The expensive cost of modelling and computation makes it difficult to establish physical-process-based models with fine-enough spatial and temporal resolutions for real-time urban flood forecasting in many cities in Southeast Asia.

The recent studies on physical-process-based hydraulic models for a real-time flood forecasting system are summarized in Table 1. Existing research on urban hydraulic drainage models may be divided into several categories, including realistic modelling, accuracy improvement, and computation time. Adding new topographic data improved the accuracy of small-scale areas in urban hydraulic drainage models. It retained impact and significance with high-resolution topographic data in 1D and 2D modelling. Existing research on real-time urban flood forecasting using hydraulic models has indicated significant efforts for real-time rainfall data from multiple sources, such as rain gauges, radar, and satellites. The 1D and 2D models are generally connected using the integration of independent software. The most severe barrier to employing physical-process-based models for a real-time flood forecasting system is computation time constraints. They usually give a short lead time, not enough for real-time systems. Several enhancements resulted in a reduced computation time, increased output frequency, and longer lead time.

Table 1. Recent research on physical-process-based models for real-time urban flood forecasting systems.

Country	Area (km^2)	Hydraulic Modelling Method	References
Germany	1.68	SWMM	[108]
France	0.3	SWMM	[109]
Australia	45	MIKE urban	[110]
USA	696.7	SWMM	[111]
Denmark	1.48	MIKE urban	[112]
Brazil	0.11	SWMM	[113]
China	0.017	SWMM, MIKE 21	[114]
USA	0.11	RBC SWMM	[115]
China	141.09	SWMM	[116]
UK	N/A	Shetran	[117]
Nepal	51.94	PCSWMM	[118]
Thailand	11	PCSWMM	[80]

4. Data-Driven Artificial Intelligence (AI) Models for Real-Time Urban Flood Forecasting Systems

4.1. The Limitation of the Existing Modelling Approach

Flood prediction is critical in decision-making and operational strategies, particularly when human lives are at stake [119,120]. The biggest obstacle to building up real-time urban flood forecasting systems is a lead time duration influenced by the rainfall forecasting process [120,121]. To describe the probability space of upcoming rainfall events, short-term and detailed ensemble prediction systems are increasingly being used [122]. The approaches for predicting the location and timing of floods in urban settings are hampered by competing goals such as lead time, model accuracy, application of results, and computing complexity [121]. The application of two-dimensional hydrodynamic models has a critical bottleneck due to the computational time [123]. This issue becomes important according to the fine-scale surface of the urban terrain [124]. Different solutions have been proposed, such as lowering their dimensionality or ignoring the momentum equation's inertial and advection elements [125,126]. However, hydrodynamic models are unsuitable for wider regions since the needed resolution is insufficient for real-time simulations [127], mainly when linked 1D–2D models are utilized [128].

4.2. Data-Driven AI Models

Several data-driven and operational processing approaches increasingly use machine learning [129]. Recently, flood prediction research has begun to employ machine learning algorithms to reduce the computation time required for hydrodynamic calculation [130]. Consequently, machine learning models strive to mimic physical-process-based simulations by learning the target systems independently with their physical links [131]. Artificial neural networks (ANNs) have shown considerable promise in simulating flood-related problems. They demonstrated a decent approximation of non-linear correlations [132]. Also, they provided excellent time series processing utilizing recurrent neural networks (RNNs) [133]. Fully connected ANNs, in particular, were used to forecast flooding characteristics at single coordinates using statistical and topographic inputs such as slope, aspect, and curvatures [134,135]. Another strategy for using fully connected ANNs was proposed by Berkhahn et al. [136]. These algorithms were used to forecast pluvial floods using uniform rainfall occurrences as training inputs.

The basic drawback of fully connected ANNs is the exponential increase in layers and parameters on high-resolution input, presenting severe processing challenges when connecting millions of neighboring raster cells in large 2D simulations [137]. Against this context, deep learning has been favored in recent years, addressing flood-related challenges. A deep learning approach concentrating on river flooding prediction is demonstrated by Zhou et al. [138]. They employed long-short-term memory (LSTM) architectures with a spatial reduction strategy to model time series and eliminate information redundancy in flood inundation data. Convolutional neural networks (CNNs) have shown promise in this application due to their capacity to (i) interpret raw input in picture format and (ii) minimize the number of parameters through the use of partly connected convolution layers and weight sharing [139]. Recent research has addressed the application of CNNs for flood area mapping using aerial or street view images [140] and flood susceptibility mapping using changeable topographic features obtained from raw elevation data [140,141]. The CNNs trained on the outputs of 2D hydraulic models were used by Kabir et al. [142] to forecast the inundation depth induced by river floods.

Existing research on data-driven AI models for real-time urban flood forecasting systems, as listed in Table 2, demonstrated significant reduction in processing time. Forecasting urban floods with a longer lead time offers a practical prediction for a longer time of forecasting, and it is beneficial for operation and mitigation. In addition, geospatial information and historical flood records are used to train the AI model, being a comprehensive urban flood forecasting model. Also, the output of hydraulic models is fundamental for AI model training because they can simulate inundation in very small sub-catchments in metropolitan areas such as streets, buildings, and social communities. CNNs turn high-dimensional data into low-dimensional ones, ensuring quick training procedures and avoiding time-consuming computation modelling.

Table 2. Recent research on data-driven AI models for real-time urban flood forecasting systems.

Country	Area (km^2)	Hydraulic Modelling Method	Data-Driven AI Model	References
France	0.3	SWMM	ANN	[109]
Belgium	0.3	-	ANN	[143]
Norway	15	-	LSTM, GRU	[144]
UK	N/A	-	LASSO, ANN	[145]
South Korea	6.8	-	CNM	[146]
Germany	N/A	-	CNM, I-Tree Canopy method	[147]
UK	21.80	-	URMOD	[148]
Germany	4.0	MIKE 21	GAN	[149]

4.3. Future Challenges for Urban Areas in Southeast Asia

The urban cities in Southeast Asia are facing severe flash flood inundation disasters under climate change and the potential for social infrastructure development. Urbanization and the future development of social infrastructure in the urban areas of Southeast Asia require adjusting the real-time urban flood forecasting systems accordingly. However, the data-driven AI models do not have the flexibility for these changing environments due mainly to the lack of the physical processes of flash floods. On the other hand, the physical-process-based hydraulic models can adjust to the changing circumstances with appropriate topographic and rainfall information from state-of-the-art technologies in geospatial science. In addition, it should be noted again that intense pluvial flash floods due to climate change need real-time forecasting systems with shorter computation and longer lead time. In this sense, the data-driven AI models show an advantage in time management compared with physical-process-based hydraulic models. Thus, one of the future challenges is the development of methodology-blending techniques of these two models for real-time flash flood forecasting systems.

5. Discussion

Weather radar observations offer immense promise in hydrological applications, while they have been underutilized in hydraulic applications. As reviewed in this paper, the current urban real-time flood forecasting systems are implemented through hydroinformatics input data, rainfall forecasting, physical-process-based hydraulic models, and data-driven AI models. This paper's primary focus is making effective use of weather radar's high temporal and spatial resolutions. It is assumed that the desired resolution for small urban catchments should be up to 1 km in space and 5 min in time. Therefore, weather radar nowcasting and forecasting have a high potential capability of providing such high-resolution rainfall input in time and space for implementing real-time flood forecasting systems.

The physical-process-based hydraulic models are essential for a flow analysis based on the 1D drainage network. A 1D hydraulic model is an unquestionably valuable tool for understanding and managing the operations of drainage networks, and its implementation is quick and stable. However, the 1D model indicates its limit when the drainage network overflows onto the urban surface. When such overflows occur, a 2D model should represent inundation flows. The 2D hydraulic models have proved effective and realistic for representing urban floods. Moreover, a 1D–2D coupling approach is considered the most realistic to represent the flow interactions between the drainage network and the urban surface. Nevertheless, the intrinsic peculiarities of urban areas need high-resolution representation, resulting in a considerable computation time unacceptable for real-time systems. Recent research suggests promising advancements in speeding up 2D models [100,103,104], hoping for real-time 2D model applications shortly.

Recent research shows that the finest topographic data provide high accuracy of inundation depths and their locations in an urban area. The grid size of input data is also a significant factor in improving forecasting accuracy. Therefore, real-time urban flood forecasting systems mainly recommend using a higher resolution and frequency of rainfall input data.

The physical-process-based hydraulic models are beneficial in terms of flexibility from future urban reconstruction, drainage infrastructure development, and land use changes. In contrast, longer computation time and limited resources restrict the usefulness of the physical-process-based hydraulic models. Therefore, data-driven AI models could be a promising approach that provides a shorter computational time and requires fewer PC resources. Moreover, the infrastructure data in urban areas of Southeast Asia have not yet been fully available and barely updated. Therefore, data-driven AI models have an advantage over traditional physical-process-based models.

In Southeast Asia, hydrological measurement has a high potential for rainfall forecasting development. On the other hand, the availability of social infrastructure data in Southeast Asia could be challenging for developing real-time urban flood forecasting systems. Therefore, a blended approach of a physical-process-based hydraulic model with a data-driven AI model is recommended for real-time urban flood forecasting systems in most of the urban cities in Southeast Asia, as shown in Figure 3.

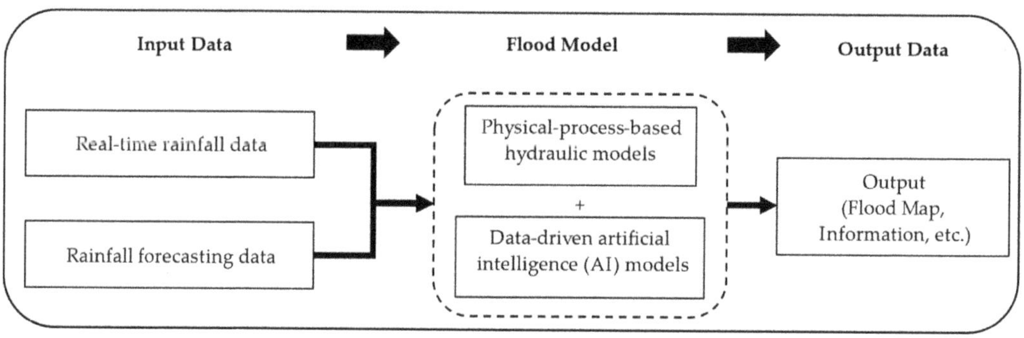

Figure 3. Conceptual diagram of the blended approach of a physical-process-based hydraulic model with a data-driven AI model for real-time urban flood forecasting systems.

Moreover, the accuracy of the real-time system is influenced by two main factors: the accuracy of rainfall forecasting and hydraulic modelling. In addition, the primary in situ records could be beneficial, e.g., the IoT (Internet of Things) water level measurement. Further developments using these techniques are of fundamental importance as a pivotal future path, expecting their approaches to be utilized to a greater extent.

6. Conclusions

This study provided an overview of current cutting-edge modelling tools and their advantages and disadvantages for real-time urban flood forecasting systems. This paper described a standard design for real-time urban flood forecasting systems in terms of three main categories, i.e., hydroinformatics for rainfall data and urban topographic information, physical-process-based hydraulic models for inundation simulations, and data-driven AI models for real-time systems. As for the input from hydroinformatics, the influence of geospatial information and temporal rainfall resolution on the accuracy of urban flood forecasting is highly recognized as an essential study field for urban flood modelling. The state-of-the-art review for real-time urban flood forecasting systems shows that the model should be chosen based on the available data, the urban flood environment, and real-time forecasting demands. Historical statistics and basic black-box models can provide quick and fascinating results when observational data are insufficient. Moreover, it could lead to implementing a data-driven AI model for the future development of real-time urban flood forecasting systems.

In urban areas in Southeast Asia, the future development of real-time urban flood forecasting systems is necessary due to the extreme flash flood trends under climate change and further social infrastructure development. The physical-process-based hydraulic models require extensive data for monitoring and operation, such as gates and pumps for the drainage network infrastructure. Moreover, the data-driven AI models significantly provide a faster output and a shorter computation time. Thus, the hybrid blended approach of physical-process-based and data-driven models have a high potential for the future best practice in developing real-time urban flood foresting systems.

Author Contributions: Conceptualization, D.C. and H.M.; methodology, D.C. and H.M.; investigation, D.C.; data curation, D.C.; writing—original draft preparation, D.C.; writing—review and editing, D.C. and H.M.; visualization, D.C. and H.M.; supervision, H.M. All authors have read and agreed to the published version of the manuscript.

Funding: This research received no external funding.

Institutional Review Board Statement: Not applicable.

Informed Consent Statement: Not applicable.

Data Availability Statement: Not applicable.

Conflicts of Interest: The authors declare no conflict of interest.

References

1. Hammond, M.J.; Chen, A.S.; Djordjević, S.; Butler, D.; Mark, O. Urban Flood Impact Assessment: A State-of-the-Art Review. *Urban Water J.* **2015**, *12*, 14–29. [CrossRef]
2. Schmitt, T.G.; Thomas, M.; Ettrich, N. Analysis and Modeling of Flooding in Urban Drainage Systems. *J. Hydrol.* **2004**, *299*, 300–311. [CrossRef]
3. Teng, J.; Jakeman, A.J.; Vaze, J.; Croke, B.F.W.; Dutta, D.; Kim, S. Flood Inundation Modelling: A Review of Methods, Recent Advances and Uncertainty Analysis. *Environ. Model. Softw.* **2017**, *90*, 201–216. [CrossRef]
4. Avila, L.; Ávila, H.; Sisa, A. A Reactive Early Warning Model for Urban Flash Flood Management. In *World Environmental and Water Resources Congress*; ASCE: Reston, VA, USA, 2017; pp. 372–382. [CrossRef]
5. Ben-Daoud, A.; Ben-Daoud, M.; Moroșanu, G.A.; M'Rabet, S. The Use of Low Impact Development Technologies in the Attenuation of Flood Flows in an Urban Area: Settat City (Morocco) as a Case. *Environ. Chall.* **2022**, *6*, 100403. [CrossRef]
6. Irvine, K.; Sovann, C.; Suthipong, S.; Kok, S.; Chea, E. Application of PCSWMM to Assess Wastewater Treatment and Urban Flooding Scenarios in Phnom Penh, Cambodia: A Tool to Support Eco-City Planning. *J. Water Manag. Model.* **2015**, *23*, C389. [CrossRef]
7. Loc, H.H.; Babel, M.S.; Weesakul, S.; Irvine, K.N.; Duyen, P.M. Exploratory Assessment of SUDS Feasibility in Nhieu Loc-Thi Nghe Basin, Ho Chi Minh City, Vietnam. *Int. J. Environ. Clim. Chang.* **2015**, *5*, 91–103. [CrossRef]
8. Pinos, J.; Quesada-Román, A. Flood Risk-Related Research Trends in Latin America and the Caribbean. *Water* **2022**, *14*, 10. [CrossRef]
9. Sidek, L.M.; Chua, L.H.C.; Azizi, A.S.M.; Basri, H.; Jaafar, A.S.; Moon, W.C. Application of PCSWMM for the 1-D and 1-D–2-D Modeling of Urban Flooding in Damansara Catchment, Malaysia. *Appl. Sci.* **2021**, *11*, 9300. [CrossRef]
10. Nkwunonwo, U.C.; Whitworth, M.; Baily, B. A Review of the Current Status of Flood Modelling for Urban Flood Risk Management in the Developing Countries. *Sci. Afr.* **2020**, *7*, e00269. [CrossRef]
11. Irvine, K. Climate Change and Urban Hydrology: Research Needs in the Developed and Developing Worlds. *J. Water Manag. Model.* **2013**, *21*, R246-11. [CrossRef]
12. Rivard, G.; Rinfret, L.A.; Davidson, S.; Morin, P.L.; Corrales, M.V.; Kompaniets, S. Applying Stormwater Management Concepts in Tropical Countries. *J. Water Manag. Model.* **2006**. [CrossRef]
13. Chaudhary, S.; Chua, L.H.C.; Kansal, A. Modeling Washoff in Temperate and Tropical Urban Catchments. *J. Hydrol.* **2021**, *603*, 126951. [CrossRef]
14. Petheram, C.; Rustomji, P.; Chiew, F.H.S.; Vleeshouwer, J. Rainfall–Runoff Modelling in Northern Australia: A Guide to Modelling Strategies in the Tropics. *J. Hydrol.* **2012**, *462–463*, 28–41. [CrossRef]
15. van Kempen, G.; van der Wiel, K.; Melsen, L.A. The Impact of Hydrological Model Structure on the Simulation of Extreme Runoff Events. *Nat. Hazards Earth Syst. Sci.* **2021**, *21*, 961–976. [CrossRef]
16. Bedient, P.B.; Holder, A.; Benavides, J.A.; Vieux, B.E. Radar-Based Flood Warning System Applied to Tropical Storm Allison. *J. Hydrol. Eng.* **2003**, *8*, 308–318. [CrossRef]
17. García, L.; Barreiro-Gomez, J.; Escobar, E.; Téllez, D.; Quijano, N.; Ocampo-Martinez, C. Modeling and Real-Time Control of Urban Drainage Systems: A Review. *Adv. Water Resour.* **2015**, *85*, 120–132. [CrossRef]

18. Georgakakos, K.P. On the Design of National, Real-Time Warning Systems with Capability for Site-Specific, Flash-Flood Forecasts. *Bull. Am. Meteorol. Soc.* **1986**, *67*, 1233–1239. [CrossRef]
19. Liang, R.; Thyer, M.A.; Maier, H.R.; Dandy, G.C.; Di Matteo, M. Optimising the Design and Real-Time Operation of Systems of Distributed Stormwater Storages to Reduce Urban Flooding at the Catchment Scale. *J. Hydrol.* **2021**, *602*, 126787. [CrossRef]
20. Martin, C.; Russell, S.; Amico, C.; Wagner, L.; James, R.; Vogelsang, T. Integrated Web-Based Flow Monitoring and Hydraulic Modeling in Erie County, New York. *J. Water Manag. Model.* **2008**, *16*, R228-08. [CrossRef]
21. Qi, W.; Ma, C.; Xu, H.; Chen, Z.; Zhao, K.; Han, H. A Review on Applications of Urban Flood Models in Flood Mitigation Strategies. *Nat. Hazards* **2021**, *108*, 31–62. [CrossRef]
22. Maiolo, M.; Palermo, S.A.; Brusco, A.C.; Pirouz, B.; Turco, M.; Vinci, A.; Spezzano, G.; Piro, P. On the Use of a Real-Time Control Approach for Urban Stormwater Management. *Water* **2020**, *12*, 2842. [CrossRef]
23. Parkinson, J.; Mark, O. *Urban Stormwater Management in Developing Countries*; IWA Publishing: London, UK, 2006.
24. Duncan, A.; Chen, A.S.; Keedwell, E.; Djordjevic, S.; Savic, D. *Urban Flood Prediction in Real-Time from Weather Radar and Rainfall Data Using Artificial Neural Networks*; International Association of Hydrological Sciences: Wallingford, UK, 2011.
25. Savina, M.; Schäppi, B.; Molnar, P.; Burlando, P.; Sevruk, B. Comparison of a Tipping-Bucket and Electronic Weighing Precipitation Gage for Snowfall. *Atmos. Res.* **2012**, *103*, 45–51. [CrossRef]
26. Influence of Small Scale Rainfall Variability on Standard Comparison Tools between Radar and Rain Gauge Data—Ecole Des Ponts ParisTech. Available online: https://hal-enpc.archives-ouvertes.fr/hal-00913286 (accessed on 23 October 2022).
27. Ochoa-Rodriguez, S.; Wang, L.-P.; Willems, P.; Onof, C. A Review of Radar-Rain Gauge Data Merging Methods and Their Potential for Urban Hydrological Applications. *Water Resour. Res.* **2019**, *55*, 6356–6391. [CrossRef]
28. Qiu, Q.; Liu, J.; Tian, J.; Jiao, Y.; Li, C.; Wang, W.; Yu, F. Evaluation of the Radar QPE and Rain Gauge Data Merging Methods in Northern China. *Remote Sens.* **2020**, *12*, 363. [CrossRef]
29. Sinclair, S.; Pegram, G. Combining Radar and Rain Gauge Rainfall Estimates Using Conditional Merging. *Atmos. Sci. Lett.* **2005**, *6*, 19–22. [CrossRef]
30. Zhang, Y.; Liu, L.; Wen, H.; Yu, B.; Wang, H.; Zhang, Y. Combined Radar Quality Index for Quantitative Precipitation Estimation of Heavy Rainfall Events. *Remote Sens.* **2022**, *14*, 3154. [CrossRef]
31. Song, Y.; Han, D.; Rico-Ramirez, M.A. High Temporal Resolution Rainfall Information Retrieval from Tipping-Bucket Rain Gauge Measurements. *Procedia Eng.* **2016**, *154*, 1193–1200. [CrossRef]
32. Sokol, Z.; Mejsnar, J.; Pop, L.; Bližňák, V. Probabilistic Precipitation Nowcasting Based on an Extrapolation of Radar Reflectivity and an Ensemble Approach. *Atmos. Res.* **2017**, *194*, 245–257. [CrossRef]
33. Ruzanski, E.; Chandrasekar, V.; Wang, Y. The CASA Nowcasting System. *J. Atmos. Ocean. Technol.* **2011**, *28*, 640–655. Available online: https://journals.ametsoc.org/view/journals/atot/28/5/2011jtecha1496_1.xml (accessed on 23 October 2022). [CrossRef]
34. Johnson, J.T.; MacKeen, P.L.; Witt, A.; Mitchell, E.D.W.; Stumpf, G.J.; Eilts, M.D.; Thomas, K.W. The Storm Cell Identification and Tracking Algorithm: An Enhanced WSR-88D Algorithm. *Weather. Forecast.* **1998**, *13*, 263–276. [CrossRef]
35. Dixon, M.; Wiener, G. TITAN: Thunderstorm Identification, Tracking, Analysis, and Nowcasting—A Radar-Based Methodology. *J. Atmos. Ocean. Technol.* **1993**, *10*, 785–797. [CrossRef]
36. Novak, P. The Czech Hydrometeorological Institute's Severe Storm Nowcasting System. *Atmos. Res.* **2007**, *83*, 450–457. [CrossRef]
37. Reyniers, M. *Quantitative Precipitation Forecasts Based on Radar Observations: Principles, Algorithms and Operational Systems*; RMI Publication: Tampa, FL, USA, 2016.
38. Sokol, Z.; Pesice, P. Nowcasting of Precipitation—Advective Statistical Forecast Model (SAM) for the Czech Republic. *Atmos. Res.* **2012**, *103*, 70–79. [CrossRef]
39. Haiden, T.; Kann, A.; Wittmann, C.; Pistotnik, G.; Bica, B.; Gruber, C. The Integrated Nowcasting through Comprehensive Analysis (INCA) System and Its Validation over the Eastern Alpine Region. *Weather. Forecast.* **2011**, *26*, 166–183. [CrossRef]
40. Sokol, Z.; Zacharov, P. Nowcasting of Precipitation by an NWP Model Using Assimilation of Extrapolated Radar Reflectivity. *Q. J. R. Meteorol. Soc.* **2012**, *138*, 1072–1082. [CrossRef]
41. Foresti, L.; Panziera, L.; Mandapaka, P.V.; Germann, U.; Seed, A. Retrieval of Analogue Radar Images for Ensemble Nowcasting of Orographic Rainfall. *Meteorol. Appl.* **2015**, *22*, 141–155. [CrossRef]
42. Bližňák, V.; Sokol, Z.; Zacharov, P. Nowcasting of Deep Convective Clouds and Heavy Precipitation: Comparison Study between NWP Model Simulation and Extrapolation. *Atmos. Res.* **2017**, *184*, 24–34. [CrossRef]
43. Schmid, W.; Mecklenburg, S.; Joss, J. Short-Term Risk Forecasts of Severe Weather. *Phys. Chem. Earth Part B Hydrol. Ocean. Atmos.* **2000**, *25*, 1335–1338. [CrossRef]
44. Pierce, C.E.; Hardaker, P.J.; Collier, C.G.; Haggett, C.M. GANDOLF: A System for Generating Automated Nowcasts of Convective Precipitation. *Meteorol. Appl.* **2000**, *7*, 341–360. [CrossRef]
45. Roberts, R.D.; Burgess, D.; Meister, M. Developing Tools for Nowcasting Storm Severity. *Weather. Forecast.* **2006**, *21*, 540–558. [CrossRef]
46. Liu, X.; Zhou, K.; Lan, Y.; Mao, X.; Trapp, R.J. On the Construction Principle of Conceptual Models for Severe Convective Weather Forecasting Operations in China. *Weather. Forecast.* **2020**, *35*, 299–308. [CrossRef]
47. Mueller, C.; Saxen, T.; Roberts, R.; Wilson, J.; Betancourt, T.; Dettling, S.; Oien, N.; Yee, J. NCAR Auto-Nowcast System. *Weather. Forecast.* **2003**, *18*, 545–561. [CrossRef]

48. Ba, M.; Xin, L.; Crockett, J.; Smith, S.B. Evaluation of NCAR's AutoNowCaster for Operational Application within the National Weather Service. *Weather. Forecast.* **2017**, *32*, 1477–1490. [CrossRef]
49. Department of Drainage and Sewerage Department of Drainage and Sewerage. Available online: https://dds.bangkok.go.th/aboutus/ (accessed on 26 October 2022).
50. Thai Meteorological Department Thai Meteorological Department. Available online: https://www.tmd.go.th/en (accessed on 26 October 2022).
51. The Meteorological Service Singapore (MSS). Available online: http://www.weather.gov.sg/weather-forecast-24hrforecast/ (accessed on 26 October 2022).
52. Laman Web Rasmi Jabatan Meteorologi Malaysia. Available online: https://www.met.gov.my/ (accessed on 26 October 2022).
53. Mignot, E.; Li, X.; Dewals, B. Experimental Modelling of Urban Flooding: A Review. *J. Hydrol.* **2019**, *568*, 334–342. [CrossRef]
54. Warsta, L.; Niemi, T.J.; Taka, M.; Krebs, G.; Haahti, K.; Koivusalo, H.; Kokkonen, T. Development and Application of an Automated Subcatchment Generator for SWMM Using Open Data. *Urban Water J.* **2017**, *14*, 954–963. [CrossRef]
55. Vojinovic, Z.; Tutulic, D. On the Use of 1D and Coupled 1D-2D Modelling Approaches for Assessment of Flood Damage in Urban Areas. *Urban Water J.* **2009**, *6*, 183–199. [CrossRef]
56. Bulti, D.T.; Abebe, B.G. A Review of Flood Modeling Methods for Urban Pluvial Flood Application. *Model. Earth Syst. Environ.* **2020**, *6*, 1293–1302. [CrossRef]
57. Guo, K.; Guan, M.; Yu, D. Urban Surface Water Flood Modelling—A Comprehensive Review of Current Models and Future Challenges. *Hydrol. Earth Syst. Sci.* **2021**, *25*, 2843–2860. [CrossRef]
58. Rosenzweig, B.R.; Herreros Cantis, P.; Kim, Y.; Cohn, A.; Grove, K.; Brock, J.; Yesuf, J.; Mistry, P.; Welty, C.; McPhearson, T.; et al. The Value of Urban Flood Modeling. *Earth's Future* **2021**, *9*, e2020EF001739. [CrossRef]
59. Mustafa, A.; Szydłowski, M. Application of Different Building Representation Techniques in HEC-RAS 2-D for Urban Flood Modeling Using the Toce River Experimental Case. *PeerJ* **2021**, *9*, e11667. [CrossRef]
60. Bermúdez, M.; Zischg, A.P. Sensitivity of Flood Loss Estimates to Building Representation and Flow Depth Attribution Methods in Micro-Scale Flood Modelling. *Nat. Hazards J. Int. Soc. Prev. Mitig. Nat. Hazards* **2018**, *92*, 1633–1648. [CrossRef]
61. Arrighi, C.; Campo, L. Effects of Digital Terrain Model Uncertainties on High-Resolution Urban Flood Damage Assessment. *J. Flood Risk Manag.* **2019**, *12*, e12530. [CrossRef]
62. Ramsauer, S.; Leandro, J.; Lin, Q. Inclusion of Narrow Flow Paths between Buildings in Coarser Grids for Urban Flood Modeling: Virtual Surface Links. *Water* **2021**, *13*, 2629. [CrossRef]
63. De Almeida, G.; Bates, P.; Ozdemir, H. Modeling Urban Floods at Submeter Resolution: Challenges or Opportunities for Flood Risk Management? *J. Flood Risk Manag.* **2016**, *11*, S855–S865. [CrossRef]
64. Yalcin, E. Assessing the Impact of Topography and Land Cover Data Resolutions on Two-Dimensional HEC-RAS Hydrodynamic Model Simulations for Urban Flood Hazard Analysis. *Nat. Hazards* **2020**, *101*, 995–1017. [CrossRef]
65. Horritt, M.S.; Bates, P.D. Effects of Spatial Resolution on a Raster Based Model of Flood Flow. *J. Hydrol.* **2001**, *253*, 239–249. [CrossRef]
66. Altenau, E.H.; Pavelsky, T.M.; Bates, P.D.; Neal, J.C. The Effects of Spatial Resolution and Dimensionality on Modeling Regional-Scale Hydraulics in a Multichannel River. *Water Resour. Res.* **2017**, *53*, 1683–1701. [CrossRef]
67. Lim, N.J.; Brandt, S.A. Flood Map Boundary Sensitivity Due to Combined Effects of DEM Resolution and Roughness about Model Performance. *Geomat. Nat. Hazards Risk* **2019**, *10*, 1613–1647. [CrossRef]
68. Sharma, V.C.; Regonda, S.K. Two-Dimensional Flood Inundation Modeling in the Godavari River Basin, India—Insights on Model Output Uncertainty. *Water* **2021**, *13*, 191. [CrossRef]
69. Leitão, J.P.; de Sousa, L.M. Towards the Optimal Fusion of High-Resolution Digital Elevation Models for Detailed Urban Flood Assessment. *J. Hydrol.* **2018**, *561*, 651–661. [CrossRef]
70. Annis, A.; Nardi, F.; Petroselli, A.; Apollonio, C.; Arcangeletti, E.; Tauro, F.; Belli, C.; Bianconi, R.; Grimaldi, S. UAV-DEMs for Small-Scale Flood Hazard Mapping. *Water* **2020**, *12*, 1717. [CrossRef]
71. Kim, S.; Shen, H.; Noh, S.; Seo, D.-J.; Welles, E.; Pelgrim, E.; Weerts, A.; Lyons, E.; Philips, B. High-Resolution Modeling and Prediction of Urban Floods Using WRF-Hydro and Data Assimilation. *J. Hydrol.* **2021**, *598*, 126236. [CrossRef]
72. Yu, D.; Lane, S.N. Urban Fluvial Flood Modelling Using a Two-Dimensional Diffusion-Wave Treatment, Part 1: Mesh Resolution Effects. *Hydrol. Process.* **2006**, *20*, 1541–1565. [CrossRef]
73. Md Ali, A.; Solomatine, D.P.; Di Baldassarre, G. Assessing the Impact of Different Sources of Topographic Data on 1-D Hydraulic Modelling of Floods. *Hydrol. Earth Syst. Sci.* **2015**, *19*, 631–643. [CrossRef]
74. Talchabhadel, R.; Nakagawa, H.; Kawaike, K.; Yamanoi, K.; Thapa, B.R. Assessment of Vertical Accuracy of Open Source 30m Resolution Space-Borne Digital Elevation Models. *Geomat. Nat. Hazards Risk* **2021**, *12*, 939–960. [CrossRef]
75. Muthusamy, M.; Casado, M.R.; Butler, D.; Leinster, P. Understanding the Effects of Digital Elevation Model Resolution in Urban Fluvial Flood Modelling. *J. Hydrol.* **2021**, *596*, 126088. [CrossRef]
76. Jung, H.C.; Jasinski, M.F. Sensitivity of a Floodplain Hydrodynamic Model to Satellite-Based DEM Scale and Accuracy: Case Study—The Atchafalaya Basin. *Remote Sens.* **2015**, *7*, 7938–7958. [CrossRef]
77. Gallegos, H.A.; Schubert, J.E.; Sanders, B.F. Two-Dimensional, High-Resolution Modeling of Urban Dam-Break Flooding: A Case Study of Baldwin Hills, California. *Adv. Water Resour.* **2009**, *32*, 1323–1335. [CrossRef]

78. Shen, J.; Tan, F. Effects of DEM Resolution and Resampling Technique on Building Treatment for Urban Inundation Modeling: A Case Study for the 2016 Flooding of the HUST Campus in Wuhan. *Nat. Hazards* **2020**, *104*, 927–957. [CrossRef]
79. Xing, Y.; Chen, H.; Liang, Q.; Ma, X. Improving the Performance of City-Scale Hydrodynamic Flood Modelling through a GIS-Based DEM Correction Method. *Nat. Hazards* **2022**, *112*, 2313–2335. [CrossRef]
80. Chitwatkulsiri, D.; Miyamoto, H.; Irvine, K.N.; Pilailar, S.; Loc, H.H. Development and Application of a Real-Time Flood Forecasting System (RTFlood System) in a Tropical Urban Area: A Case Study of Ramkhamhaeng Polder, Bangkok, Thailand. *Water* **2022**, *14*, 1641. [CrossRef]
81. David, A.; Schmalz, B. A Systematic Analysis of the Interaction between Rain-on-Grid-Simulations and Spatial Resolution in 2D Hydrodynamic Modeling. *Water* **2021**, *13*, 2346. [CrossRef]
82. Akoh, R.; Ishikawa, T.; Kojima, T.; Tomaru, M.; Maeno, S. High-Resolution Modeling of Tsunami Run-up Flooding: A Case Study of Flooding in Kamaishi City, Japan, Induced by the 2011 Tohoku Tsunami. *Nat. Hazards Earth Syst. Sci.* **2017**, *17*, 1871–1883. [CrossRef]
83. Gayer, G.; Leschka, S.; Nöhren, I.; Larsen, O.; Günther, H. Tsunami Inundation Modelling Based on Detailed Roughness Maps of Densely Populated Areas. *Nat. Hazards Earth Syst. Sci.* **2010**, *10*, 1679–1687. [CrossRef]
84. Shen, J.; Tan, F.; Zhang, Y. Improved Building Treatment Approach for Urban Inundation Modeling: A Case Study in Wuhan, China. *Water* **2018**, *10*, 1760. [CrossRef]
85. Schubert, J.E.; Sanders, B.F.; Smith, M.J.; Wright, N.G. Unstructured Mesh Generation and Landcover-Based Resistance for Hydrodynamic Modeling of Urban Flooding. *Adv. Water Resour.* **2008**, *31*, 1603–1621. [CrossRef]
86. Schubert, J.E.; Sanders, B.F. Building Treatments for Urban Flood Inundation Models and Implications for Predictive Skill and Modeling Efficiency. *Adv. Water Resour.* **2012**, *41*, 49–64. [CrossRef]
87. Chen, A.S.; Evans, B.; Djordjević, S.; Savić, D.A. Multi-Layered Coarse Grid Modelling in 2D Urban Flood Simulations. *J. Hydrol.* **2012**, *470–471*, 1–11. [CrossRef]
88. Cao, X.; Lyu, H.; Ni, G.; Tian, F.; Ma, Y.; Grimmond, C.S.B. Spatial Scale Effect of Surface Routing and Its Parameter Upscaling for Urban Flood Simulation Using a Grid-Based Model. *Water Resour. Res.* **2020**, *56*, e2019WR025468. [CrossRef]
89. Ghosh, I.; Hellweger, F.L. Effects of Spatial Resolution in Urban Hydrologic Simulations. *J. Hydrol. Eng.* **2012**, *17*, 129–137. [CrossRef]
90. Ozdemir, H.; Sampson, C.C.; de Almeida, G.A.M.; Bates, P.D. Evaluating Scale and Roughness Effects in Urban Flood Modelling Using Terrestrial LIDAR Data. *Hydrol. Earth Syst. Sci.* **2013**, *17*, 4015–4030. [CrossRef]
91. Hossain Anni, A.; Cohen, S.; Praskievicz, S. Sensitivity of Urban Flood Simulations to Stormwater Infrastructure and Soil Infiltration. *J. Hydrol.* **2020**, *588*, 125028. [CrossRef]
92. Molinari, D.; De Bruijn, K.M.; Castillo-Rodríguez, J.T.; Aronica, G.T.; Bouwer, L.M. Validation of Flood Risk Models: Current Practice and Possible Improvements. *Int. J. Disaster Risk Reduct.* **2019**, *33*, 441–448. [CrossRef]
93. Macchione, F.; Costabile, P.; Costanzo, C.; De Santis, R. Moving to 3-D Flood Hazard Maps for Enhancing Risk Communication. *Environ. Model. Softw.* **2019**, *111*, 510–522. [CrossRef]
94. Scotti, V.; Giannini, M.; Cioffi, F. Enhanced Flood Mapping Using Synthetic Aperture Radar (SAR) Images, Hydraulic Modelling, and Social Media: A Case Study of Hurricane Harvey (Houston, TX). *J. Flood Risk Manag.* **2020**, *13*, e12647. [CrossRef]
95. Fernández-Pato, J.; García-Navarro, P. An Efficient GPU Implementation of a Coupled Overland-Sewer Hydraulic Model with Pollutant Transport. *Hydrology* **2021**, *8*, 146. [CrossRef]
96. Hu, R.; Fang, F.; Salinas, P.; Pain, C.C. Unstructured Mesh Adaptivity for Urban Flooding Modelling. *J. Hydrol.* **2018**, *560*, 354–363. [CrossRef]
97. Geuzaine, C.; Remacle, J.-F. Gmsh: A 3-D Finite Element Mesh Generator with Built-in Pre- and Post-Processing Facilities. *Int. J. Numer. Methods Eng.* **2009**, *79*, 1309–1331. [CrossRef]
98. Dewals, B.; Bruwier, M.; Pirotton, M.; Erpicum, S.; Archambeau, P. Porosity Models for Large-Scale Urban Flood Modelling: A Review. *Water* **2021**, *13*, 960. [CrossRef]
99. Kwon, S.H.; Kim, J.H. Machine Learning and Urban Drainage Systems: State-of-the-Art Review. *Water* **2021**, *13*, 3545. [CrossRef]
100. Jamali, B.; Löwe, R.; Bach, P.M.; Urich, C.; Arnbjerg-Nielsen, K.; Deletic, A. A Rapid Urban Flood Inundation and Damage Assessment Model. *J. Hydrol.* **2018**, *564*, 1085–1098. [CrossRef]
101. Craninx, M.; Hilgersom, K.; Dams, J.; Vaes, G.; Danckaert, T.; Bronders, J. Flood4castRTF: A Real-Time Urban Flood Forecasting Model. *Sustainability* **2021**, *13*, 5651. [CrossRef]
102. Kim, H.I.; Keum, H.J.; Han, K.Y. Real-Time Urban Inundation Prediction Combining Hydraulic and Probabilistic Methods. *Water* **2019**, *11*, 293. [CrossRef]
103. Burger, G.; Sitzenfrei, R.; Kleidorfer, M.; Rauch, W. Parallel Flow Routing in SWMM 5. *Environ. Model. Softw.* **2014**, *53*, 27–34. [CrossRef]
104. Kalyanapu, A.J.; Shankar, S.; Pardyjak, E.R.; Judi, D.R.; Burian, S.J. Assessment of GPU Computational Enhancement to a 2D Flood Model. *Environ. Model. Softw.* **2011**, *26*, 1009–1016. [CrossRef]
105. Bates, P.D. Flood Inundation Prediction. *Annu. Rev. Fluid Mech.* **2022**, *54*, 287–315. [CrossRef]
106. Guidolin, M.; Chen, A.S.; Ghimire, B.; Keedwell, E.C.; Djordjević, S.; Savić, D.A. A Weighted Cellular Automata 2D Inundation Model for Rapid Flood Analysis. *Environ. Model. Softw.* **2016**, *84*, 378–394. [CrossRef]

107. Wolfs, V.; Willems, P. A Data Driven Approach Using Takagi–Sugeno Models for Computationally Efficient Lumped Floodplain Modelling. *J. Hydrol.* **2013**, *503*, 222–232. [CrossRef]
108. Müller, H.; Haberlandt, U. Temporal Rainfall Disaggregation Using a Multiplicative Cascade Model for Spatial Application in Urban Hydrology. *J. Hydrol.* **2018**, *556*, 847–864. [CrossRef]
109. Abou Rjeily, Y.; Abbas, O.; Sadek, M.; Shahrour, I.; Hage Chehade, F. Flood Forecasting within Urban Drainage Systems Using NARX Neural Network. *Water Sci. Technol.* **2017**, *76*, 2401–2412. [CrossRef]
110. Thrysøe, C.; Arnbjerg-Nielsen, K.; Borup, M. Identifying Fit-for-Purpose Lumped Surrogate Models for Large Urban Drainage Systems Using GLUE. *J. Hydrol.* **2019**, *568*, 517–533. [CrossRef]
111. Wang, X.; Kinsland, G.; Poudel, D.; Fenech, A. Urban Flood Prediction under Heavy Precipitation. *J. Hydrol.* **2019**, *577*, 123984. [CrossRef]
112. Lund, N.S.V.; Madsen, H.; Mazzoleni, M.; Solomatine, D.; Borup, M. Assimilating Flow and Level Data into an Urban Drainage Surrogate Model for Forecasting Flows and Overflows. *J. Environ. Manag.* **2019**, *248*, 109052. [CrossRef] [PubMed]
113. Silva, C.D.M.; da Silva, G.B.L. Cumulative Effect of the Disconnection of Impervious Areas within Residential Lots on Runoff Generation and Temporal Patterns in a Small Urban Area. *J. Environ. Manag.* **2020**, *253*, 109719. [CrossRef]
114. Yin, D.; Evans, B.; Wang, Q.; Chen, Z.; Jia, H.; Chen, A.S.; Fu, G.; Ahmad, S.; Leng, L. Integrated 1D and 2D Model for Better Assessing Runoff Quantity Control of Low Impact Development Facilities on Community Scale. *Sci. Total Environ.* **2020**, *720*, 137630. [CrossRef]
115. Li, J. A Data-Driven Improved Fuzzy Logic Control Optimization-Simulation Tool for Reducing Flooding Volume at Downstream Urban Drainage Systems. *Sci. Total Environ.* **2020**, *732*, 138931. [CrossRef]
116. Liu, J.; Shao, W.; Xiang, C.; Mei, C.; Li, Z. Uncertainties of Urban Flood Modeling: Influence of Parameters for Different Underlying Surfaces. *Environ. Res.* **2020**, *182*, 108929. [CrossRef]
117. Birkinshaw, S.J.; O'Donnell, G.; Glenis, V.; Kilsby, C. Improved Hydrological Modelling of Urban Catchments Using Runoff Coefficients. *J. Hydrol.* **2021**, *594*, 125884. [CrossRef]
118. Kc, S.; Shrestha, S.; Ninsawat, S.; Chonwattana, S. Predicting Flood Events in Kathmandu Metropolitan City under Climate Change and Urbanisation. *J. Environ. Manag.* **2021**, *281*, 111894. [CrossRef]
119. Borga, M.; Anagnostou, E.N.; Blöschl, G.; Creutin, J.-D. Flash Flood Forecasting, Warning and Risk Management: The HYDRATE Project. *Environ. Sci. Policy* **2011**, *14*, 834–844. [CrossRef]
120. René, J.-R.; Djordjević, S.; Butler, D.; Madsen, H.; Mark, O. Assessing the Potential for Real-Time Urban Flood Forecasting Based on a Worldwide Survey on Data Availability. *Urban Water J.* **2014**, *11*, 573–583. [CrossRef]
121. Jasper-Tönnies, A.; Hellmers, S.; Einfalt, T.; Strehz, A.; Fröhle, P. Ensembles of Radar Nowcasts and COSMO-DE-EPS for Urban Flood Management. *Water Sci. Technol.* **2018**, *2017*, 27–35. [CrossRef] [PubMed]
122. Zanchetta, A.D.L.; Coulibaly, P. Recent Advances in Real-Time Pluvial Flash Flood Forecasting. *Water* **2020**, *12*, 570. [CrossRef]
123. Henonin, J.; Russo, B.; Mark, O.; Gourbesville, P. Real-Time Urban Flood Forecasting and Modelling—A State of the Art. *J. Hydroinform.* **2013**, *15*, 717–736. [CrossRef]
124. Leandro, J.; Schumann, A.; Pfister, A. A Step towards Considering the Spatial Heterogeneity of Urban Key Features in Urban Hydrology Flood Modelling. *J. Hydrol.* **2016**, *535*, 356–365. [CrossRef]
125. Bates, P.D.; Horritt, M.S.; Fewtrell, T.J. A Simple Inertial Formulation of the Shallow Water Equations for Efficient Two-Dimensional Flood Inundation Modelling. *J. Hydrol.* **2010**, *387*, 33–45. [CrossRef]
126. de Almeida, G.A.M.; Bates, P.; Freer, J.E.; Souvignet, M. Improving the Stability of a Simple Formulation of the Shallow Water Equations for 2-D Flood Modeling. *Water Resour. Res.* **2012**, *48*, W05528. [CrossRef]
127. Courty, L.G.; Pedrozo-Acuña, A.; Bates, P.D. Itzï (Version 17.1): An Open-Source, Distributed GIS Model for Dynamic Flood Simulation. *Geosci. Model Dev.* **2017**, *10*, 1835–1847. [CrossRef]
128. Leandro, J.; Chen, A.S.; Djordjević, S.; Savić, D.A. Comparison of 1D/1D and 1D/2D Coupled (Sewer/Surface) Hydraulic Models for Urban Flood Simulation. *J. Hydraul. Eng.* **2009**, *135*, 495–504. [CrossRef]
129. Shen, C. A Transdisciplinary Review of Deep Learning Research and Its Relevance for Water Resources Scientists. *Water Resour. Res.* **2018**, *54*, 8558–8593. [CrossRef]
130. Chang, F.-J.; Hsu, K.; Chang, L.-C. *Flood Forecasting Using Machine Learning Methods*; MDPI: Basel, Switzerland, 2019; ISBN 978-3-03897-548-9.
131. Chu, H.; Wu, W.; Wang, Q.J.; Nathan, R.; Wei, J. An ANN-Based Emulation Modelling Framework for Flood Inundation Modelling: Application, Challenges and Future Directions. *Environ. Model. Softw.* **2020**, *124*, 104587. [CrossRef]
132. Lin, Q.; Leandro, J.; Gerber, S.; Disse, M. Multistep Flood Inundation Forecasts with Resilient Backpropagation Neural Networks: Kulmbach Case Study. *Water* **2020**, *12*, 3568. [CrossRef]
133. Tealab, A. Time Series Forecasting Using Artificial Neural Networks Methodologies: A Systematic Review. *Future Comput. Inform. J.* **2018**, *3*, 334–340. [CrossRef]
134. Kim, H.I.; Han, K.Y. Urban Flood Prediction Using Deep Neural Network with Data Augmentation. *Water* **2020**, *12*, 899. [CrossRef]
135. Tien Bui, D.; Hoang, N.-D.; Martínez-Álvarez, F.; Ngo, P.-T.T.; Hoa, P.V.; Pham, T.D.; Samui, P.; Costache, R. A Novel Deep Learning Neural Network Approach for Predicting Flash Flood Susceptibility: A Case Study at a High Frequency Tropical Storm Area. *Sci. Total Environ.* **2020**, *701*, 134413. [CrossRef]

136. Berkhahn, S.; Fuchs, L.; Neuweiler, I. An Ensemble Neural Network Model for Real-Time Prediction of Urban Floods. *J. Hydrol.* **2019**, *575*, 743–754. [CrossRef]
137. Guo, Z.; Leitão, J.P.; Simões, N.E.; Moosavi, V. Data-Driven Flood Emulation: Speeding up Urban Flood Predictions by Deep Convolutional Neural Networks. *J. Flood Risk Manag.* **2021**, *14*, e12684. [CrossRef]
138. Zhou, Y.; Wu, W.; Nathan, R.; Wang, Q.J. A Rapid Flood Inundation Modelling Framework Using Deep Learning with Spatial Reduction and Reconstruction. *Environ. Model. Softw.* **2021**, *143*, 105112. [CrossRef]
139. LeCun, Y.; Bengio, Y.; Hinton, G. Deep Learning. *Nature* **2015**, *521*, 436–444. [CrossRef]
140. Moy de Vitry, M.; Kramer, S.; Wegner, J.D.; Leitão, J.P. Scalable Flood Level Trend Monitoring with Surveillance Cameras Using a Deep Convolutional Neural Network. *Hydrol. Earth Syst. Sci.* **2019**, *23*, 4621–4634. [CrossRef]
141. Wang, Y.; Fang, Z.; Hong, H.; Peng, L. Flood Susceptibility Mapping Using Convolutional Neural Network Frameworks. *J. Hydrol.* **2020**, *582*, 124482. [CrossRef]
142. Kabir, S.; Patidar, S.; Xia, X.; Liang, Q.; Neal, J.; Pender, G. A Deep Convolutional Neural Network Model for Rapid Prediction of Fluvial Flood Inundation. *J. Hydrol.* **2020**, *590*, 125481. [CrossRef]
143. Bermúdez, M.; Ntegeka, V.; Wolfs, V.; Willems, P. Development and Comparison of Two Fast Surrogate Models for Urban Pluvial Flood Simulations. *Water Resour. Manag.* **2018**, *32*, 2801–2815. [CrossRef]
144. Zhang, F.; Wang, X.; Guan, J.; Wu, M.; Guo, L. RN-Net: A Deep Learning Approach to 0–2 Hour Rainfall Nowcasting Based on Radar and Automatic Weather Station Data. *Sensors* **2021**, *21*, 1981. [CrossRef] [PubMed]
145. Zhao, W.; Beach, T.H.; Rezgui, Y. Automated Model Construction for Combined Sewer Overflow Prediction Based on Efficient LASSO Algorithm. *IEEE Trans. Syst. Man Cybern. Syst.* **2019**, *49*, 1254–1269. [CrossRef]
146. Dao, D.A.; Kim, D.; Kim, S.; Park, J. Determination of Flood-Inducing Rainfall and Runoff for Highly Urbanized Area Based on High-Resolution Radar-Gauge Composite Rainfall Data and Flooded Area GIS Data. *J. Hydrol.* **2020**, *584*, 124704. [CrossRef]
147. Xu, C.; Rahman, M.; Haase, D.; Wu, Y.; Su, M.; Pauleit, S. Surface Runoff in Urban Areas: The Role of Residential Cover and Urban Growth Form. *J. Clean. Prod.* **2020**, *262*, 121421. [CrossRef]
148. Fidal, J.; Kjeldsen, T.R. Accounting for Soil Moisture in Rainfall-Runoff Modelling of Urban Areas. *J. Hydrol.* **2020**, *589*, 125122. [CrossRef]
149. Hofmann, J.; Schüttrumpf, H. FloodGAN: Using Deep Adversarial Learning to Predict Pluvial Flooding in Real Time. *Water* **2021**, *13*, 2255. [CrossRef]

water

MDPI

Review

Frequency Analysis of Hydrological Data for Urban Floods—Review of Traditional Methods and Recent Developments, Especially an Introduction of Japanese Proper Methods

Chiharu Mizuki [1,2,3,*] and Yasuhisa Kuzuha [1,2]

1 Disaster Mitigation Research Center, Mie University, 1577 Kurima-Machiya, Tsu 514-8507, Japan; kuzuha@crc.mie-u.ac.jp

2 Mie Disaster Mitigation Center, Mie Prefecture and Mie University, 1577 Kurima-Machiya, Tsu 514-8507, Japan

3 Graduate School of Regional Innovation Studies, Mie University, 1577 Kurima-Machiya, Tsu 514-8507, Japan

* Correspondence: mizkichi@crc.mie-u.ac.jp

Abstract: Frequency analysis has long been an important theme of hydrology research. Although meteorological techniques (physical approaches) such as radar nowcasting, remote sensing, and forecasting heavy rainfall events using meteorological simulation models are quite effective for urban disaster prevention, statistical and stochastic theories that include frequency analysis, which are usually used in flood control plans, are also valuable for flood control plans for disaster prevention. Master plans for flood control projects in urban areas often use the concept of T-year hydrological values with a T-year return period. A flood control target is a "landside area that is safe against heavy rainfall or floods with a return period of T years". This review emphasizes discussions of parameter estimation of stochastic models and selection of optimal statistical models, which include evaluation of goodness-of-fit techniques of statistical models. Based on those results, the authors criticize Japanese standard procedures recommended by the central government. Consistency between parameter estimation and evaluation of goodness-of-fit is necessary. From this perspective, we recommend using the maximum likelihood method and AIC, both of which are related to Kullback–Leibler divergence. If one prefers using SLSC, we recommend not SLSC itself but SLSC's non-exceedance probability. One important purpose of this review is the introduction of well-used Japanese methods. Because some techniques that are slightly different from the international standard have been used for many years in Japan, we introduce those in the review article.

Keywords: AIC; frequency analysis; goodness-of-fit; maximum likelihood method; parameter estimation; stochastic model; statistical test

Citation: Mizuki, C.; Kuzuha, Y. Frequency Analysis of Hydrological Data for Urban Floods—Review of Traditional Methods and Recent Developments, Especially an Introduction of Japanese Proper Methods. *Water* **2023**, *15*, 2490. https://doi.org/10.3390/w15132490

Academic Editor: Renato Morbidelli

Received: 10 April 2023
Revised: 19 June 2023
Accepted: 4 July 2023
Published: 7 July 2023

1. Introduction

First, we would like to emphasize that parameter-estimation processes and processes for selecting the optimal probability distribution are the most important processes in hydrological frequency analyses. Therefore, we focus on only these techniques in this review article. Nevertheless, we have used similar techniques for the past three or four decades. We cannot propose optimal and decisive techniques that numerous researchers think are the optimal techniques.

For preventing water-related disasters, flood control plans are usually made for large rivers. In Japan, main rivers are designated as "Class-1 rivers" in principle, managed by the central government, or as "Class-2 rivers" managed by local governments. A certain numerical goal is set in a flood control plan, for which the jurisdictional government has a responsibility to protect people, residences, and other properties in the river basin.

According to Nakamura [1], such numerical goals are set using two methods. Japan, the Netherlands, the Philippines, and other countries have adopted stochastic goals: T-year hydrological values with a return period of T years. The United States, China, and other nations have adopted historical maximum values. This review specifically examines the former case. For the former case, the government estimates T-year hydrological values. The estimation processes are divisible mainly into methods of two kinds: non-parametric and parametric methods. Takara [2] described that non-parametric methods can be adopted when the sample size is sufficiently large.

Non-parametric methods are likely to be superior to parametric methods because they use no specific probability distribution: neither a parameter nor an optimal probability distribution need to be selected. By contrast, using the parametric methods, one must estimate parameters and select the optimal probability distribution. Selecting parameters and probability distribution processes include subjective judgments. If one uses parametric methods, subjective judgment must be eliminated to the greatest degree possible. The "Japanese MLIT (Ministry of Land, Infrastructure, Transport and Tourism) flow chart" described later includes some subjective judgment. Therefore, the authors are critical of the method. One should not refer to the flow chart by Japanese MLIT.

The next chapter briefly presents international standard procedures used for hydrological frequency analysis. Because some techniques used in Japan are slightly different from the international standard, we introduce those in Section 3. Techniques described in Sections 2 and 3 are those which have been used for many years. Section 4 presents other techniques developed in recent years. Subsequently, we introduce some future perspectives.

2. International Standard Procedure

The World Meteorological Organization (WMO) published its "Guide to Hydrological Practices (WMO-No. 168 fifth edition)" [3] in 1994. One chapter has the title "Frequency analysis (Chapter 27)". The chapter includes the statement that "hydrological phenomena that are commonly described by frequency analysis are storm precipitation and annual flood maxima". They present 16 probability distributions that are commonly used in hydrology. These include a lognormal distribution, Pearson type three distribution, Gumbel distribution, general extreme value distribution, and others, which have been used for hydrologically extreme values. The sixth edition of the guide [4] was published later, and the Kolmogorov–Smirnov test, the probability plot correlation test, AIC, and BIC were introduced. Those are related to the goodness-of-fit test. Moreover, the L-moment method was also mentioned in the guide.

In the "Handbook of Hydrology" [5], one chapter has the title "Frequency Analysis of Extreme Events". As a parameter estimation method, the authors first introduced the method of moments (MOM), the method of L-moments, and maximum likelihood. They describe that maximum likelihood estimators (MLEs) have very good statistical properties for large samples. Experience has shown that they generally perform well with data from records available from hydrology studies, but experience has also shown that MLEs often cannot be reduced to simple formulas. Regarding the selection of the optimal probability distribution, the authors described goodness-of-fit tests and L-moment diagrams. The textbook introduces the Kolmogorov–Smirnov test, the probability plot correlation coefficient test, L-moment diagrams [6], and ratio tests.

Rao and Hamed [7] explicitly described the selection of distributions. After reviewing many reports of the literature, including reports by Hazen [8], Markovic [9], Gupta [10], McCuen and Rawls [11], McCuen [12], Campbell and Sidel [13], Turkman [14], Vogel [15], Vogel and McMartin [16], Haktanir [17], Bobee et al. [18], and Onoz and Bayazit [19], they expounded the chi-square test, Kolmogorov–Smirnoff test, and Akaike's Information Criterion (AIC) [20]. Using these three methods, they described that probability distributions for flood frequency analysis had been selected.

3. Japanese History of Estimating T-Year Hydrological Value

As described in Section 2, Akaike proposed the information criterion—AIC [20]. Moreover, many researchers developed their own statistical hydrological theories. We have an impression that some hydrological procedures used in Japan differ somewhat from international standard procedures. Some effective theories might not be known worldwide because they have been published only in Japanese-language journals.

In Japan, the main class-1 rivers are managed by MLIT. An organization related to MLIT published some manuals [21–23] in which they explained river plan production.

3.1. Iwai Method for Parameter Estimation of a Three-Parameter Lognormal Distribution

Iwai [24] proposed their method, which belongs to the "quantile method" type and is used for parameter estimation of the three parameters lognormal distribution. The so-called "Slade type [25] of lognormal distribution" has a bounded probability distribution function. Iwai used "Slade type II" described below. His method consists of the estimation of parameters of the three-parameter lognormal distribution. First, one can define a cumulative distribution function $F(x)$ as explained below. In Equation (1), ξ is designated as "reduced variate" (Equation (2)).

$$F(x) = \frac{1}{\sqrt{\pi}} \int_{-\infty}^{\xi} \exp\left(-t^2\right) dt \quad (1)$$

$$\xi = \alpha \log_{10} \frac{x+b}{x_0+b} \quad (2)$$

This function for the lognormal distribution has three parameters: α, x_0, and b. Additionally, −b is a lower bound ($x > -b$). After Kadoya [26] proposed a modification of the original Iwai method, the modified Iwai method has come to be used in most cases. Therefore, we intend to present the "modified Iwai method" herein.

Presuming that there are extreme data with sample size n, then they are annual maxima data. We present these samples as x_n (i = 1, 2, 3, ... , n), which is the ascending order of statistics.

A. Approximation of x_0

First, we use Equation (3) to estimate x_g(approximation of x_0)

$$\log_{10} x_g = \frac{\sum_{i=1}^{n} \log_{10} x_i}{n} \quad (3)$$

B. Estimation of b and x_0

First, we produce $b_{(s)i}$ $(i = 1, 2, 3, \ldots m)$ values (Equation (4)). Integer m is the nearest integer to $n/10$.

$$b_{(s)i} = \frac{x_i x_{n-i+1} - {x_g}^2}{2x_g - (x_i + x_{n-i+1})} \quad (4)$$

Then, b is estimated using the following equation.

$$\hat{b} = \frac{1}{m} \sum_{i=1}^{m} b_{(s)i} \quad (5)$$

By defining $X_i = \log_{10}(x_i + b)$, $\hat{x_0}$ can be estimated by solving the following Equation (6).

$$\log_{10}(\hat{x_0} + b) = \frac{1}{n} \sum_{i=1}^{n} \log_{10}(x_i + b) = \frac{1}{n} \sum_{i=1}^{n} X_i \quad (6)$$

In this Equation, $\hat{x_0}$ is the estimate of x_0; $\hat{b}$ is obtained using Equation (5), and is substituted for b in Equation (6).

C. Final process: Estimation of α.

α is estimated by solving the following equation.

$$\frac{1}{\hat{\alpha}} = \sqrt{\frac{2n}{n-1}}\sqrt{\overline{X^2} - (\overline{X})^2} \tag{7}$$

In Equation (7), $\overline{X} = \frac{1}{n}\sum_{i=1}^{n} X_i$ and $\overline{X^2} = \frac{1}{n}\sum_{i=1}^{n} X_i^2$.

3.2. Ishihara–Takase Method for Parameter Estimation of Three-Parameter Lognormal Distribution

Ishihara and Takase proposed their method [27], which belongs to the "moment method" type. Their method, similar to the Iwai method, estimates the parameters of the three-parameter lognormal distribution. Although a natural logarithm can be used instead of a common logarithm, we use Equation (2) for assigning priority to uniformity with the Iwai method described above.

First, we calculate the sample average $\overline{x}$, standard deviation s, and coefficient of skewness C_{S1}. These are estimated using Equations (8)–(10) presented below.

$$\overline{x} = \frac{1}{n}\sum_{i=1}^{n} x_i \tag{8}$$

$$s^2 = \frac{1}{n-1}\sum_{i=1}^{n}(x_i - \overline{x})^2 = \frac{n}{n-1}\left(\overline{x^2} - (\overline{x})^2\right), s = \sqrt{s^2} \tag{9}$$

$$C_{S1} = \sum_{i=1}^{n}(x_i - \overline{x})^3 / s^3 / (n-1) \tag{10}$$

The parameters that must be estimated are α, b, and x_0. Ishihara and Takase concluded that α is estimated using the following Equation (11).

$$\begin{aligned} k &= 1/\left(\sqrt{2}\sqrt{\ln\left[-1 + \frac{2^{1/3}}{\left(2+C_S{}^2+\sqrt{4C_S{}^2+C_S{}^4}\right)^{1/3}} + \frac{\left(2+C_S{}^2+\sqrt{4C_S{}^2+C_S{}^4}\right)^{1/3}}{2^{1/3}}\right]}\right) \\ \alpha &= k \cdot \ln 10 \end{aligned} \tag{11}$$

The reason for using k is that their original paper adopted natural instead of common logarithms in Equation (2); k is a parameter for the case of using natural logarithms. Furthermore, C_S is not C_{S1} in Equation (10). Therein, C_{S1} is biased; C_S is corrected when using correction factor F_{C_S} in Equation (12).

$$C_S = C_{S1}(1 + F_{C_S}) \tag{12}$$

As for the correction factor F_{C_S}, Ishihara and Takase showed it using a figure. One can obtain F_{C_S}, which is a function of sample size n and C_{S1}, using their figure, which is well-known as Ishihara–Takase's figure. However, calculating F_{C_S} by PC can be performed more easily than ever using the following procedure. Therefore, we recommend that analysts calculate F_{C_S} by themselves.

A. Estimating tentative k and α using Equation (11)

First, we estimate k and α. In Equation (11), C_{S1} is substituted into C_S. C_{S1} is calculated using Equation (10); sample x_i represents observed data.

B. Generating ξ_i $(i = 1, 2, 3, \ldots. n)$

According to Hazen's plotting position formula (as for plotting position formula, see [8,28,29]), $F_i(i = 1, 2, 3, \ldots. n$, which is the probability of non-exceedance) is calculated.

Additionally, ξ_i is calculated by the inverse function of Equation (1) as $\xi(F)$. Hazen's plotting position formula is the following, where i is the order of ascending-order statistics:

$$F_i = \frac{2i-1}{2n} \tag{13}$$

The method for obtaining $\xi(F)$ using the inverse function depends on the software used. Equation (1) can be written as $F(\xi) = \{1 + Erf(\xi)\}/2$, where $Erf(\cdot)$ is the error function. Therefore, an inverse function of it can be expressed as the following Equation (14).

$$\xi(F) = Erf^{-1}(2F-1) \tag{14}$$

y_i is obtained using an inverse function of Equation (2): —Equation (15). Then we can use $x_0 = 1$, $b = 0$ for simplicity of calculation.

$$\begin{aligned} y(\xi) &= 10^{\xi/\alpha}(x_0+b) - b \\ &= e^{\xi/k}(x_0+b) - b \end{aligned} \tag{15}$$

C. Calculating $C^*{}_{S1\_y}$ and $C^*{}_{S\_y}$ of samples

We can calculate $C^*{}_{S1\_y}$ using y_i, which is the coefficient of skewness not of x_i but of y_i by Equation (10). Then, we obtain the theoretical coefficient of skewness $C^*{}_{S\_y}$ using the following Equation (16). When one calculates $C^*{}_{S\_y}$, k is the value estimated by using sample x_i first.

$$C^*{}_{S\_y} = \frac{\exp(9/(4k^2)) - 3\exp(5/(4k^2)) + 2\exp(3/(4k^2))}{(\exp(1/(k^2)) - \exp(1/(2k^2)))^{3/2}} \tag{16}$$

As a result, $F_{C_S} = C^*{}_{S\_y}/C^*{}_{S1\_y} - 1$ is obtained.

D. Calculating three parameters

Using the corrected coefficient of skewness, k or α is obtained. Then b and x_0 are estimated using the following equations (Iwai and Ishiguro [30]).

$$\begin{aligned} \lambda &= \exp(1/(4k^2)) \\ b &= \tfrac{1}{\sqrt{\lambda^2-1}}\sigma - \overline{x} \\ x_0 &= \overline{x} - \tfrac{\lambda-1}{\lambda\sqrt{\lambda^2-1}}\sigma \end{aligned} \tag{17}$$

In Equation (17), $\overline{x}$ and σ, respectively, denote the average and standard deviation of the observed sample x_i.

3.3. Etoh's Distribution

Etoh et al. proposed the probability distribution for extreme values (Etoh et al. [31]). The cumulative distribution function of Etoh's distribution, which has the two parameters a and b, is the following. This probability density function has a heavy tail.

$$F(x) = \begin{cases} \exp\left\{-a\left(1+\sqrt{bx}\right)\exp\left(-\sqrt{bx}\right)\right\} & (x \geq 0) \\ 0 & (x < 0) \end{cases} \tag{18}$$

Although the following probability density function (Equation (19)) has been used, $F(0)$ (left-hand limit) is e^{-a}. It is not zero in accordance with Equation (18). Therefore,

Hayashi et al. [32] proposed the modified function as Equation (20), where $\delta(x)$ represents Dirac's delta.

$$f(x) = \frac{ab}{2}\exp\left\{-\sqrt{bx} - a\left(1+\sqrt{bx}\right)\exp\left(-\sqrt{bx}\right)\right\}\ (x \geq 0) \tag{19}$$

$$f(x) = \begin{cases} \frac{ab}{2}\exp\left\{-\sqrt{bx} - a\left(1+\sqrt{bx}\right)\exp\left(-\sqrt{bx}\right)\right\} + \delta(x)\exp(-a) & (x \geq 0) \\ 0 & (x < 0) \end{cases} \tag{20}$$

Because $\exp(-a)$ is usually small, however, the use of Equation (19) is adequate.

As a parameter estimation method, we usually use the maximum likelihood method. Etoh et al. [31] and Hoshi [33] recommend the following procedure. The log-likelihood of this probability distribution is presented as Equation (21).

$$\begin{aligned} L(a,b) &= \sum_{j=1}^{N} \ln f(x_j) \\ &= N\ln a + N\ln b - N\ln 2 - \sum_{j=1}^{N}\sqrt{bx_j} \\ &-a\left[\sum_{j=1}^{N}\exp\left(-\sqrt{bx_j}\right) + \sum_{j=1}^{N}\sqrt{bx_j}\exp\left(-\sqrt{bx_j}\right)\right] \end{aligned} \tag{21}$$

By solving $\frac{\partial L}{\partial b} = 0$, we can obtain a, which is a function of b, as the following Equation (22), which is referred to as a_1.

$$a_1 = \frac{\sum_{j=1}^{N}\sqrt{bx_j} - 2N}{\sum_{j=1}^{N} bx_j\exp\left(-\sqrt{bx_j}\right)} \tag{22}$$

Then, substituting a_1, obtained by Equation (22) into Equation (21), $L(a,b)$ is modified to $L(b)$. Finally, we seek the largest $L(b)$—the optimal b in some way. Kubota [34] proposed the following procedure. Solving $\frac{\partial L}{\partial a} = 0$, one can obtain a (designated as "a_2"), which is a function of b, from Equation (23).

$$a_2 = \frac{N}{\sum_{j=1}^{N}\exp\left(-\sqrt{bx_j}\right) + \sum_{j=1}^{N}\sqrt{bx_j}\exp\left(-\sqrt{bx_j}\right)} \tag{23}$$

The solution of a can be obtained by minimizing $h(b) = |a_1 - a_2|$ [33], which can be performed easily using software such as Mathematica [34]. In Japan, Etoh's distribution is thought to be appropriate for extreme values data. Kuzuha and Mizuki [35] applied several probability distributions to 42,500 pieces of annual maximum one-hour rainfall data whose sample size is 60. They reported that Etoh's distribution was most appropriate for 37% of the 42,500 data. Two-parameters lognormal distribution was most appropriate for 42%, and the Gumbel distribution was most appropriate for 14%.

3.4. *Approach Proposed by Tsuchiya and Takeuchi*

Although Etoh's distribution is quite an effective probability distribution, the L-moment solution has not been known. This probability distribution was not described by Hosking and Wallis [6] because this probability distribution is not well-known internationally.

Tsuchiya et al. [36] (see also Kuzuha [37]) presented the PWM solution of this probability distribution as follows. Their solution was obtained using numerical procedures, but the method is simple.

Specifically, we can estimate the parameters using the following procedure.

$$\beta_r = \int_0^1 x(F)F^r dF = \int_0^\infty x(F)F^r f(x)dx \tag{24}$$

$$\begin{aligned} M_{1,0,0} &= \beta_0 = \textstyle\int_0^\infty xf(x)dx = \frac{ab}{2}\int_0^\infty x\exp\left\{-\sqrt{bx} - a\left(1+\sqrt{bx}\right)\exp\left(-\sqrt{bx}\right)\right\}dx \\ M_{1,1,0} &= \beta_1 = \textstyle\int_0^\infty xF(x)f(x)dx \\ &= \textstyle\frac{ab}{2}\int_0^\infty x\exp\left\{-\sqrt{bx} - 2a\left(1+\sqrt{bx}\right)\exp\left(-\sqrt{bx}\right)\right\}dx \end{aligned} \tag{25}$$

Equations (24) and (25) indicate the first-order and second-order probability weighted moments. Equation (26) presents the sample probability weighted moments.

$$\begin{aligned} \widehat{M}_{1,0,0} &= b_0 = \tfrac{1}{n}\sum_{i=1}^{n} x_i \\ \widehat{M}_{1,1,0} &= b_1 = \tfrac{1}{n}\sum_{i=1}^{n} x_i \tfrac{i-1}{n-1} \end{aligned} \tag{26}$$

As Tsuchiya and Takeuchi reported [36], $M_{1,1,0}/M_{1,0,0}$ is independent of b; it is a function of only a. Therefore, we can ignore b and can set $b = 1$ as the following Equation (27).

$$\frac{\int_0^\infty x\exp\{-\sqrt{x} - 2a(1+\sqrt{x})\exp(-\sqrt{x})\}dx}{\int_0^\infty x\exp\{-\sqrt{x} - a(1+\sqrt{x})\exp(-\sqrt{x})\}dx} = \frac{b_1}{b_0} \tag{27}$$

a is obtained by numerically solving Equation (27).

Finally, estimation of b is obtained by numerical solution of Equation (28) after substituting the $\widehat{a}$ obtained into a.

$$\widehat{M}_{1,0,0} = \frac{ab}{2}\int_0^\infty x\exp\left\{-\sqrt{bx} - a\left(1+\sqrt{bx}\right)\exp\left(-\sqrt{bx}\right)\right\}dx \tag{28}$$

Furthermore, we would like to mention the following facts. Takeuchi and Tsuchiya reported the PWM solution of the normal distribution [38], a lognormal distribution, and Pearson type three distribution [39]. Because their findings were published in a Japanese journal, they have not become well-known internationally, but they found their solution ahead of the international hydrological community.

3.5. Ueda–Kawamura's Criterion for Evaluating Goodness-of-Fit

Ueda and Kawamura [40] proposed a criterion to evaluate the goodness-of-fit of a probability model. Although many textbooks have recommended the evaluation of the validity of a probability model based on probability studies, it is difficult to evaluate their validity quantitatively. They sought to quantitively evaluate the probability model's goodness-of-fit, as explained below.

A. Presuming sample data with size n and that have ascending order statistics, then using the plotting position formula, the non-exceedance probability $F_P(x_i)$ is estimated. Several plotting position formulas are expressed as Equation (29).

$$F_P(x_i) = \frac{i-\alpha}{n+1-2\alpha-\beta} \tag{29}$$

For example, for Cunnan's formula [29], α is 0.4 and β is 0.

B. If the cumulative distribution function of the probability model is $F(x)$, then, of course, the non-exceedance probability is $F(x_i)$.

C. Ueda and Kawamura plot $(F(x_i), F_P(x_i))$ on a graph with the normal axis. The minimum and maximum of both axes are 0 and 1. From the viewpoint of goodness-of-fit, the data shown are near the line of $y = x$.

Ueda and Kawamura proposed the use of the χ^2 test as a goodness-of-fit test. As a result, the χ^2 value of each probability distribution is a candidate "fair criterion" when choosing a probability distribution.

3.6. Takasao–Takara's SLSC for Evaluating Goodness-of-Fit

Takasao et al. [41] proposed the standard least-squares criterion for goodness of fit (SLSC). This criterion evaluates goodness-of-fit by linearity on a probability plotting paper. The SLSC is expressed as the following Equation (30).

$$\mathrm{SLSC} = \frac{\sqrt{\sum_{i=1}^{n} (s_i - s^*{}_i)^2 / n}}{|s_{0.99} - s_{0.01}|} \tag{30}$$

In Equation (30), s is a reduced variate and calculated according to Equation (31), where ξ and α, respectively, denote the location and scale parameter.

$$s = (x - \xi)/\alpha \tag{31}$$

$$x_i{}^* = x(F(x_i)) \tag{32}$$

The value of $x_i{}^*$ is calculated using Equation (32); in addition, $s^*{}_i$ is transformed from $x_i{}^*$ by Equation (31). One can assume a probability plotting paper with a horizontal axis x and vertical axis s. (x_i, s_i) is on a linear line because of the definition. However, $(x_i, s^*{}_i)$ is plotted nearly as a straight line but not on the line: SLSC is the mean distance between a straight line and $(x_i, s^*{}_i)$. That is, SLSC evaluates the mean distance which is the degree of separation of the probability model from the sample, not by vision but by values.

Takasao et al. used a denominator of the right side of Equation (30) to maintain the fairness of the criterion. They regarded vertical scales of the probability plotting paper of each probability distribution as corrected to the same scale, divided by the denominator. As Kuzuha [42] and Hayashi et al. [43] found and Kuzuha et al. [35] [44–46] later examined in detail, however, it is not true. That point is explained in the next section.

3.7. Procedure for Parameter Estimation and Choosing the Probability Distribution of the Japan Ministry of Land, Infrastructure, Transport, and Tourism

For estimating long-term stochastic hydrological values (e.g., 100-year precipitation whose return period is 100 years), THE MLIT used their own flow chart for parameter estimation and for choosing an optimal probability distribution [21]. Recently, Kuzuha and Mizuki criticized the flow chart. The flow chart has several shortcomings, but it is wholly inappropriate for three main reasons:

(1) The most important process of the MLIT flow chart is THE evaluation of goodness-of-fit by SLSC and the evaluation of variability by resampling technique for each probability distribution: some candidates are first chosen for the optimal probability distribution by SLSC. However, the authors found that SLSC is not valid from the perspective of fairness among probability distributions. An unfair referee should not judge the match.

(2) In the MLIT flow chart, the probability distribution with the smallest variability is thought to be the optimal one among the candidates selected above. They regard the probability distribution having the least variability of T-year values as optimal. They use three criteria in the flow chart: "at parameter-estimation process", "at process of selecting the optimal", and "at evaluating variability".

(3) The criteria of the least likelihood method for parameter estimation and AIC are related to Kullback–Leibler divergence [47]. If they use the L-moment method (or the

"conventional" moment method), they assign importance to the coincidence of the L-moment (moment) between the model and data.

(4) Work by Tanaka and Takara [48] probably affected the MLIT flow chart the most. Tanaka and Takara mentioned that "if SLSC is less than 0.04, we regard that the probability distribution's goodness-of-fit as sufficient. If using 0.03 for the threshold, most probability distributions are evaluated as inappropriate from the viewpoint of the goodness-of-fit. Then, we use 0.04 as the threshold". The authors have criticized this rationale as it is not scientific. It is for the convenience of administration—the Japan MLIT.

3.8. Current Best Practice

We think that consistency is extremely important between the processes of parameter estimation and choosing the probability distribution. In this context, "consistency" means using the same or similar criterion for parameter-estimation and evaluation of goodness-of-fit. Moreover, we believe that "evaluating variability in MLIT flow chart" is not necessary. Let us explain the reason in detail. The most important is that the criterion for evaluating the goodness-of-fit is a fair criterion from the perspective of comparing probability distributions. Because we compare a goodness-of-fit-measure of each probability distribution and select the optimal probability distribution, fairness is most important. From this perspective, SLSC is not a fair measure at all.

Suppose that an analyst uses the maximum likelihood method for parameter estimation and that they estimate parameters of an A-probability distribution and a B-probability distribution. Moreover, suppose that the analyst chooses Takasao–Takara's criterion (SLSC) for selecting the optimal probability distribution. Parameters are selected to maximize the likelihood. Then, the A-probability distribution and B-probability distribution are compared. If the SLSC of the A-probability distribution is smaller than that of the B-probability distribution, the A-probability is selected as the optimal distribution. This poses a big problem since there is a possibility that other parameter sets are selected, and the B-probability distribution is selected as the optimal distribution if parameters are selected to minimize SLSC. This is the reason why we insist that the consistency of measure for parameter-estimation and evaluating goodness-of-fit is quite important.

According to the arguments presented above, using the maximum likelihood method for parameter estimation and using AIC for testing goodness-of-fit are recommended procedures. The main reason is that both are related to Kullback–Leibler divergence [46]. As described in Section 3.7, Tanaka and Takara's explanation [47] for the threshold (=0.04) is inappropriate. However, one can understand the difficulty of policymakers in government agencies in changing their methods quickly to align with an academic perspective. Therefore, we presented some issues related to the conventional method in earlier reports [35,44,45].

A. We recommend using the maximum likelihood method and AIC (or TIC, etc.).

B. If an analyst prefers using SLSC, then we recommend not using SLSC itself but SLSC's non-exceedance probability $F(SLSC)$. For calculating $F(SLSC)$, one must know SLSC's probability distribution function. Hayashi et al. [43] and Kuzuha and Mizuki [35,44] demonstrated how to obtain the SLSC's probability density function using Monte Carlo simulation.

C. If an analyst uses the SLSC's non-exceedance probability, then they can evaluate the goodness-of-fit of each probability distribution, even if SLSC is not a fair criterion. That procedure can be applied to any criterion, even if the criterion is not a fair one from the viewpoint of comparing the degrees of goodness-of-fit.

4. Novel Techniques and Future Perspectives

In 2004, Gelder [49] described some well-known techniques for parameter estimation: the method of moments (MOM), maximum likelihood estimation (MLE), least squares, Bayesian estimation, minimum cross-entropy, probability weighted moments (PWMs), and

L-moments. More recent reports, such as that by Yuan et al. (2018) [50], described the adoption of the so-called MOM for parameter estimation. Langat et al. (2019) [51] adopted MLE after reviewing some techniques. Those are MOM, L-moments, LH moments [52], and the expected moments algorithm (EMA). Anghel and Ilinca (2022) [53] used both MOM and L-moments for parameter estimation.

Coles [54] and Hayashi et al. [43] considered non-stationary hydrological models. Hayashi et al. discussed non-stationary hydrological frequency models introducing time-dependent parameters. Their report recommended the use of MLE for parameter estimation. Langat et al. commented on the method of Bayesian estimation: "although there are drawbacks of complexity in its implementation in present time, it might become a useful non-stationarity flood frequency analysis model in the future, with advancements in technology".

Yuan et al. (2018) [50] described that "the choice of an appropriate PDF is still one of the major issues in engineering practice because there is no general agreement as to which distribution could be used for the frequency analysis of extreme rainfalls". They adopted the chi-square test for selecting the optimal probability distribution. Langat et al. [51] introduced the Kolmogorov–Smirnov, Anderson–Darling, and Cramer–Von Mises tests in addition to the chi-square test.

Most techniques described above have a long history; quite attractive and novel techniques that have become a new international standard have not been proposed in recent years. Nevertheless, because hydrological frequency analyses that use non-stationary hydrological data have become increasingly important in light of drastic climate change, non-stationary analyses have become ever more necessary. Some techniques are useful for non-stationary analyses. The maximum likelihood method and AIC, TIC, or BIC, which are related to Kullback–Leibler divergence [47], are expected to be crucially important in the research area. In addition, the method of Bayesian estimation might be particularly effective.

5. Conclusions

We reviewed statistical hydrological studies, especially those conducted in Japan. Many Japanese government analysts often use procedures developed in Japan, which have been recommended by Japanese MLIT. We criticized the use of those procedures. Some consistency between parameter estimation and evaluation of goodness-of-fit is necessary. From this perspective, we recommend using the maximum likelihood method and AIC, both of which are related to Kullback–Leibler divergence. If one prefers using SLSC, we recommend not SLSC itself but SLSC's non-exceedance probability.

Techniques for parameter estimation and selecting the optimal probability distribution should be discussed from an international viewpoint. Some techniques related to Kullback–Leibler divergence or Bayesian estimation might be candidates for the solution of non-stationary flood frequency analyses.

Author Contributions: Conceptualization, C.M. and Y.K.; methodology, C.M. and Y.K.; investigation, C.M. and Y.K.; writing, C.M.; supervision, Y.K.; project administration, Y.K.; funding acquisition, Y.K. All authors have read and agreed to the published version of the manuscript.

Funding: This research was partly supported by JSPS Grants-in-Aid for Scientific Research. The grant Number is JP19K04613.

Institutional Review Board Statement: Not applicable.

Informed Consent Statement: Not applicable.

Data Availability Statement: Not applicable.

Conflicts of Interest: The authors declare no conflict of interest.

References

1. Nakamura, S. *Floods and Probability;* University of Tokyo Press: Tokyo, Japan, 2021; 195p. (In Japanese)
2. Takara, K. Frequency Analysis of Larger Samples of Hydrologic Extreme-Value Data—How to estimate the T-year quantile for samples with a size of more than the return period T. *Annu. Disaster Prev. Res. Inst. Kyoto Univ.* **2006**, *49B*, 7–12. (In Japanese)

3. World Meteorological Organization (WMO). *Guide to Hydrological Practices*, 5th ed.; World Meteorological Organization (WMO): Geneva, Switzerland, 1994. Available online: http://www.innovativehydrology.com/WMO-No.168-1994.pdf (accessed on 28 December 2022).
4. World Meteorological Organization (WMO). *Guide to Hydrological Practices*, 6th ed.; World Meteorological Organization (WMO): Geneva, Switzerland, 2009. Available online: https://www.hydrology.nl/images/docs/hwrp/WMO_Guide_168_Vol_II_en.pdf (accessed on 28 December 2022).
5. Stedinger, J.R.; Vogel, R.M.; Foufoula-Georgiou, D. Frequency Analysis of Extreme Events. In *Handbook of Hydrology*; Maidment, D.R., Ed.; McGraw-Hill: New York, NY, USA, 1992; pp. 18.1–18.66.
6. Hosking, J.R.M.; Wallis, J.R. *Regional Frequency Analysis*; Cambridge University Press: New York, NY, USA, 1977; 224p.
7. Rao, R.R.; Hamed, K.H. *Flood Frequency Analysis*; CRC Press: Boca Raton, FL, USA, 2000; 350p.
8. Hazen, A. Storage to be Provided Impounding Resevoirs for Municipal Water Supply. *Trans. ASCE* **1914**, *77*, 1308.
9. Markovic, R.D. *Probability Functions of the Best Fit to Distributions of Annual Precipitation and Runoff*; Hydrology paper No. 8; Colorado State University: Fort Collins, CO, USA, 1965.
10. Gupta, V.L. Selection of Frequency Distribution Models. *Water Resour. Res.* **1970**, *6*, 1193–1198. [CrossRef]
11. McCuen, R.H.; Rawls, W.J. Classification of Evaluation of Flood Flow Frequency Estimation Techniques. *Water Resour. Bull.* **1979**, *15*, 88–93. [CrossRef]
12. McCuen, R.H. Statistical Terminology: Definitions and Interpretation for Flood Peak Estimation. *Water Resour. Bull.* **1979**, *15*, 1106–1116. [CrossRef]
13. Campbell, A.J.; Sidel, R.C. Prediction of Peak Flows on Small Watersheds in Oregon for Use in Culvert Design. *Water Resour. Bull.* **1984**, *20*, 9–14. [CrossRef]
14. Turkman, K.F. The choice of extremal models by Akaike's information criterion. *J. Hydrol.* **1985**, *82*, 307–315. [CrossRef]
15. Vogel, R.M. The Probability Plot Correlation Coefficient Test for the Normal, Lognormal, and Gumbel Distributional Hypotheses. *Water Resour. Res.* **1986**, *22*, 587–590. [CrossRef]
16. Vogel, R.W.; McMartin, D.E. Probability Plot Goodness-of-Fit and Skewness Estimation Procedures for the Pearson Type 3 Distribution. *Water Resour. Res.* **1991**, *27*, 3149–3158. [CrossRef]
17. Haktanir, T. Comparison of various flood frequency distributions using annual flood peaks data of rivers in Anatolia. *J. Hydrol.* **1992**, *136*, 1–31. [CrossRef]
18. Bobée, B.; Cavadias, G.; Ashkar, F.; Bernier, J.; Rasmussen, P. Towards a systematic approach to comparing distributions used in flood frequency analysis. *J. Hydrol.* **1993**, *142*, 121–136. [CrossRef]
19. Önöz, B.; Bayazit, M. Best-fit distributions of largest available flood samples. *J. Hydrol.* **1995**, *167*, 195–208. [CrossRef]
20. Akaike, H. Information Theory and an Extension of the Maximum Likelihood Principle. In Proceedings of the 2nd International Symposium on Information Theory, Tsahkadsor, Armenia, 2–8 September 1973; pp. 267–281.
21. Committee to Discuss on River Plan Design for Small and Medium-Sized Rivers. *Guide for River Plan Design for Small and Medium-Sized Rivers*; Japan Institute of Countryology and Engineering: Tokyo, Japan, 1999; 243p. (In Japanese)
22. Japan Institute of Countryology and Engineering. *Guide for Discussion on High-Water Plan*; Japan Institute of Countryology and Engineering: Tokyo, Japan, 2007; 43p. (In Japanese)
23. International Centre for Water Hazard and Risk Management (ICHARM). User's Manual of Hydrological Statistics Utility. Available online: https://www.pwri.go.jp/icharm/special_topic/20171013_manual_en_hsu/english_manual_for_hydrological_statistics_utility.pdf (accessed on 28 December 2022).
24. Iwai, S. Some Estimating Methods of Probable Flood and Their Application to Japanese Rivers. *Bull. Math. Stat.* **1949**, *2*, 21–36. (In Japanese)
25. Slade, J.J.J. An asymmetric probability function. *Trans. ASCE* **1936**, *101*, 35–61. [CrossRef]
26. Kadoya, M. On the Applicable Ranges and Parameters of Logarithmic Normal Distributions of Slade-Type. *J. Irrig. Eng. Rural Plan.* **1962**, *3*, 12–16. (In Japanese)
27. Ishithara, T.; Takase, N. The Logarithmic-Normal Distribution and its Solution Based on Moment Method. *Trans. JSCE* **1957**, *47*, 18–23. (In Japanese)
28. Barnett, V. Probability Plotting Methods and Order Statistics. *J. R. Stat. Soc.* **1975**, *24*, 95–108. [CrossRef]
29. Cunnane, C. Unbiased plotting positions—A review. *J. Hydrol.* **1978**, *37*, 205–222. [CrossRef]
30. Iwai, S.; Ishiguro, M. *Applied Hydrological Statistics*; Morikita Publishing: Tokyo, Japan, 1970; 370p. (In Japanese)
31. Etoh, T.; Murota, A.; Yonetani, T.; Kinoshita, T. Frequency of Record-breaking Large Precipitation. *Proc. JSCE* **1986**, *369*, 165–174. (In Japanese) [CrossRef]
32. Hayashi, H.; Tachikawa, Y.; Shiiba, M. Non-Stationary Hydrologic Frequency Analysis Using Time Dependent Parameters and Its Model Selection. *J. Jpn. Soc. Civ. Eng. Ser B1* **2015**, *71*, 28–42.
33. Hoshi, K. Hydrological Statistical Analysis. *Month. Rep. Civil Eng. Res. Inst.* **1998**, *540*, 31–63. (In Japanese)
34. Kubota, K. On Probability Distribution and Method for Estimating Statistics. Available online: http://civilyarou.web.fc2.com/WANtaroHP_html5_win/f90_ENGI/dir_HFA/suimon.pdf (accessed on 28 December 2022). (In Japanese).
35. Kuzuha, Y.; Mizuki, C. T-year Hydrological Event Estimation Using the Akaike Information Criterion and Some Considerations. *J. Jpn. Soc. Hydol. Water Resour.* **2022**, *35*, 134–147. (In Japanese) [CrossRef]

36. Tsuchiya, K.; Takeuchi, K. Application of PWM Method to SQRT-ET-max Distribution. In Proceedings of the 42th Annual Conference of the Japan Society of Civil Engineers (Division 2); 1987; pp. 34–35. (In Japanese).
37. Kuzuha, Y. L-moment Solution of Etoh's Distribution. *J. JSCE* **2023**, *unpublished manuscript*.
38. Takeuchi, K.; Tsuchiya, K. A PWM Solution for Parameters of Normal Distribution. *Annu. J. Hydraul. Eng.* **1987**, *31*, 191–196. (In Japanese)
39. Takeuchi, K.; Tsuchiya, K. PWM Solutions to Nomal, Lognormal and Pearson-III Distributions. *Proc. JSCE* **1988**, *393/II-9*, 95–101. (In Japanese)
40. Ueda, T.; Kawamura, A. A New Graphical Method of Testing the Goodness of fit of Data to Probability Distributions. *Proc. JSCE* **1985**, *357/II-3*, 243–246. (In Japanese)
41. Takasao, T.; Takara, K.; Shimizu, A. A Basic Study on Frequency Analysis of Hydrologic Data in The Lake Biwa Basin. *Annu. Disaster Prev. Res. Inst. Kyoto Univ.* **1986**, *29B-2*, 157–171. (In Japanese)
42. Kuzuha, Y. Considerations of Statistical Method in Flood-control planning -SLSC and Cost Benefit Analysis. *J. JSCE* **2010**, *66*, 66–75. (In Japanese) [CrossRef]
43. Hayashi, H.; Tachikawa, Y.; Shiiba, M.; Yorozu, K.; Sunmin, K. Introducing a Statistical Hypothesis Testing into SLSC Goodness-of -fit Evaluation for Hydrological Frequency Analysi Models. *J. JSCE* **2012**, *68*, I_1381–I_1386.
44. Kuzuha, Y.; Mizuki, C. Estimating T-year Hydrological Event and Issues of Conventional Methods—Improved Standard Least Squares Criterion (SLSC) Method for Goodness-of-fit Evaluation. *J. Jpn. Soc. Hydol. Water Resour.* **2021**, *34*, 283–302. (In Japanese) [CrossRef]
45. Kuzuha, Y.; Mizuki, C. Proof of Problems in SLSC—Simple Explanation Using the Mean of SLSC Derived by Semi-analytic Method. *J. JSCE* **2022**, *78*, I-481–I-486. (In Japanese) [CrossRef]
46. Kuzuha, Y.; Mizuki, C. Some Issues Related to SLSC Method and Guideline for Prameter Estimation and Goodness-of-fit Test. *J. JSCE* **2022**, *78*, I-487–I-92. (In Japanese) [CrossRef]
47. Kullback, S.; Leibler, R.A. On information and sufficiency. *Ann. Math. Stat.* **1951**, *22*, 79–86. [CrossRef]
48. Tanaka, S.; Takara, K. Goodness-of-fit and Stability Assessment in Flood Frequency Analysis. *Annu. J. Hydraul. Eng.* **1999**, *43*, 127–132. (In Japanese) [CrossRef]
49. Van Gelder, P.H.A.J.M. Statistical Estimation Methods in Hydrological Engineering. In *Proceedings International Scientific Seminar*; Korytny, L.M., Luxemburg, W.M., Eds.; Publishing House of the Institute of Geography: Irkutsk, Russia, 2004; pp. 11–57.
50. Yuan, J.; Emura, K.; Farnham, C.; Alam, M.A. Frequency analysis of annual maximum hourly precipitation and determination of best fit probability distribution for regions in Japan. *Urban Clim.* **2018**, *24*, 276–286. [CrossRef]
51. Langat, P.K.; Kumar, L.; Koech, R. Identification of the Most Suitable Probability Distribution Models for Maximum, Minimum, and Mean Streamflow. *Water* **2019**, *11*, 734. [CrossRef]
52. Wang, Q.J. Lh moments for statistical analysis of extreme events. *Water Resour. Res.* **1997**, *33*, 2841–2848. [CrossRef]
53. Anghel, C.G.; Ilinca, C. Parameter Estimation for Some Probability Distributions Used in Hydrology. *Appl. Sci.* **2022**, *12*, 12588. [CrossRef]
54. Coles, S. *An Introduction to Statistical Modeling of Extreme Values*; Springer: London, UK, 2001; 223p.

water

MDPI

Review

Application of Porous Concrete Infiltration Techniques to Street Stormwater Inlets That Simultaneously Mitigate against Non-Point Heavy Metal Pollution and Stormwater Runoff Reduction in Urban Areas: Catchment-Scale Evaluation of the Potential of Discrete and Small-Scale Techniques

Shigeki Harada

Department of Agroenvironmental Sciences, Faculty of Food and Agricultural Sciences, Fukushima University, Fukushima 960-1296, Japan; harada.shigeki@agri.fukushima-u.ac.jp

Abstract: The expansion of pervious areas is an essential and common concept in mitigating non-point pollution runoff in urban areas. In this review, literature related to the expansion of pervious areas is introduced. In addition, the potential application of porous concrete as a medium for constructing the bottom and side walls of street stormwater inlets is investigated. The effectiveness of this medium in reducing (i) the stormwater runoff volume via porous concrete by exfiltrating from the bottom and the wall, and (ii) the heavy metal pollution runoff loads via infiltration through the porous concrete is assessed using data obtained by the author and published in the literature. The urban hydrological model Infoworks ICM (Innovyze) was used to estimate the exfiltration rates through the porous concrete plates set at the bottom and side walls of the street stormwater inlets. The exfiltration rates used in the pre-reported literature varied depending on the methods used. In the present study, sensitivity tests were performed by changing the exfiltration rates. The results of this study indicated that porous concrete used at only the bottom and side walls of the street stormwater inlets is suitable for reducing the runoff volume and removing any heavy metals from stormwater at a catchment scale.

Keywords: non-point pollution; stormwater drainage systems; infiltration technique; storage of runoff water; quantity and quality of stormwater; porous concrete; heavy metals; urbanized areas

Citation: Harada, S. Application of Porous Concrete Infiltration Techniques to Street Stormwater Inlets That Simultaneously Mitigate against Non-Point Heavy Metal Pollution and Stormwater Runoff Reduction in Urban Areas: Catchment-Scale Evaluation of the Potential of Discrete and Small-Scale Techniques. *Water* **2023**, *15*, 1998. https://doi.org/10.3390/w15111998

Academic Editor: Domenico Cicchella

Received: 3 April 2023
Revised: 10 May 2023
Accepted: 18 May 2023
Published: 24 May 2023

1. Introduction

1.1. Overview and Review Objectives

The potential and realized effects of non-point sources of pollution originating from runoff from urban areas, such as stormwater drains adjacent to roadways, have received increased attention in recent years. Comprehensive reviews of state-of-art remediation techniques and their development have been reported [1,2]. The impacts of non-point sources of pollution have been demonstrated [3–5] and the runoff characteristics have been described [6,7]. Major non-point pollutants, including organic materials (chemical oxygen demand (COD), biochemical oxygen demand (BOD), total organic carbon (TOC)), nutrients (various forms of nitrogen and phosphorus), heavy metals, suspended solids, PAHs [8], microplastics [9–11], and water-soluble aerosols [12], have all been found in the runoff from urbanized areas and their surrounding areas. Of these pollutants, heavy metal pollutions from urbanized areas [13–24] are the primary focus of this review owing to its high toxicity [25–29]. In this review, studies on heavy metal pollution in road effluents [13–15] will be cited and the data therein will be compared to the author's own data.

Since non-point pollutants, including heavy metals, in runoff water can be captured using infiltration techniques (or infiltration unit processes) [30–35], the author focused on the application of porous concrete in stormwater drainage systems. The aim of the study was the reduction of both non-point source heavy metals in runoff and surface runoff in urbanized areas. The author focused on the application of porous concrete plates set at the bottom [36–40] and sides of street stormwater inlets. The author intended to demonstrate that the placement of relatively small porous concrete plates at the bottom and sides of street stormwater inlets is sufficient to capture the heavy metals within the runoff and facilitate the efficient drainage of the stormwater itself in impermeable urban catchment areas. Specifically, the author examines the following points:

- Heavy metal concentrations in urban road runoff.
- Adsorption of heavy metals by porous concrete.
- Estimating the amount of runoff that can be treated in stormwater drains fitted with porous concrete filters.

1.2. Features and Definitions of Non-Point Pollution Sources and the Countermeasures Required for Their Reduction in Runoff

The information provided on the homepage of the Japan Society for Water Environment (JSWE) and the US Environmental Protection Agency (EPA) is discussed in order to explore the similarities and differences in the approaches adopted by the two agencies to expand pervious areas and control non-point sources of pollution.

The JSWE (https://www.jswe.or.jp/eng/index.html (accessed on 10 May 2023)) has based its description of non-point sources of pollution in urban areas on the features of these sources and the actions that need to be implemented (http://jswe-nonpoint.com/1/documents.html (accessed on 10 May 2023)). The definitions employed by the EPA are based on basic information, such as "non-point vs. point sources" and "what we can do first, etc.".

JSWE describes the actions that are required to remediate non-point pollution runoff by emphasizing the need to understand runoff behavior and develop pollution control measures.

The US EPA published a fact sheet entitled, "Protecting Water Quality from Urban Runoff" (https://nepis.epa.gov/Exe/ZyPDF.cgi/20004PP1.PDF?Dockey=20004PP1.PDF (accessed on 10 May 2023)), which shows "How Urbanized Areas Affect Water Quality" in terms of increasing runoff and pollutant loads. Of particular interest were the roles of porous and pervious areas in natural landscapes, such as forests, wetlands, and grasslands, at trapping rainwater and snowmelt and how they promote water filtration into the ground. The roles of pervious areas as countermeasures to non-point pollution, and the ways in which pervious areas can be expanded include the following [41]:

- Infiltration practice
- Infiltration basins
- Infiltration trenches
- Pervious or porous pavements
- Vegetated open channel practices
- Filtering practices
- Detention ponds or vaults
- Retention ponds
- Wetlands
- Other practices, including water quality inlets.

These could be included in the BMPs of water managers [42]. These infiltration measures are similar to those employed in the experimental sewer system (ESS) in Japan [43]. The system, which is described in [41], could be considered to include the whole catchment area, while the porous concrete plates at the bottoms and sides of the street stormwater inlets could be considered as small discrete points.

2. Runoff Behavior from Non-Point Sources in Urban Areas

2.1. Road Runoff Water Quality Assessments in Sendai City, Miyagi Prefecture, Japan

2.1.1. Materials and Methods

In Wakabayashi ward, Sendai City in Miyagi Prefecture, the author deployed a water collection device in a street stormwater inlet. The device, which consisted of a glass bottle with a floating ball, can collect 1 L samples of stormwater runoff. When the ball rises to the top of the bottle, further inflow is prevented. Therefore, the collected samples correspond to the initial stormwater runoff from 1.5 mm rainfall events (Figures 1–3). This estimate is based on the assumption that the area of inflow into the inlet was 5 m by 10 m, the initial loss was 0.75 mm, the runoff ratio of the area was 0.8, and the proportion of water entering the sample collecting vessel was 0.05 (Figure 3).

Figure 1. Study site and the street stormwater inlet.

Figure 2. Inside the street stormwater inlet.

Figure 3. Water collection device.

The daily rainfall volumes during the periods 1985–1990 and 2015–2019 in Sendai obtained from the Japan Meteorological Agency showed daily rainfall events of <1.5 mm, which accounted for 34.1% and 31.2% of the rainy days during the 1985–1990 and 2015–2019 periods, respectively. The histograms show that daily rainfall events ranged between 0.5 and 10 mm during these two periods (Figures 4 and 5), while daily rainfall events of 0.5 mm were the highest in frequency. In addition, despite the changes in climatological conditions over these periods, the shapes of the histograms are generally similar. The figures show that the sample collecting system employed in this study is well suited to collecting runoff from the most frequent rainfall events and, also, for collecting samples for some of the larger rainfall events. Whether collecting runoff for the first <1.5 mm rainfall events can capture the first flush effects should be clarified by further monitoring [6,44], literature reviews [2,45,46], and analyses using EMC [47].

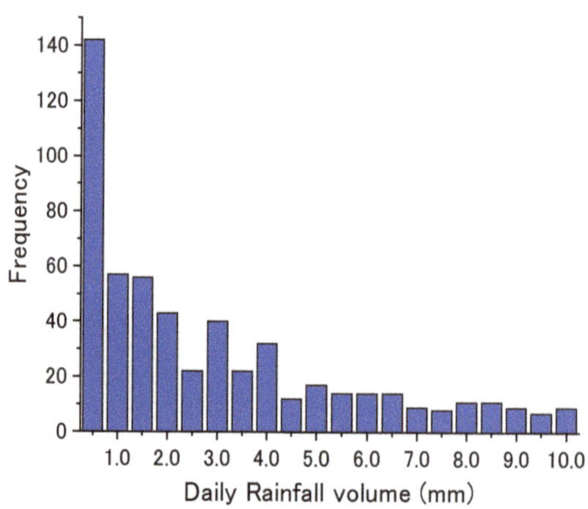

Figure 4. Frequency of the daily rainfall volume focusing on 0.5–10 mm events in Sendai (1985–1990).

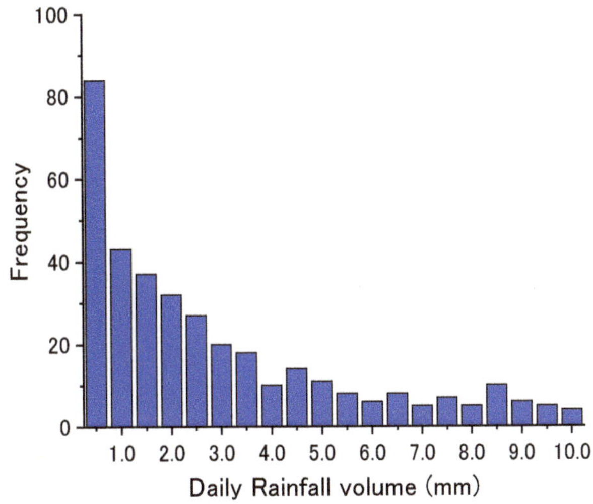

Figure 5. Frequency of the daily rainfall volume focusing on 0.5–10 mm events in Sendai (2015–2019).

In 2019, the author collected 1 L samples at intervals of 1–2 weeks (Table 1).

Table 1. Features of the samples collected 18 May to 19 December in 2019 and qualities (SS, particulate and dissolved heavy metals).

Sampling Date	Settled Date	Antecedent Dry Weather Days	SS (mg/L)	GF/B Filtration	Cr (ppb)	Cu (ppb)	Zn (ppb)	Cd (ppb)	Pb (ppb)
18 May	-------	-------------	0.289	None	29.88	49.27	395.47	0.22	17.35
				Done	16.32	6.37	73.47	0.04	0.94
25 May	18 May	3	0.051	None	18.30	16.94	178.23	0.08	3.00
				Done	15.17	12.36	109.46	0.04	0.83
7 Jun.	25 May	4	0.414	None	21.60	55.88	578.17	0.30	38.38
				Done	14.70	8.38	116.56	0.05	1.36
9 Jun.	7 Jun.	0	0.122	None	19.87	31.03	246.51	0.15	12.78
				Done	14.89	16.10	107.69	0.06	2.16
29 Jun.	9 Jun.	2	0.382	None	19.57	23.19	337.92	0.18	11.21
				Done	14.53	5.41	92.99	0.05	0.19
6 Jul.	29 Jun.	0	0.187	None	4.07	18.60	169.89		7.02
				Done	0.95	8.04	85.29		1.10
19 Jul.	6 Jul.	0	0.162	None		32.93	374.60	0.49	8.92
				Done		10.32	162.52	0.32	0.71
28 Jul.	19 Jul.	0	0.837	None	12.05	46.10	914.82	0.79	27.26
				Done	0.92	6.99	87.31	0.75	1.29
30 Jul.	28 Jul.	0	0.511	None	20.34	58.15	615.98	0.67	31.01
				Done	2.56	4.60	29.34	0.17	1.10

Table 1. *Cont.*

Sampling Date	Settled Date	Antecedent Dry Weather Days	SS (mg/L)	GF/B Filtration	Cr (ppb)	Cu (ppb)	Zn (ppb)	Cd (ppb)	Pb (ppb)
10 Aug.	30 Jul.	9	0.151	None		88.85	487.80	1.94	14.50
				Done		6.53	240.23	0.17	1.78
23 Aug.	10 Aug.	2	0.128	None		38.47		0.71	13.25
				Done		7.92		0.18	0.38
31 Aug.	23 Aug.	4	0.872	None	28.27	131.21	1385.67	2.04	69.69
				Done	3.26	7.20	138.00	0.21	0.13
26 Sep.	31 Aug.	1	0.187	None	10.73	40.34	413.30	0.44	20.73
				Done	2.22	5.44	84.48	0.07	0.99
10 Oct.	26 Sep.	3	0.731	None	15.97	51.42	806.82	0.65	
				Done	2.54	5.52	249.62	0.13	
6 Dec.	10 Oct.	23	1.097	None	28.68	81.86	833.19	0.57	
				Done	2.70	8.48	76.93	0.05	
19 Dec.	6 Dec.	11	0.334	None	22.36	53.44	632.71	0.50	18.80
				Done	14.79	17.99	334.95	0.16	0.44

The collected water samples were initially stored in a cool dark space before being transferred to a refrigerator (4 °C) until filtration.

The collected water was pre-filtered using a 2 mm mesh filter followed by filtration using a glass fiber filter (Whatman GF/B, 47φ, pore size: 1 μm). Suspended solids (SS, see also Figures 5 and 6) on the glass fiber filters were measured after drying the filter for two hours at 105 °C. The filtrates were used to measure heavy metals (Cr, Cu, Zn, Cd, and Pb) by ICP–MS. Separately, pre-filtrated waters were used to measure the total heavy metals (Cr, Cu, Zn, Cd, and Pb) by ICP–MS to determine the proportion of the particulate heavy metals.

2.1.2. Results

A list of the water samples and their physicochemical parameters is shown in Table 1. High SS values were often observed, which were consistent with the visual observations (Figure 6).

SS was observed in the samples, even when the antecedent dry weather days were zero. The author observed road sediment residual, even after rainfall events, and the SS values obtained, when antecedent dry weather days were zero, were consistent with this observation. Table 1 shows similar trends among SS, particulate heavy metals, and dissolved heavy metals alongside antecedent dry days. However, the correlation factor between the antecedent dry days and SS was not high (0.475) suggesting that more precise analyses, which consider the build-up and wash-off mechanisms, are needed to quantitatively characterize the effects of the antecedent dry weather days on the variations in water quality.

Table 2 shows that the runoff contained high levels of heavy metals, which was consistent with the visible abundance of SS (Figure 6), and suggests an important potential role for porous concrete in the removal of particulate heavy metals during the passage of runoff water via porous concrete because the particles are expected to be trapped.

Figure 6. Collected water after shaking to resuspend the particulate matter.

Table 2. Proportions of particulate heavy metal to total heavy metal in the samples shown in the Table 1.

Sampling Date	Cr (%)	Cu (%)	Zn (%)	Cd (%)	Pb (%)
18 May	45.38	87.07	81.42	81.15	94.60
26 May	17.11	27.02	38.58	44.72	72.40
7 Jun.	31.94	85.01	79.84	81.63	96.45
9 Jun.	25.03	48.11	56.32	58.67	83.11
29 Jun.	25.74	76.67	72.48	71.03	98.28
6 Jul.	76.57	56.81	49.80		84.28
19 Jul.		68.66	56.62	34.18	92.07
28 Jul.	92.36	84.84	90.46	4.39	95.28
30 Jul.	87.42	92.10	95.24	75.02	96.44
10 Aug.		92.65	50.75	91.03	87.70
23 Aug.		79.41		75.07	97.15
31 Aug.	88.46	94.51	90.04	89.58	
26 Sep.	79.28	86.50	79.56	83.19	95.20
1 Nov.	84.10	89.26	69.06	79.86	
6 Dec.	90.57	89.65	90.77	90.62	
19 Dec.	33.86	66.33	47.06	68.60	97.67
Average	59.83	76.54	69.87	68.58	91.59

2.2. *Comparison of Dissolved Heavy Metal Concentrations with Previous Studies*

Compared to previous studies [13–15], the heavy metal concentrations in the collected samples, shown in Table 1, were high; the results of the three studies are reviewed in Murakami [48] (Table 2.11). Specifically, the concentration of Cr was similar to those in Sansalone and Buchberger [14] and Pitt et al. [15], while it was one to two orders higher than in Shinya et al. [13]. The concentration of Cu was one order lower than in Shinya et al. [13] and Sansalone and Buchberger [14] and similar to those found by Pitt et al. [15].

The concentration of Zn was one to two orders lower than in Shinya et al. [13] and Sansalone and Buchberger [14] and similar to those found by Pitt et al. [15]. The Cd concentration was higher than in Shinya et al. [13], one order lower than in Sansalone and Buchberger [14], and similar to those in Pitt et al. [15]. The Pb concentration was one order lower than those identified in the three aforementioned studies [13–15].

The results of these comparisons suggest that the heavy metals levels in all four studies (including this study) varied markedly, presumably reflecting the differences in the environmental conditions at the four sites. For Zn, the environmental standards in Japan are set at 0.03 mg/L for rivers and lakes; this level was exceeded only once at the study site (19 December 2019). The mean heavy metal concentrations reported by Shinya et al. [13] and Sansalone and Buchberger [14] were one order higher than the environmental standards, suggesting that there is a need to reduce non-point Zn in Japan, and presumably the other heavy metals in stormwater runoff in urban areas.

A study by Flores-Rodriguez et al. [49] measured Pb, Zn, and Cd concentrations in stormwater samples collected at eight sites. The Pb and Cd concentrations were one order higher than most of the values obtained in this study, while the Zn levels were more varied, yet tended to be one order lower than the values shown in Table 1. In a study by Mikkelsen et al. [50], the Pb, Zn, Cd, and Cu concentrations were measured in different types of urban runoff. They found that the Cd concentrations were higher and Cu concentrations were lower than the values shown in Table 1, while the concentrations of Pb and Zn were mostly similar. Numerous factors are considered to affect non-point sources of heavy metals, as demonstrated in the studies by Ozaki et al. [51,52].

3. Control of Non-Point Sources of Pollution and Sewage Systems

3.1. *Non-Point Source Pollution and Sewage Systems*

The present study focused on reducing non-point heavy metal runoff. The benefits of using infiltration techniques to decrease heavy metals in runoff are (i) to avoid constructing water treatment facilities; (ii) to facilitate multipurpose uses for the surfaces used for infiltration during fine weather (i.e., these surfaces could be used for other activities); (iii) to retain the water that can be recycled for uses other than drinking.

Owing to the high ratio of paved (i.e., impervious) to non-paved (i.e., pervious) areas in urban centers, high runoff volumes and peak discharge rates are observed. Thus, mitigation measures employing infiltration and storage are very important in urban areas because they prevent flooding and/or inundation.

Furthermore, infiltration techniques are both indirectly and directly effective for water quality control, as described below.

Direct methods of control include methods such as those that the author is attempting to implement. In these cases, the infiltration site (i.e., the bottom and the sides of the street stormwater inlets) could be said to achieve two aims: water volume control and quality control. These two aims could be met by water penetration and these sites could be regarded as high-performance facilities.

3.2. *Indirect and Direct Means of Reducing Non-Point Pollution Runoff Loads*

Indirect control methods promote penetration of the water at the surface of the infiltration stratum and the subsequent retention therein. Percolation of the water in the infiltration stratum is referred to as temporal retention, which is different from the water inside the retention ponds because it is exfiltrated into the natural base soil below the stratum and/or through drainage pipes [53–55]. This method reduces the volume of runoff and decreases the volume and frequency of combined sewage overflow (CSO), as well as the sweep flow inside stormwater pipes, gutters, etc. Originally, the idea of runoff volume reduction developed from the viewpoint of flood control. The frequent occurrence of heavy rain events around the world in recent years has required catchment managers to re-evaluate runoff volume reduction, i.e., to consider the mutual benefits of flood control and runoff load reduction.

Direct control methods to control non-point pollution runoff aim at controlling (i.e., trapping and adsorbing) the pollution in the stormwater inside the infiltration stratum. In the case of porous concrete, the mechanisms of pollution control were clarified based on laboratory experiments (see Section 5). The economic aspects of the infiltration technique have been previously analyzed [36]. We intended to develop direct methods for runoff reduction into separate sewage systems.

4. Infiltration as a Direct Pollution Control Method in a Separate System

4.1. Porous Concrete as An Infiltration Medium

Porous concrete is produced using large aggregates, which ensures that permeability is maintained with only a minor decrease in hardness compared to normal cement. The author used an aggregate called Gmax15, which is composed of gravel measuring less than 15 mm.

Porous concrete columns (Figure 7) and porous concrete cubes (Figure 8) were used in the laboratory experiments. Columns were prepared using a mixture of cement (0.3 kg), gravel (Gmax15, 1.55 kg), water (8.05 kg), and admixture (high-range water-reducing admixture, 0.003 kg) [38].

Figure 7. Experimental porous concrete column measuring 10 cm in diameter and 10 cm in depth.

Figure 8. Porous concrete cube with each side measuring 4 cm.

The saturated hydraulic conductivity of the column was measured using the constant water level method and estimated to be 1800 mm/h.

4.2. Deployment of Porous Concrete Plate at the Bottom of Street Stormwater Inlets

Using porous concrete on the bottom of the street and not at the ground surface of stormwater inlets (Figure 9) has been proposed by the author's research group as a means of reducing heavy metal runoff, via filtration and adsorption, and water runoff reduction, via exfiltration, into the natural base soil around the bottom of the street stormwater inlets [36–38]. The reasons why the bottoms of the street stormwater inlets were selected as sites to deploy the porous concrete plates were because (i) porous concrete has a structural weakness and cannot bear significant loads, (ii) the inflows of the stormwater during rainfall events will keep the porous composite unclogged, and (iii) the street stormwater inlets are located in the stormwater drainage networks.

Figure 9. Porous concrete plate at the bottom of the street stormwater inlet.

Figure 10 shows the behavior of the water passing through the permeable bottom plates in the stormwater inlets, although, in reality, the walls could also be permeable, as described below. Here, the behavior of the water at the bottom of the inlets is shown to illustrate the two proposed functions of the stormwater drains, i.e., the reduction of heavy metals in the runoff via filtration and adsorption reactions, and the overall reduction in water runoff, via the exfiltration into the natural base soil around the bottom of the street stormwater inlets [36–38].

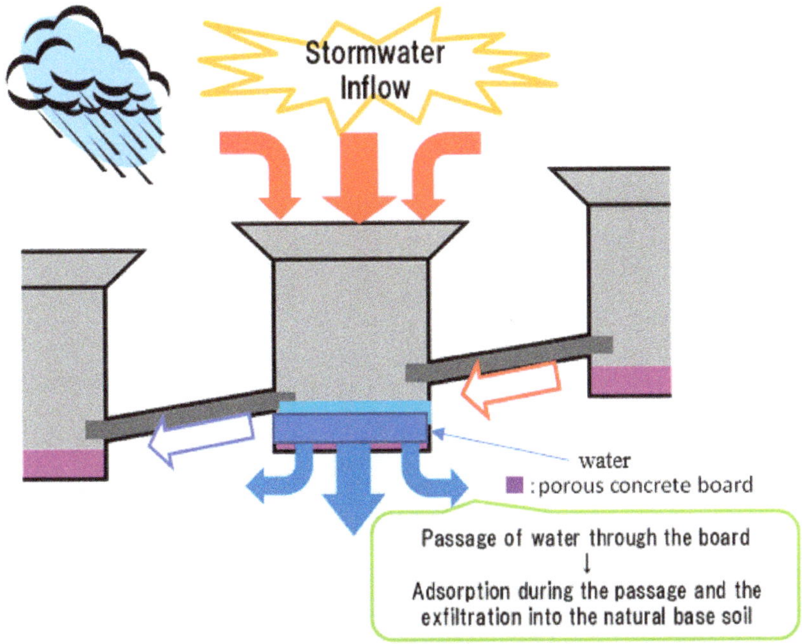

Figure 10. Schematic diagram showing the behavior of water through the porous concrete when placed only at the bottoms of street stormwater inlets.

5. Laboratory Experiments Examining the Potential Reduction in Heavy Metals in Porous Concrete Exposed to Runoff

Using artificial rainfall (mixture of Zn, Cu, and Pb), the author conducted experiments using porous concrete columns (Figure 7; diameter and depth: 10 cm) and porous concrete cubes (Figure 8; 64 cm^3). Specifically, the author investigated the adsorption rates of heavy metals using concrete columns and cubes under various conditions.

The porous concrete column samples (C series) consisted of a column prepared in February 2008 at the School of Food, Agricultural and Environmental Sciences at Miyagi University, alongside the column samples (N series) prepared outside the University in November 2016. Both columns were prepared using the same mixture, coefficient of permeability, and size.

We performed 12 experimental runs (Table 3). In each run, a single column was placed in a Petri dish and a 50 mL solution was added comprising either a mixture of the heavy metals Zn, Cu, and Pb, or Zn or Cu, or Pb individually. The concentrations used for Pb were higher than the level of dissolved Pb in Table 1, although Zn and Cu concentrations were of the same order of magnitude, while each was sprayed onto the top of the sample column using a 50 mL volumetric pipette (Figure 11). The experiment was left to run overnight. Then, the leachate that had accumulated in the glass Petri dish was collected, and the amount of the leachate and the concentrations of Pb, Zn, and Cu were measured. The runs for each experimental condition were repeated 2–13 times. The amount of leachate in each run varied from 25 to 45 mL. The adsorption rate was determined based on the relationship between the volume of leachate, the concentration of each heavy metal, the initial volume of solution added, and the concentration. In addition, the time taken for the 50 mL of the solution to flow out of the sample (about 25 s) was measured periodically to confirm that the spray intensity had not changed.

Table 3. Twelve experimental runs for the leachate experiments.

RUN	Column	Experiment Timing	Heavy Metal Concentration (ppb)		
			Pb	Zn	Cu
1-1	C	January 2009	1500	----------------	----------------
1-2	C	January 2009	----------------	580	----------------
1-3	C	January 2009	----------------	----------------	4150
2-1	C	December 2017	21	----------------	----------------
2-2	C	December 2017	----------------	53	----------------
2-3	C	December 2017	----------------	----------------	42
3-1	C	April 2018	55	----------------	----------------
3-2	C	April 2018	----------------	66	----------------
3-3	C	April 2018	----------------	----------------	22
4-1	N	April 2018	----------------	----------------	22
5	C	April 2018	10	25	7
6	N	April 2018			

Figure 11. Leachate experiments.

The variations in the proportion of Pb, Zn, and Cu adsorbed by the concrete columns in the indoor artificial rainfall experiments are shown in Table 4. The findings revealed that (i) similar proportions of Pb, Zn, and Cu were absorbed when the leachates contained mixed heavy metal solutions and when they contained individual heavy metals (Run 1-1, Run 2-1 and Run 3-1 and Run 5; Run 1-2, Run 2-2, Run 3-2 and Run 5; Run 1-3, Run 2-3 and Run 3-3 and Run 5), (ii) similar proportions of the three heavy metals were absorbed by the columns, even when the concentration of each heavy metal showed variations (Run 1-1, Run 2-1 and Run 3-1; Run 1-2, Run 2-2 and 3-2; Run 1-3, Run 2-3 and Run 3-3), (iii) similar adsorption proportions for Cu (Run 3-3 and Run 4-1) and for Zn and Pb (Run 5 and Run 6) were shown and (iv), smaller proportions were shown when the concentrations of the heavy metals were low (less than 10 ppb), such as in Runs 4-1 and 6, where a few ppb Cu elution from the column N series arose. The elution was confirmed separately by adding the ultrapure water (Millipore MQW) onto sample columns and again by the subsequent measurements of Pb, Zn, and Cu in the leachate. The results showed that a few ppb of each heavy metal leached from the recycled concrete used in the porous concrete columns. The small amount of heavy metal leaching decreased the apparent adsorption proportions. To resolve this problem, the author attempted to purify the porous concrete columns by submerging the columns in ca. 30 L pure water for approximately 2 weeks; this process was repeated three times after replacing the water with new pure water. Using this procedure, leaching from the columns decreased by 90% compared to the original concentrations of heavy metals in leachate and the porous concrete columns could be used to assess actual stormwater samples.

Table 4. Proportion of heavy metals adsorbed by the columns.

RUN	Adsorption Proportion (%)		
	Pb	**Zn**	**Cu**
1-1	83.1	----------------	----------------
1-2	----------------	66.6	----------------
1-3	----------------	----------------	69.6
2-1	84.9	----------------	----------------
2-2	----------------	77.1	----------------
2-3	----------------	----------------	66.6
3-1	87.9	----------------	----------------
3-2	----------------	63.2	----------------
3-3	----------------	----------------	69.6
4-1	----------------	----------------	70.2
5	80.5	60.2	73.0
6	85.8	64.4	52.6

6. Effectiveness of the Porous Concrete Plates Placed at the Bottom of the Inlets Based on Calculations Using Infoworks ICM (Innovyze)

The magnitude of infiltration for a 1 ha catchment at 20 discrete street stormwater inlets where the porous concrete plates were deployed was estimated using the author's own simulation results.

Firstly, the density of the stormwater inlets in the catchment needed to be estimated. In Japan, the density of the street stormwater inlets is 10–30/ha (20 in Harada and Komuro [37]); therefore, in this study, a value of 20 was used for the inlet density in 1 ha. Using the maximum adsorption capacity of Zn, by the cube (Figure 8) [37], the duration that the Zn runoff did not occur was estimated at about 41 years [37]. By obtaining the EMC values, a more precise duration could be calculated.

Assuming that the bottom plate was circular with a bottom thickness of 10 cm and side walls that were 10 cm thick with a 10 cm water level, while the average diameter of all stormwater inlets was set as 1.8 m, the area of the porous concrete at 1 inlet would correspond to 31,086 cm^2. Thus, for the 1100 inlets in the Fukumuro catchment (Figure 12, the 9 inlets are the ones selected to conduct the passage proportion of water via porous concrete, as mentioned in Section 8), the area of the permeable media was estimated as

3419.5 m^2. This area corresponds only to 0.04% of the catchment area (ca. 900 ha). Using Infoworks ICM (Innovyze) [56,57], Harada and Kim [36] showed that mitigation of the inundation occurrence happened following a 68.5 mm rainfall event in the Fukumuro catchment.

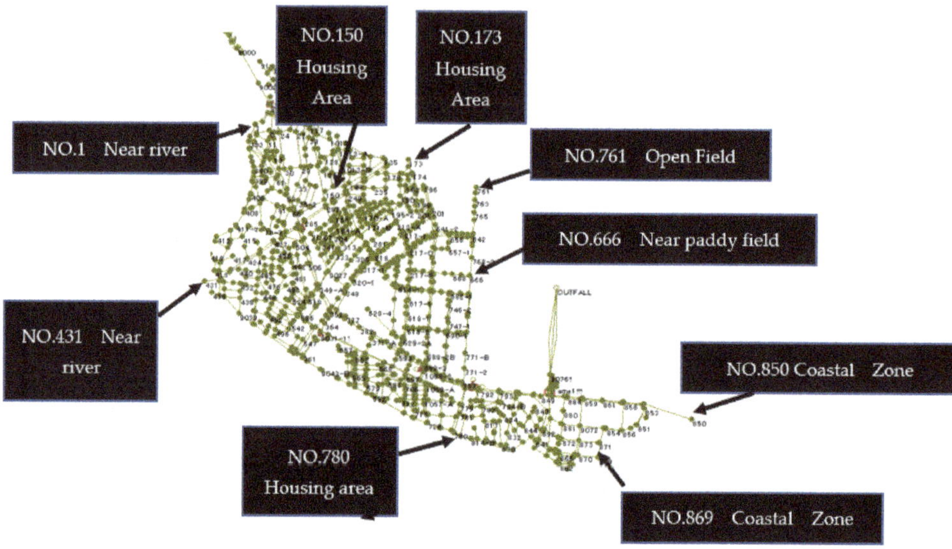

Figure 12. Fukumuro catchment and the nine sewage traps used in the study.

The analyses described here show that the ability to control stormwater runoff is quite high when small and discrete infiltration is used in a catchment.

7. Verification of Exfiltration Coefficient Obtained Using Infoworks ICM

The simulation described in Section 6 assumed that the exfiltration rate (i.e., three-dimensional water seepage into the natural base soil below and around the inlets) of every street stormwater inlet was 2000 mm/h. The magnitude of the exfiltration coefficient was quite sensitive in those analyses, thus, the suitability should be examined. This examination is of the validity of the 3D to 1D conversion coefficient. Here, the author has introduced four concepts from previous studies to clarify the suitable exfiltration rate.

Firstly, based on the observed infiltration rate in the vicinity of the street stormwater inlet in the Fukumuro catchment where the natural base soil was exposed at the surface (3600 mm/h, the 3D to 1D conversion coefficient was 36 because of the saturated hydraulic conductivity of the natural base soil (sand and silt), which was 100 mm/h) [36].

Next, referring to Herath et al. [58] and Herath and Mushiake [59], the ratio of the exfiltration rate to the hydraulic conductivity from infiltration trench (q/k_0 in the Fig.2 of Herath and Mushiake [59]) was considered according to the matric and gravity potential slopes and water pressure (the model domain used for the numerical simulation is shown in Fig.1 of Herath and Mushiake [59]). The Fig.1 of Herath and Mushiake [59] shows a symmetric analysis, thus, the width of the trench shown in the Fig.1 of Herath and Mushiake [59] could be replaced with the average radius of the inlets in the Fukumuro area, 0.9 m. Substituting the width of the trench as 0.9 m, and the water level, 10 cm (obtained via infoworks ICM calculations) into the Fig.2 of Herath and Mushiake [59], the author obtained the 3D to 1D conversion coefficient corresponds to ca. 100.

Blazejewski et al. (2018) [60] did not consider the matric potential slope and demonstrated a 3D to 1D conversion coefficient of 1.0–1.3 [60].

The author performed the 3D to 1D conversion coefficient sensitivity tests in Infoworks ICM by changing the coefficient to 0, 5, 10, and 20 [36]. The mitigation mentioned in Section 6 used a coefficient of 20.

The author assumed that a value of 10–20 was plausible. First, the coefficient should be larger than in Blazejewski et al. (2018) [60], thereby considering the matric and gravity potential slopes with the water head. Moreover, the coefficient should be smaller than that shown by the author's observed value of 36 (where the soil was dry) because the natural base soil around the street stormwater inlets should be wet, whereby the design runoff ratio of the Fukumuro area was 0.40 [36].

8. Proportion of Water Passing through the Porous Concrete Plates at the Bottom of the Street Stormwater Inlets to the Total Volume of Inflow Based on Estimates Calculated Using Infoworks ICM (Innovyze)

The author analyzed the proportion of water passage through the porous concrete at the bottom and the side walls of the street stormwater inlets, using alternative 3D to 1D conversion coefficients of 0, 5, 10, 15, and 20 and simultaneously changing the rainfall volume to 1, 3, and 5 mm. The proportions of water passage through the porous concrete column were calculated as the "volume of water exfiltrated into the soil from the street stormwater inlets" divided by "the volume of water that enters the street stormwater inlets" multiplied by 100 (%). The proportions at the 9 street stormwater inlets in the Fukumuro area (Figure 12) during 3 mm of rainfall, when the 3D to 1D conversion coefficient was 10 using Infoworks ICM (Innovyze) [56,57], is shown in Figure 13.

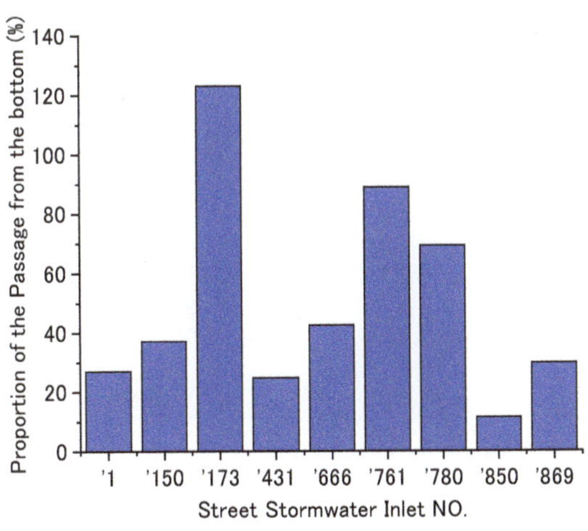

Figure 13. The proportion of the passage of the porous concrete at the bottom of the sewage traps in case the rainfall volume is 3mm and the exfiltration rate is 1500 mm/h.

Figure 13 shows the passage proportion of the water through the porous concrete at the bottom and side walls of the inlets. The proportions varied from about 10% to 88%, excluding the proportion at the inlet of 173, meaning that the calculations there showed that these numerical analyses were unsuitable. The proportions were inversely related to the increasing rainfall volume and related to the increasing 3D to 1D conversion coefficient. However, reaching 100% was uncommon, even though heavy metal removal was expected as the proportions increased. Thus, the author proposed using a hanging-type porous concrete plate, shown in Figure 14. This configuration of which has already been deployed in Sendai.

Figure 14. Hanging-type placement of the porous concrete at the same inlet as the one shown in Figure 1. The grating is open. Closing the grating results in the porous concrete board assuming a horizontal orientation.

9. Conclusions and Future Work

The present study reported the results of the author's monitoring of heavy metal concentrations in urban road runoffs. Indoor experiments were conducted to analyze the adsorption of heavy metals by porous concrete. In addition, the study also examined the amount of runoff that could be treated in stormwater drains fitted with porous concrete filters. While Section 2 presented some of the behavior of non-point heavy metals, there remains a need to explain them in terms of environmental factors using statistical analyses, as highlighted by Ozaki et al. [51,52]. It is recommended that the EMC values should be corrected, and improvements be made to the 3D/1D conversion methods.

Funding: This research was funded by Japan Science and Technology Agency grant number [JP-MJTM20BK], Fukushima University Research Fund grant number [262q006], Miyagi University Research Funds and KC Miyagi Research Fund 2019.

Data Availability Statement: Not applicable.

Conflicts of Interest: The author declares no conflict of interest.

References

1. Fletcher, T.; Duncan, H.; Poelsma, P.; Lloyd, S. *Stormwater Flow and Quality, and the Effectiveness of Non-Proprietary Stormwater Treatment Measures—A Review and Gap Analysis*; Technical Report 04/8; Cooperative Research Center for Catchment Hydrology: Melbourne, Australia, 2004.
2. Shinya, M. Non-point pollution. *Seikatsu Eisei* **2008**, *52*, 87–97. (In Japanese with English Abstract)
3. Marsalek, J.; Ng, H. Evaluation of Pollution Loadings from Urban Nonpoint Sources: Methodology and Applications. *J. Great Lakes Res.* **1989**, *15*, 444–451. [CrossRef]
4. Marsalek, J. Evaluation of Pollutant Loads from Urban Nonpoint Sources. *Water Sci. Technol.* **1990**, *22*, 23–30. [CrossRef]
5. Brett, M.T.; Arhonditsis, G.B.; Mueller, S.E.; Hartley, D.M.; Frodge, J.D.; Funke, D.E. Non-Point-Source Impacts on Stream Nutrient Concentrations Along a Forest to Urban Gradient. *Environ. Manag.* **2005**, *35*, 330–342. [CrossRef] [PubMed]
6. Wada, K.; Fujii, S. Characterization of Stormwater Runoff Processes from Urban Roadways. *J. Jpn. Soc. Water Environ.* **2006**, *29*, 699–704. (In Japanese with English Abstract) [CrossRef]
7. Wada, Y.; Miura, H. Characterization and Control of Runoff Pollution Loads in Catchment Area of the Lake. *J. Jpn. Soc. Water Environ.* **1996**, *19*, 161–169. (In Japanese with English Abstract) [CrossRef]

8. Zgheib, S.; Moileron, R.; Saad, M.; Chebbo, G. Partition of pollution between dissolved and particulate phases: What about emerging substances in urban stormwater catchments? *Water Res.* **2011**, *43*, 913–925. [CrossRef]
9. Katsumi, N.; Kusube, T.; Nagao, S.; Okochi, H. The role of coated fertilizer used in paddy fields as a source of microplastics in the marine environment. *Mar. Pollut. Bull.* **2020**, *161*, 111727. [CrossRef]
10. Katsumi, N.; Kusube, T.; Nagao, S.; Okochi, H. Accumulation of microcapsules derived from coated fer fields. *Chemosphere* **2021**, *267*, 129185. [CrossRef]
11. Katsumi, N.; Kusube, T.; Nagao, S.; Okochi, H. The input–output balance of microplastics derived from coated fertilizer in paddy fields and the timing of their discharge during the irrigation season. *Chemosphere* **2021**, *279*, 130574. [CrossRef]
12. Gjessing, E.; Lygren, E.; Berglind, L.; Gulbrandsen, T.; Skanne, R. Effect of highway runoff on lake water quality. *Sci. Total. Environ.* **1984**, *33*, 245–257. [CrossRef]
13. Shinya, M.; Tsuchinaga, T.; Kitano, M.; Yamada, Y.; Ishikawa, M. Characterization of heavy metals and polycyclic aromatic hydrocarbons in urban highway runoff. *Water Sci. Technol.* **2000**, *42*, 201–208. [CrossRef]
14. Sansalone, J.J.; Buchberger, S.G. Partitioning and First Flush of Metals in Urban Roadway Storm Water. *J. Environ. Eng.* **1997**, *123*, 134–143. [CrossRef]
15. Pitt, R.; Field, R.; Lalor, M.; Brown, M. Urban stormwater toxic pollutants assessment, sources, and treatability. *Water Environ. Res.* **1995**, *67*, 260–275. [CrossRef]
16. Murakami, M.; Fujita, M.; Furumai, H.; Kasuga, I.; Kurisu, F. Sorption behavior of heavy metal species by soakaway sediment receiving urban road runoff from residential and heavily trafficked areas. *J. Hazard. Mater.* **2009**, *164*, 707–712. [CrossRef] [PubMed]
17. Murakami, M.; Nakajima, F.; Furumai, H. The sorption of heavy metal species by sediments in soakaways receiving urban road runoff. *Chemosphere* **2008**, *70*, 2099–2109. [CrossRef]
18. Kumar, M.; Furumai, H.; Kurisu, F.; Kasuga, I. Potential mobility of heavy metals through coupled application of sequential extraction and isotopic exchange: Comparison of leaching tests applied to soil and soakaway sediment. *Chemosphere* **2013**, *90*, 796–804. [CrossRef]
19. Kumar, M.; Furumai, H.; Kurisu, F.; Kasuga, I. Tracing source and distribution of heavy metals in road dust, soil and soakaway sediment through speciation and isotopic fingerprinting. *Geoderma* **2013**, *211–212*, 8–17. [CrossRef]
20. Ozaki, H.; Watanabe, I.; Kuno, K. Investigation of the Heavy Metal Sources in Relation to Automobiles. *Water Air Soil Pollut.* **2004**, *157*, 209–223. [CrossRef]
21. Root, R.A. Lead loading of urban streets by motor vehicle wheel weights, Environ. *Health Perspect.* **2000**, *108*, 937–940. [CrossRef]
22. Westerlund, K.G. *Metal Emissions from Stockholm Traffic-Wear of Brake Linings*; Report SLB Analysis No.3; Stockholm Environment and Health Protection Administration: Stockholm, Sweden, 2001.
23. Adach, K.; Tainosho, Y. Single particle characterization of size-fractionated road sediments. *Appl. Geochem.* **2005**, *20*, 849–859. [CrossRef]
24. Gray, S.R.; Becker, N.S.C. Contaminant flows in urban residential water systems. *Urban Water* **2002**, *4*, 331–346. [CrossRef]
25. Rijstenbil, J.W.; Poortvliet, T.C.W. Copper and zinc in estuarine water: Chemical speciation in relation to bioavailability to the marine planktonic diatom Ditylum Brightwellii. *Environ. Toxicol. Chem.* **1992**, *11*, 1615–1625. [CrossRef]
26. Borgmann, U. Interactive Effects of Metals in Mixtures on Biomass Production Kinetics of Freshwater Copepods. *Can. J. Fish. Aquat. Sci.* **1980**, *37*, 1295–1302. [CrossRef]
27. Enserink, E.; Maas-Diepeveen, J.; Van Leeuwen, C. Combined effects of metals; an ecotoxicological evaluation. *Water Res.* **1991**, *25*, 679–687. [CrossRef]
28. Watanabe, H.; Nakajima, F.; Kasuga, I.; Furumai, H. Application of whole sediment toxicity identification evaluation proce-dures to road dust using a benthic ostracod Heterocypris incongruens. *Ecotoxicol. Environ. Saf.* **2013**, *89*, 245–251. [CrossRef]
29. Watanabe, H.; Nakajima, F.; Kasuga, I.; Furumai, H. Toxicity evaluation of road dust in the runoff process using a benthic ostracod Heterocypris incongruens. *Sci. Total. Environ.* **2011**, *409*, 2366–2372. [CrossRef]
30. Mikkelsen, P.S.; Jacobsen, P.; Fujita, S. Infiltration practice for control of urban stormwater. *J. Hydraul. Res.* **1996**, *34*, 827–840. [CrossRef]
31. Mikkelsen, P.; Häfliger, M.; Ochs, M.; Tjell, J.; Jacobsen, P.; Boller, M. Experimental assessment of soil and groundwater contamination from two old infiltration systems for road run-off in Switzerland. *Sci. Total Environ.* **1996**, *189–190*, 341–347. [CrossRef]
32. Bardin, J.-P.; Barraud, S.; Alfakih, E.; Dechesne, M. Performance Assessment of Stormwater Infiltration Strategies: A Multi-Indicator Approach. In Proceedings of the 9th International Conference on Urban Drainage, Portland, OR, USA, 8–13 September 2002; CD-ROM; ASCE Publications: Reston, VA, USA, 2002. [CrossRef]
33. Goebel, P.; Fach, S.; Kories, H.; Geiger, W.F.; Coldewey, W.G. Effects of Stormwater Infiltration on the Water Balance of an Urban Area. In Proceedings of the 9th International Conference on Urban Drainage, Portland, OR, USA, 8–13 September 2002; CD-ROM; ASCE Publications: Reston, VA, USA, 2002. [CrossRef]
34. Datry, T.; Malard, F.; Vitry, L.; Hervant, F.; Gibert, J. Solute dynamics in the bed sediments of a stormwater infiltration basin. *J. Hydrol.* **2003**, *273*, 217–233. [CrossRef]
35. Legret, M.; Colandini, V. Effects of a porous pavement with reservoir structure on runoff water: Water quality and fate of heavy metals. *Water Sci. Technol.* **1999**, *39*, 111–117. [CrossRef]

36. Harada, S.; Kim, J. Application of infiltration technic for residential zone in agricultural area. *Water Land Environ. Eng.* **2014**, *82*, 301–304.
37. Harada, S.; Komuro, Y. Decrease of heavy metal non-point runoff using porous concrete. *Trans. JSIDRE* **2009**, *264*, 81–82. (In Japanese)
38. Harada, S.; Komuro, Y. Decrease of non-point zinc runoff using porous concrete. *Chemosphere* **2010**, *78*, 488–491. [CrossRef]
39. Harada, S.; Yanbe, M. Adsorption by and artificial release of zinc and lead from porous concrete for recycling of adsorbed zinc and lead and of porous concrete to reduce urban non-point heavy metal runoff. *Chemosphere* **2018**, *197*, 451–456. [CrossRef]
40. Harada, S.; Kitatusji, M.; Onodera, S. Reducing Non-Point Heavy Metal Runoff from Urban Catchment Using Porous Concrete Filters in Storm Sewer Catch Basins and Manholes. In Proceedings of the DIPCON/ARC-2014, Kyoto, Japan, 3–4 September 2014.
41. United States Environmental Protection Agency. *National Management Measures to Control Nonpoint Source Pollution from Urban Area*; United States Environmental Protection Agency: Washington, DC, USA, 2005.
42. Fletcher, T.D.; Shuster, W.; Hunt, W.F.; Ashley, R.; Butler, D.; Arthur, S.; Trowsdale, S.; Barraud, S.; Semadeni-Davies, A.; Bertrand-Krajewski, J.L.; et al. SUDS, LID, BMPs, WSUD and more—The evolution and application of terminology surrounding urban drainage. *Urban Water J.* **2015**, *12*, 525–542. [CrossRef]
43. Fujita, S. Experimental sewer system for reduction of urban storm runoff. *Proc. Environ. Sani. Eng. Res.* **1986**, *22*, 175–185. (In Japanese with English Abstract)
44. Hijioka, Y.; Furumai, H. Long-Term Runoff Monitoring of Urban Nonpoint Pollution Load in a Separate Sewer System using Autosampler and Monitoring Devices. *J. Jpn. Soc. Water Environ.* **2003**, *26*, 237–242. [CrossRef]
45. Al Mamun, A.; Shams, S. Nuruzzaman Review on uncertainty of the first-flush phenomenon in diffuse pollution control. *Appl. Water Sci.* **2020**, *10*, 525–542. [CrossRef]
46. Maniquiz-Redillas, M.; Robles, M.E.; Cruz, G.; Reyes, N.J.; Kim, L.-H. First Flush Stormwater Runoff in Urban Catchments: A Bibliometric and Comprehensive Review. *Hydrology* **2022**, *9*, 63. [CrossRef]
47. Kojima, K.; Sano, S.; Furumai, H. Contributions of Road Surface Deposits and Atmospheric Deposition to Heavy-Metal and Nitrogen Contamination by Road Runoff in a Residential Area. *J. Jpn. Soc. Water Environ.* **2012**, *35*, 119–126. [CrossRef]
48. Murakami, M. Adsorption and Desorption of Heavy Metals in Infiltration Facilities Receiving Urban Runoff. Ph.D. Thesis, Tokyo University, Tokyo, Japan, 2006; p. 17.
49. Flores-Rodríguez, J.; Bussy, A.-L.; Thevenot, D. Toxic Metals in Urban Runoff: Physico-Chemical Mobility Assessment Using Speciation Schemes. *Water Sci. Technol.* **1994**, *29*, 83–93. [CrossRef]
50. Mikkelsen, P.S.; Weyer, G.; Berry, C.; Waldent, Y.; Colandini, V.; Poulsen, S.; Grotehusmann, D.; Rohlfing, R. Pollution from Urban Stormwater Infiltration. *Water Sci. Technol.* **1994**, *29*, 293–302. [CrossRef]
51. Ozaki, N.; Wada, K.; Murakami, M.; Nakajima, F.; Furumai, H. Prediction and its Reliability Evaluation of Urban Non-point Pollution Loading by Statistical Analysis of the Data of Published Researches. *J. Jpn. Soc. Water Environ.* **2017**, *40*, 115–124. [CrossRef]
52. Ozaki, N.; Wada, K.; Murakami, M.; Nakajima, F.; Furumai, H. Prediction of Urban Non-point Pollution Load by Statistical Analysis of Data of Published Research and Its Reliability Evaluation –Statistical Analysis of Mean Load and Verification and Modification of Previously Proposed Model Using Newly Obtained Data. *J. Jpn. Soc. Water Environ.* **2021**, *44*, 1–8. [CrossRef]
53. Harada, S.; Ichikawa, A. Mitigating peak discharge of urban overland surface runoff using drainage infiltration strata. In Proceedings of the 5th International Conference on Ultrafast Structural Dynamics (ICUSD), Daejeon, Republic of Korea, 25–28 June 1989; Volume 5, pp. 821–826.
54. Harada, S.; Ichikawa, A. Performance of water infiltrated within infiltration strata and Kanto loam taking into account hysteresis. *J. Nat. Disaster Sci.* **1990**, *12*, 21–35.
55. Harada, S.; Ichikawa, A. Performance of the Drainage Infiltration Strata: Statistical and Numerical Analyses. *Water Sci. Technol.* **1994**, *29*, 255–265. [CrossRef]
56. Harada, S.; Endo, K. Application of urban hydrological model for prediction of runoff from agriculture dominated area. *Inter. J. Latest Res. Sci. Technol.* **2013**, *2*, 44–48.
57. Harada, S.; Sato, I. Evaluation of Exfiltration Rates at the Bottom of Manholes at an Urbanized Area Using Infoworks. *Inter. J. Latest Res. Sci. Technol.* **2015**, *4*, 134–137.
58. Herath, S.; Musiake, K.; Hironaka, S. Field Estimation of Saturated Conductivity Using Borehole Test: Effect of Unsaturated Flow and Soil Anisotropy. *Annu. J. Hydraul. Eng. (JSCE)* **1992**, *36*, 435–440. [CrossRef]
59. Herath, S.; Musiake, K. Simulation of Basin Scale Runoff Reduction by Infiltration Systems. *Water Sci. Technol.* **1994**, *29*, 267–275. [CrossRef]
60. Blazejewski, R.; Nieć, J.; Murat-Błażejewska, S.; Zawadzki, P. Comparison of infiltration models with regard to design of rectangular infiltration trenches. *Hydrol. Sci. J.* **2018**, *63*, 1707–1716. [CrossRef]

water

MDPI

Article

Groundwater Quality and Potential Pollution in the Southern Shimabara Peninsula, Japan

Kei Nakagawa [1,*], Hiroki Amano [2], Zhi-Qiang Yu [3] and Ronny Berndtsson [4]

1 Institute of Integrated Science and Technology, Nagasaki University, 1-14 Bunkyo-machi, Nagasaki 852-8521, Japan
2 Undergraduate School of Humanities and Science, Tokai University, 9-1-1 Toroku, Higashi-ku, Kumamoto 862-8652, Japan
3 Graduate School of Fisheries and Environmental Sciences, Nagasaki University, 1-14 Bunkyo-machi, Nagasaki 852-8521, Japan
4 Division of Water Resources Engineering & Centre for Advanced Middle Eastern Studies, Lund University, P.O. Box 118, SE-221 00 Lund, Sweden
* Correspondence: kei-naka@nagasaki-u.ac.jp

Abstract: Nitrate pollution in groundwater is a severe problem in Shimabara Peninsula, Nagasaki Prefecture, Japan. Previous studies have investigated water quality characteristics in the northern part of the peninsula and shown serious effects of nitrate pollution in the groundwater. The present study aimed to investigate the groundwater quality in the southern areas of the peninsula for improved understanding of the water quality status for the entire peninsula. Groundwater samples were collected at 56 locations in Minami-Shimabara City from 28 July to 4 August 2021. The spatial distribution of water quality constituents was assessed by Piper-trilinear and Stiff diagrams for major ion concentrations. One agricultural area in the western parts exceeded Japanese recommended standards for water. According to the Piper-trilinear diagram, 44 sampling sites (78.6%) were classified as alkaline earth carbonate type, nine sites (16.1%) as alkaline earth non-carbonate type, and three sites (5.3%) as alkaline carbonate type. Stiff diagrams displayed Ca-HCO_3 water type for most of the sites. Na-HCO_3 and Mg-HCO_3 types were found in coastal areas. Principal component analyses showed that the first component corresponded to dissolved constituents in groundwater and denitrification, the second effects of ion exchange and low nitrate pollution, and the third effects of severe nitrate pollution. Hierarchical cluster analysis was used to classify the groundwater into five groups. The first group included sites with relatively high nitrate concentration. The second group had relatively low ion concentration, distributed from center to eastern parts. The third group included intermediate ion concentration, distributed at lower altitudes along the coastal line. The fourth and fifth groups had a higher ion concentration, especially characterized by high sodium and bicarbonate concentration.

Keywords: groundwater; principal component analysis; hierarchical cluster analysis

Citation: Nakagawa, K.; Amano, H.; Yu, Z.-Q.; Berndtsson, R. Groundwater Quality and Potential Pollution in the Southern Shimabara Peninsula, Japan. *Water* **2022**, *14*, 4106. https://doi.org/10.3390/w14244106

Academic Editor: Cesar Andrade

Received: 9 November 2022
Accepted: 14 December 2022
Published: 16 December 2022

Publisher's Note: MDPI stays neutral with regard to jurisdictional claims in published maps and institutional affiliations.

1. Introduction

Groundwater is a globally important resource used for domestic, industrial, and agricultural purposes. In 2018, groundwater dependency for these uses in Japan was 20.1%, 27.3%, and 5.4%, respectively [1]. In some areas, the dependence on groundwater for domestic use is 100%. Therefore, groundwater quality assessment is essential to protect residents' health and preserve their living environment. In Japan, assessments have continuously been carried out for 28 water quality parameters (e.g., cadmium, lead, arsenic, dichloromethane, carbon tetrachloride, benzene, nitrate+nitrite-nitrogen (NO_3+NO_2-N), fluoride, and boron) for which environmental standards of groundwater pollution are established under the Basic Environment Law. In 2020, an investigation was conducted for 3103 wells to assess the overall groundwater quality. In total, 10 parameters were found to exceed environmental standards [2]. NO_3+NO_2-N had the highest exceedance at 3.3%. This was followed by arsenic at 2.1% and fluoride at 0.8%. NO_3+NO_2-N has continued to present the highest exceedance rates since it was added to the criteria in 1999, and this is typified throughout Japan. Nitrate pollution in groundwater is the most prevalent type of anthropogenic pollution [3]. In many cases, nitrate pollution in groundwater is caused by fertilizer and livestock wastewater in agricultural areas [3]. Other pollutants, such as arsenic and fluoride stem from natural sources originally contained in rocks [3]. The release and fate of these elements in groundwater are controlled by water-rock interaction processes in hydrogeological paths of the water [4,5].

The Shimabara Peninsula in Nagasaki Prefecture, Japan, depends on groundwater for most of its water supply due to the small amounts of surface water [6]. However, the peninsula is experiencing severe nitrate pollution in groundwater. Water quality monitoring has shown that the number of tap water source (groundwater) wells above environmental standards for NO_3+NO_2-N ranged from 5.1% to 10.0% during 2005 to 2020 [7]. Although the general pollution level is slowly decreasing, some wells are still showing increasing concentrations and require continuous monitoring. The Shimabara Peninsula consists of three cities: Shimabara, Unzen, and Minami-Shimabara. In 2011, we started an investigation on groundwater quality in Shimabara City, where pollution is particularly severe. We found concentrations up to 26.6 mg/L NO_3-N related to intensive agricultural activities (fertilizer and livestock production) [8]. Further investigation displayed pollution spread with depth in groundwater [9] and that there is a strong relationship between pollution in soil and groundwater [10]. In 2018, the study area was expanded to Unzen City. We found maximum NO_3-N concentrations of 19.9 mg/L and that the pollution is related to agriculture as in Shimabara City [11]. Although the assessment of groundwater pollution has been established for Shimabara and Unzen Cities; groundwater chemistry including nitrate pollution has not been studied in the Minami-Shimabara City. For this reason, we evaluated spatial characteristics of groundwater chemistry in the Minami-Shimabara City for improved understanding of the status of nitrate pollution in this area and the hydrochemical characteristics of groundwater in vulnerable aquifers compared to the hydrochemistry of the Shimabara and Unzen Cities.

2. Materials and Methods

2.1. Study Area

Minami-Shimabara City is located in the south of Shimabara Peninsula, Nagasaki Prefecture (Figure 1) with an area of approximately 17,000 ha. In 2017, the population was 44,200 with a water supply coverage of 92.8% [12]. There are 5 surface reservoirs, a main river, 60 groundwater sources, and 2 springs supplying water to the city [12]. Groundwater has the highest ratio of water withdrawal, reaching 86.1% [12]. The land use map is shown in Figure 1a [13]. Forests cover 35.6% of the total area [14]. The agricultural land area is 4730 ha (paddy fields 1650 ha; upland fields 3070 ha) [14], equivalent to 27.8% of the city area. This ratio is higher than that of Nagasaki Prefecture as a whole and is characterized by agriculture use in upland fields. Mainly cultivated crops are rice, potatoes, fodder crops, and leaf tobacco. Nitrogen application by chemical fertilizer and manure for the 10 major crops is estimated at 1715 kg/day [7]. Livestock production is thriving, with 223 livestock facilities (49 dairies, 138 beef cattle, 14 pork, 5 egg, and 17 broiler production units) in 2019 [7]. This corresponds to 35,950 cattle, 79,000 pigs, and 2,900,000 chicken [7]. The resulting nitrogen generation from livestock waste is estimated at 6625 kg/day [7].

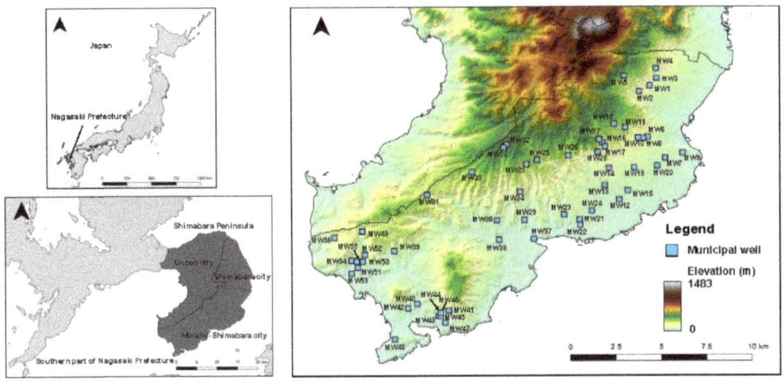

Figure 1. Study area and groundwater sampling locations in the Minami-Shimabara City.

Figure 2b shows the geology of the study area [15]. In the southern part of the area, the Kuchinotsu formation, which consists mainly of freshwater sediment and shallow-marine sediment of middle Pliocene to early Pleistocene age [16], is exposed. Pre-Unzen volcanic rocks (4 Ma–500 ka) are partially distributed above the Kuchinotu formation. Pre-Unzen volcanic rocks and Kuchinotsu formation are overlain by the Unzen volcanic rocks (500 ka–present), which are widely distributed in the Shimabara Peninsula [16]. Unzen volcanic rocks are observed near the southern border and in the center of the peninsula. Alluvial deposits are found at lower elevation areas in the north. Pre-unzen volcanic rocks are mainly composed of olivine basalt and amphibole andesite, while Unzen volcanic rocks contain hornblende andesite to dacite rich in plagioclase, hornblende, and biotite mottling [16]. Figure 2c shows the hydrogeological map of the study area [17]. Groundwater levels are only evident in a part of the study area. There are many faults in this area, and the groundwater levels differ greatly due to these. Especially, in the south and east, groundwater levels are complex due to faults.

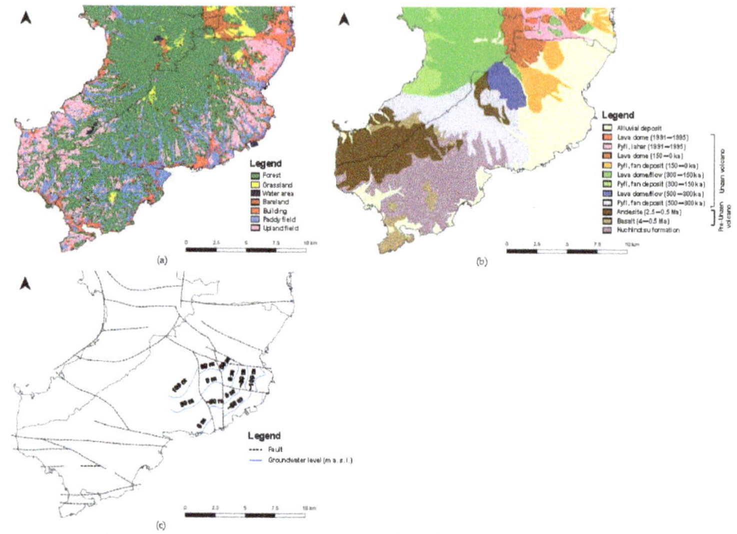

Figure 2. Land use and geology, (**a**) vegetation, (**b**) geology, (**c**) hydrogeology; (Pyfl) Pyroclastic flow. Vegetation map is based on data collected by Biodiversity Center of Japan [13]. Geological map 1/200,000 scale is based on data collected by Geological Survey of Japan [15]. Hydrogeological map is based on the groundwater investigation by Murakami [17].

The climate is humid subtropical. Mean temperature and mean annual precipitation are 17.3 °C and 1836 mm, respectively (1992 to 2021; data from Kuchinotsu observatory, 32°36.7′ N, 130°11.6′ E) [18].

2.2. Sampling and Analyses

Groundwater samples were taken from 56 municipal wells (Figure 1) during July to August 2021. All samples were stored in pre-washed bottles. Stagnant water in the well pipes was removed before water sampling. pH, electrical conductivity (EC), oxidant redox potential (ORP), and dissolved oxygen (DO) were measured on-site by using hand-held instruments (HORIBA D-51 and D-54, and HACH HQ30d). Bicarbonate ion (HCO_3^-) was determined by the titration method with 0.1 N HCl on-site. All water samples were filtered by 0.45 μm membrane filter on-site. Major anions (Cl^-, NO_3^-, SO_4^{2-}) and cations (Na^+, K^+, Mg^{2+}, Ca^{2+}) were measured by ion-chromatography (DKK-TOA, IA-300) in the laboratory. The validity of the analytical results was confirmed by converting ion concentrations from mg/L to $mmol_c$/L and then calculating the charge balance errors (CBE) using:

$$CBE = \left(\frac{\sum cations - \sum anions}{\sum cations + \sum anions} \right) \times 100, \tag{1}$$

According to Rahman et al. [19], CBE < ±5 is good, and CBE < ±10 is acceptable. In this study, the CBEs of the samples ranged from −3.6 to 7.3, with <±5 in 41 samples and <±10 in 15 samples.

Principal component analysis (PCA) and hierarchical cluster analysis (HCA) for the hydrochemical parameters were used to interpret processes controlling water chemistry (mineral dissolution, anthropogenic input, sea water intrusion, ion exchange, evapotranspiration, etc.) and grouping the water samples e.g., [20–24]. These methods were also used previously for the Shimabara and Unzen groundwater [8,11]. PCA and HAC were performed based on major ion concentrations (Cl^-, NO_3^-, SO_4^{2-}, HCO_3^-, Na^+, K^+, Mg^{2+}, and Ca^{2+}). The principal components were extracted based on the Kaiser criterion to only retain components with eigenvalues greater than 1. The HCA was performed based on

Ward's method. All analyses were performed by using the statistical software JMP Pro 13 (SAS Institute Inc., 100 SAS Campus Drive, Cary, NC 27513-2414, USA).

3. Results and Discussion

3.1. Water Chemistry

The analysis results of hydrochemical parameters are shown in Table 1 and Figure 3. pH ranged from 6.3 to 8.7, indicating that groundwater in the study is weakly acidic to weakly alkaline. EC ranged from 74 to 563 μS/cm. The range of pH and EC was generally close to the Shimabara and Unzen groundwater [8,11]. ORPs ranged from −21 to 738 mV, meaning that groundwater is oxic or anoxic. The latter was observed at two sampling points (MW51 and MW55). The redox conditions are strong indicators of contaminants that might be present at elevated concentrations [25]. The nitrate concentration is more likely to exceed the recommended limits for oxic conditions [25]. On the other hand, microbially driven reduction of nitrate to nitrogen gas occurs only under anoxic conditions [25]. Gillham and Cherry [26] reported that denitrification processes can occur when the DO in groundwater is less than 2 mg/L. In our study, DO ranged from 1.7 to 10.2 mg/L, with two sampling points (MW44 and MW50) showing less than 2.0 mg/L. Thus, at sampling points (MW44, MW50, MW51, and MW55) with low DO or ORP, NO_3^- concentrations between 0.1 to 1.9 mg/L indicate that denitrification is likely to occur. These conditions were not observed in Shimabara and Unzen [8,11]. The total dissolved solids (TDS) were estimated from EC by the following equation [27].

$$TDS = 640 \times EC, \tag{2}$$

where TDS and EC are in mg/L and dS/m, respectively. TDS ranged from 47.4 to 360.3 mg/L with a mean of 134.4 mg/L. The samples were characterized by HCO_3^- followed by SO_4^{2-}, Cl^-, and NO_3^- for anions (Figure 3). Cations were characterized by Ca^{2+} and Na^+ followed by Mg^{2+} and K^+.

Table 1. Descriptive statistics of hydrochemical components of groundwater samples.

	Cl^-	NO_3^-	SO_4^{2-}	HCO_3^-	Na^+	K^+	Mg^{2+}	Ca^{2+}	pH	EC	ORP	DO
	mg/L	mg/L	mg/L	mg/L	mg/L	mg/L	mg/L	mg/L		μS/cm	mV	mg/L
Min	3.3	0.1	1.4	11.4	6.3	3.1	1.6	3.9	6.3	74	−21	1.7
Max	49.2	55.7	43.5	299.7	134.3	7.5	18.2	37.0	8.7	563	738	10.2
Mean	10.0	7.9	12.7	81.5	17.9	4.9	6.3	14.6	7.4	210	235	6.8
S.D. [1]	7.7	11.2	11.2	60.3	23.8	1.0	4.0	8.7	0.5	126	106	2.5

Note: [1] S.D. = standard deviation.

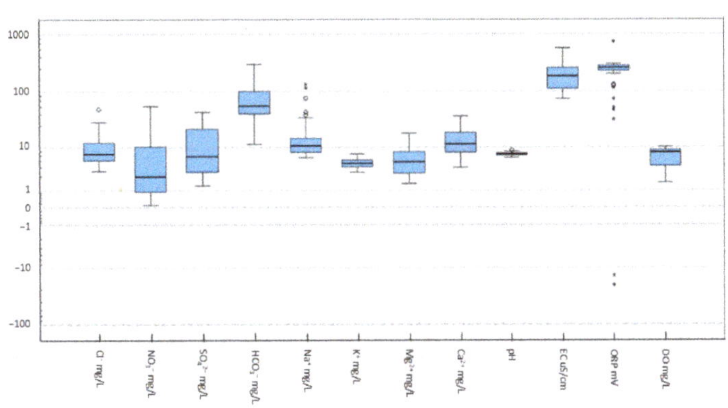

Figure 3. Box plots of hydrochemical components of groundwater samples. A circle is a far outlier. Starred point is a "far out" outlier.

Groundwater samples were plotted using trilinear Piper diagram for easy visual understanding the characteristics of water chemistry (Figure 4). Forty-four sampling points (78.6%) were classified as alkaline earth carbonate type, which is common in shallow groundwater in Japan. Nine points (16.1%) showed alkaline earth non-carbonate type, which is classified as groundwater with high nitrate concentration in Shimabara and Unzen [8,11]. Three samples (5.3%) were classified as alkaline carbonate type. This type of water is commonly found in deep groundwater with long residence time and was not identified in Shimabara and Unzen [8,11]. To consider salinity, Total Ionic Salinity (TIS) is shown in Figure 5 [28]. According to the figure groundwater in Minami-Shimabara has low TIS (<10 $mmol_c/L$).

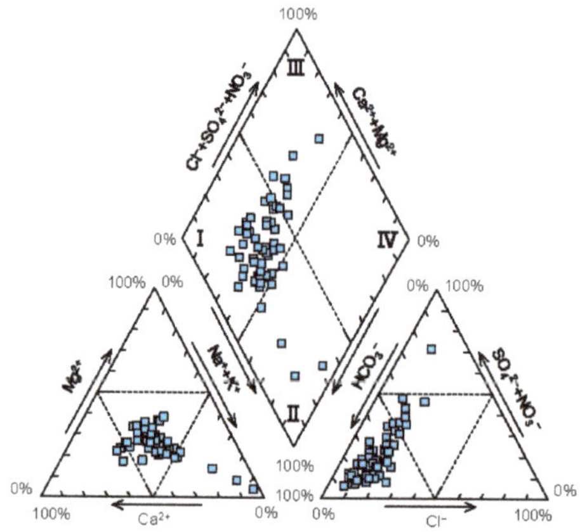

Figure 4. Trilinear Piper diagram of groundwater in Minami-Shimabara City.

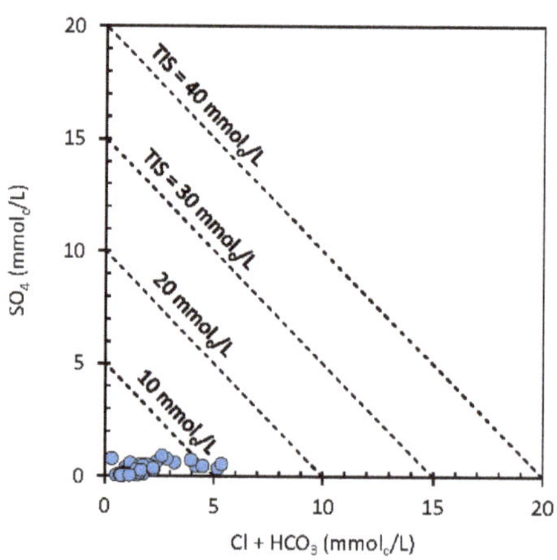

Figure 5. Total Ionic Salinity of groundwater in Minami-Shimabara City.

The concentration of major dissolved ions is visualized using Stiff diagram in Figure 6. Most of the samples represent the $Ca\text{-}HCO_3$ type. The $Ca\text{-}(SO_4+NO_3)$, $Ca\text{-}SO_4$, $Na\text{-}HCO_3$, and $Mg\text{-}HCO_3$ types were observed in a small number of wells. $Na\text{-}HCO_3$ and $Mg\text{-}HCO_3$ were found in the coastal area and thus, have larger arrows than other samples. To decipher mechanisms controlling groundwater chemistry, Gibbs diagram [29–31] is shown in Figure 7. Gibbs diagram is described by the ratio of cation and anion endmembers ($Na^+/(Na^+ + Ca^{2+})$) and ($Cl^-/(Cl^- + HCO_3^-)$), respectively, as a function of TDS. This indicates the functional sources of dissolved chemical constituents, such as precipitation, rock weathering, or evaporation dominance. The diagram shows that most groundwater samples are affected by rock weathering processes, suggesting that longer residence time, large contribution of mineral dissolution, and the influence of ion exchange form water chemistry in the study area. A few samples were influenced by precipitation. The difference of these factors appears as the difference of ion concentrations and water type. The water samples in precipitation dominance have lower concentration. Higher concentrations and water types of Na-HCO_3 and Mg-HCO_3 are dominated by rock weathering processes. Influence of seawater was mainly found in the south part of Minami-Shimabara City [17], but no Na-Cl type samples were identified. Figure 8 shows the distribution of groundwater samples in the Hydrochemical Facies Evolution (HFE) diagram [32] by using Excel Macro [33]. HFE-diagram reveals that 70% of the groundwater samples correspond to freshening phase and 30% correspond to intrusion phase. Since the study area faces to the sea, it is assumed that groundwater samples, which is characterized by high concentration and ratio of Na^+, such as MW42, MW43, and MW44 in the left upper corner of freshening phase, are affected by seawater intrusion.

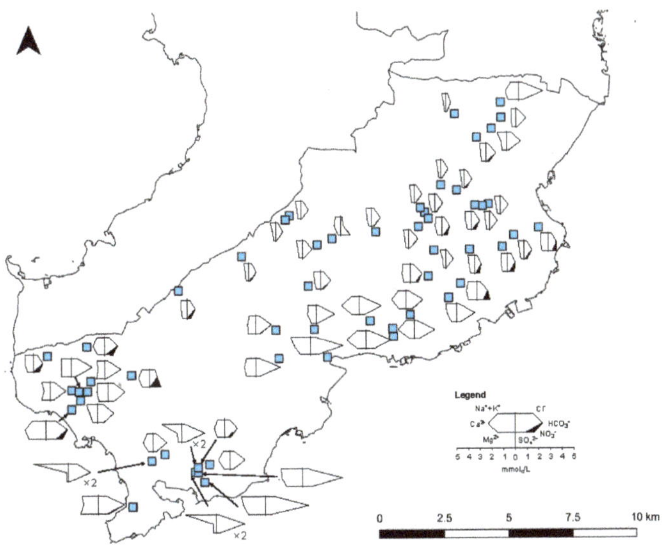

Figure 6. Major ion variation in Minami-Shimabara City groundwater using Stiff diagrams.

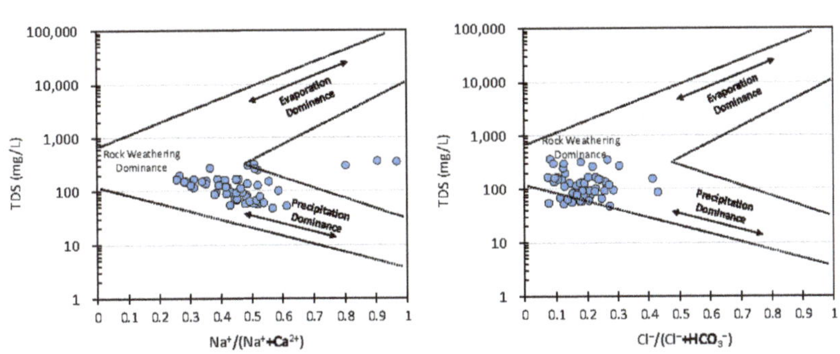

Figure 7. Gibbs diagram of groundwater in Minami-Shimabara City.

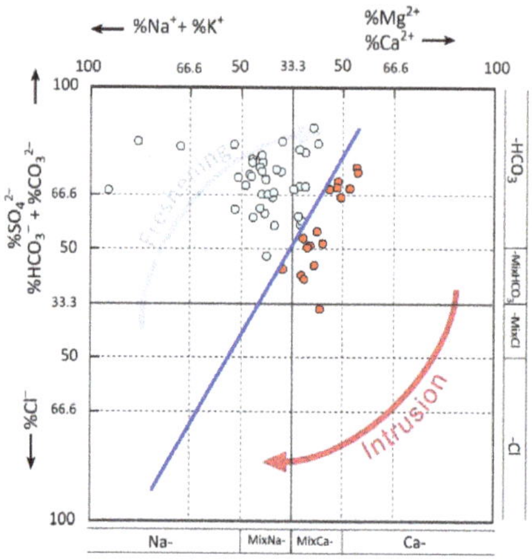

Figure 8. Hydrochemical Facies Evolution (HFE) of groundwater in Minami-Shimabara City.

The southern part of the city has little surface water and relies on groundwater for irrigation. However, as mentioned above, groundwater with high Na^+ concentrations were observed. Therefore, the suitability of groundwater for irrigation was evaluated by plotting a USSL diagram (Figure 9). The USSL diagram is drawn based on the salinity (EC) and sodium hazard (Sodium Adsorption Ratio: SAR) [34]. The SAR is calculated by [34]:

$$SAR = \frac{Na^+}{\sqrt{\left(Ca^{2+} + Mg^{2+}\right)/2}}, \quad (3)$$

where the concentrations of Na^+, Ca^{2+}, and Mg^{2+} are all expressed in $mmol_c/L$. Most of the groundwater samples belong to C1-S1 (low salinity hazard and low sodium hazard) class or C2-S1 (medium salinity hazard and low sodium hazard). Only two samples (MW42 and MW43) were classified in the C2-S2 (medium salinity hazard and medium sodium hazard) class. The following is a summary of each class [34]. Low salinity water (C1) can be used for irrigation with most crops on most soils with little likelihood that soil salinity will develop. Some leaching is required, but this occurs under normal irrigation practices except in soils of extremely low permeability. Medium salinity water (C2) can be used if a moderate amount of leaching occurs. Plants with moderate salt tolerance can be grown in most cases without species practices for salinity control. Low salinity water (S1) can be used for irrigation on almost all soils with little danger of the development of harmful levels of exchangeable sodium. However, sodium sensitive crops, such as stone-fruit trees and avocados may accumulate harmful concentrations of sodium. Medium sodium water (S2) will present an appreciable sodium hazard in fine textured soils having high cation exchange capacity, especially under low leaching conditions, unless gypsum is present in the soil. Consequently, Figure 9 shows that groundwater in the study area can be used for irrigation, but attention must be paid to the use of groundwater belonging to the C2-S2 class. Further evaluation of water quality for irrigation was conducted based on Food and Agriculture Organization (FAO) guidelines [35] (Table 2). According to the FAO guidelines, although most of the groundwater samples has no potential irrigation problem, some samples have elevated EC, SAR, Na, NO_3-N, and HCO_3 or even potentially severe problem in view of SAR. In addition, some of the samples have a pH somewhat out of range. Therefore, caution should be exercised in some places when using groundwater. Since most tree crops and woody plants are sensitive to sodium [35], groundwater having high SAR should be avoided. These results are generally consistent with the caution given from the USSL diagram.

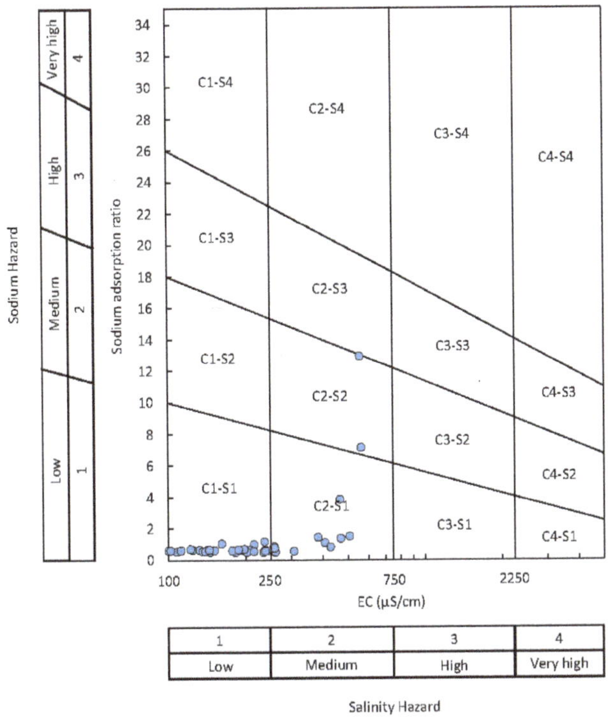

Figure 9. USSL diagram based on EC and SAR.

Table 2. Guidelines for interpretations of water quality for irrigation.

	Unit	Degree of Restriction on Use		
		None	Slight to Moderate	Severe
EC	dS/m	<0.7	0.7–3.0	>3.0
TDS	mg/L	<450	450–2000	>2000
Na (surface irrigation)	SAR	<3	3–9	>9
Na (sprinkler irrigation)	meq/L	<3	>3	
Cl (surface irrigation)	meq/L	<4	4–10	>10
Cl (sprinkler irrigation)	meq/L	<3	>3	
NO_3-N	mg/L	<5	5–30	>30
HCO_3	meq/L	<1.5	1.5–8.5	>8.5
pH			Normal range 6.5–8.4	

3.2. Nitrate Pollution

The nitrate (NO_3-N) concentration ranged between 0.02 and 12.6 mg/L, with an average of 1.8 mg/L. The maximum concentration was found at MW39, the only point where NO_3-N concentration exceeded the Japanese drinking water standard (10 mg/L). The drinking water standard of World Health Organization [36] for NO_3^- concentration (50 mg/L) was exceeded in one point. The maximum nitrate concentration in Shimabara City was 26.6 mg/L [8] and that in Unzen City 19.9 mg/L [11]. Compared to these values, the pollution level in Minami-Shimabara City is lower. Only one sampling point exceeded the standard limits for NO_3-N concentration. This is lower than the 38% in Shimabara City [8] and the 12% in Unzen City [11]. Even if the standard limits were not exceeded, relatively high NO_3-N concentrations were observed, e.g., 8.4 mg/L at MW49, 7.3 mg/L at MW53, and 6.5 mg/L at MW15. Nitrate can occur naturally, but nitrate concentrations greater than 1 mg/L are likely to indicate effects of human activities [37]. Under this criterion, 21 sampling points (38%: except for the smallest blue circle in Figure 10) can be considered polluted by nitrate. As mentioned above, denitrification may occur in the groundwater in this area, and it is considered that at some sampling points the nitrate concentration has decreased below 1 mg/L due to denitrification. Sampling points with nitrate concentration exceeding the criteria of 1.0 mg/L [37] were observed in the southwestern and eastern parts of the study area (Figure 10), and the land use here is upland agriculture (Figure 2). This suggests that nitrate pollution of groundwater in this area is related to agricultural activities.

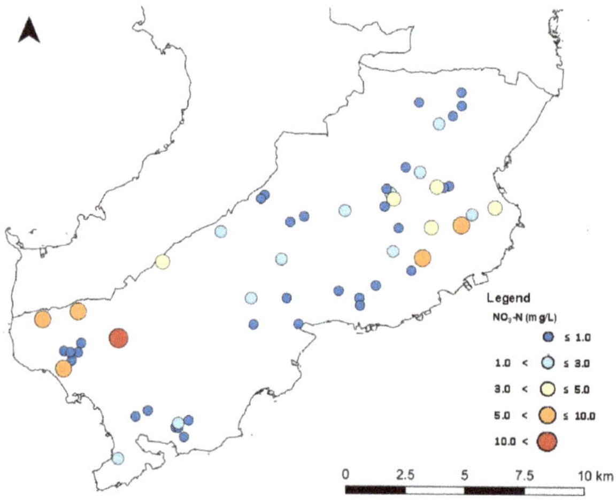

Figure 10. Distribution of nitrate concentration in groundwater.

Correlation among the eight major dissolved ion components is shown in Table 3. In the investigation of Shimabara and Unzen [8,11], NO_3^- showed a high positive correlation with Cl^-, SO_4^{2-}, and K^+, suggesting that nitrate is derived from chemical fertilizers ($(NH_4)_2SO_4$), manure, and livestock waste. However, in Minami-Shimabara, NO_3^- is not correlated with Cl^-, SO_4^{2-}, and K^+. Focusing on Cl^-, SO_4^{2-}, and K^+ concentrations in MW39, where NO_3-N concentration exceeded the Japanese drinking water standard, the relationship is as follows: Cl^- 16.9 mg/L > K^+ 5.1 mg/L > SO_4^{2-} 4.2 mg/L. For MW49 with NO_3-N of 8.4 mg/L, which has the next highest NO_3-N concentration, a relationship of Cl^- 12.3 mg/L > K^+ 6.4 mg/L > SO_4^{2-} 2.7 mg/L was observed. The SO_4^{2-} concentration at these points is similar to or lower than those at other points with low NO_3-N concentration (Figure 6). At the sampling points where such a relationship between ion concentrations is found, SO_4^{2-} concentration is low, and manure and/or livestock waste are supposed to be a dominant nitrate source. Analysis of stable nitrogen oxygen isotopes NO_3^- is required for a more valid assessment of the nitrate source.

Table 3. Correlation matrix for eight dissolved ions (n = 56).

	Cl^-	NO_3^-	SO_4^{2-}	HCO_3^-	Na^+	K^+	Mg^{2+}	Ca^{2+}
Cl^-	1.00	0.15	0.50	0.65	0.74	−0.10	0.52	0.39
NO_3^-	-	1.00	−0.05	−0.26	−0.13	0.24	0.14	0.10
SO_4^{2-}	-	-	1.00	0.57	0.38	0.12	0.71	0.72
HCO_3^-	-	-	-	1.00	0.84	0.07	0.56	0.52
Na^+	-	-	-	-	1.00	−0.10	0.19	0.10
K^+	-	-	-	-	-	1.00	0.20	0.32
Mg^{2+}	-	-	-	-	-	-	1.00	0.83
Ca^{2+}	-	-	-	-	-	-	-	1.00

3.3. Multivariate Analysis

A principal component analysis summarized the 12 hydrochemical parameters into three principal components based on the Kaiser index (Table 4). The eigenvalues of principal component 1 (PC1), 2 (PC2), and 3 (PC3) were 5.64, 1.92, and 1.33, respectively. These explained 46.9%, 16.0%, and 11.1%, respectively, of the total variance. PC1 had positive loadings for EC and all ions except NO_3^- and K^+, indicating that it is related to dissolved constituents controlled by rock weathering and precipitation. The negative loadings of ORP and DO in PC1 are related to denitrification. PC2 had positive loadings for Ca^{2+} and negative loadings for Na^+, implying ion exchange. The positive loading of NO_3^- in PC2 represents lower nitrate pollution exceeding the criteria of 1.0 mg/L indicating effects of human activities. PC3 had positive loadings for Cl^- and NO_3^-, indicating relatively severe nitrate pollution including the exceedance of standard limits.

Table 4. Relationship between extracted principal components (PCs) and ions.

	PC 1	PC 2	PC 3
Cl^-	0.770	−0.213	0.337
NO_3^-	−0.083	0.421	0.775
SO_4^{2-}	0.793	0.278	−0.093
HCO_3^-	0.903	−0.267	0.006
Na^+	0.708	−0.599	0.235
K^+	0.104	0.543	0.286
Mg^{2+}	0.716	0.548	−0.002
Ca^{2+}	0.773	0.508	−0.029
pH	0.430	−0.560	0.227
EC	0.969	−0.082	0.182
ORP	−0.518	−0.113	0.435
DO	−0.755	−0.084	0.447
Eigenvalue	5.64	1.92	1.33
Explained variance %	46.9	16.0	11.1
Cumulative % of variance	46.9	62.9	74.0

Results of the HCA are shown in Table 5 and Figure 11. The 56 groundwater samples were classified into five groups, with the number of samples in each group ranging from 3 to 26. Figure 12 shows the scatter plot of the principal components related to groups. PC1 effectively separated groups, meaning that dissolved constituents are different for each group. Group 1 shows relatively low scores for PC1, indicating lower dissolved constituents. Mainly the sampling points indicated by smaller Stiff diagrams in Figure 6 are contained in this group. Group 1 is distributed from the center to east in the study area. Group 2 and 3 has similar scores of PC1 and PC2, representing intermediate ion concentrations. However these groups are be distinguished by PC3, meaning that nitrate pollution level are different for these groups. As shown in Table 5, water samples having higher NO_3^- concentrations were classified into Group 2. The difference between Group 2 and 3 is characterized by difference in locality (Figure 13). Group 2 is located in the western part of study area, while Group 3 is mainly located at lower altitude along the coastal line. PC1 and PC3 of Group 4 is similar to that of Group 5, but these groups are clearly distinguished by PC2. In other words, Group 4 has relatively high negative scores for PC2, indicating negative effects of ion exchange, and anthropogenic activities. Group 4 is associated with deep groundwater with high concentrations of sodium and bicarbonate ions (Table 5). The five groups can be generally summarized in three categories.PC1 and Figure 11 show that Group 4 and 5, which are characterized by high ion concentration especially Na^+ and HCO_3^-, are different from the other groups. Based on overall ion concentration including nitrate, the remaining groups can be categorized by either intermediate ion concentration (Group 3) or lower ion concentration (Group 1 and 2).

Table 5. Hydrochemical component depending on each group.

	Samples		Cl^-	NO_3^-	SO_4^{2-}	HCO_3^-	Na^+	K^+	Mg^{2+}	Ca^{2+}	NO_3-N	pH	EC	ORP	DO
			mg/L	mg/L	mg/L	mg/L	mg/L	mg/L	mg/L	mg/L	mg/L		μS/cm	mV	mg/L
Group 1	26	Min	3.3	0.1	1.4	23.2	6.3	3.1	1.6	4.2	0.0	6.8	74	242	7.2
		Max	12.1	26.1	17.9	97.5	15.6	6.1	6.4	14.7	5.9	7.9	190	738	10.1
		Mean	5.9	7.4	4.3	44.4	8.7	4.7	3.5	8.2	1.7	7.3	116	287	8.8
		S.D.	1.9	7.4	3.2	13.7	2.0	0.8	1.2	2.6	1.7	0.2	29	93	0.7
Group 2	3	Min	10.9	28.6	2.2	41.3	11.6	5.1	5.7	14.2	6.5	7.0	187	201	7.3
		Max	16.9	55.7	4.2	65.7	11.7	6.4	7.4	19.9	12.6	7.4	241	287	10.2
		Mean	13.4	40.5	3.0	52.5	11.7	5.8	6.8	17.6	9.2	7.3	222	230	8.5
		S.D.	3.1	13.8	1.0	12.3	0.1	0.7	1.0	3.0	3.1	0.2	30	49	1.5
Group 3	20	Min	5.0	0.1	6.5	11.4	7.7	3.3	2.5	9.9	0.0	6.3	136	−21	1.9
		Max	16.1	28.9	37.3	171.8	34.7	7.5	11.0	33.3	6.5	8.0	383	302	9.0
		Mean	9.6	3.8	19.3	90.4	14.1	5.3	7.6	19.3	0.9	7.3	234	204	5.1
		S.D.	3.4	6.7	8.0	33.6	5.8	1.3	2.2	6.5	1.5	0.5	52	101	2.2
Group 4	3	Min	13.8	1.9	17.3	231.3	76.3	3.4	2.6	3.9	0.4	8.1	466	74	1.7
		Max	49.2	2.5	26.6	299.7	134.3	4.8	8.2	16.5	0.6	8.7	563	231	6.2
		Mean	26.2	2.1	21.9	254.3	108.8	4.3	5.5	10.3	0.5	8.4	529	177	3.7
		S.D.	19.9	0.3	4.7	39.3	29.6	0.8	2.8	6.3	0.1	0.3	55	89	2.3
Group 5	4	Min	17.5	0.7	22.0	111.2	23.6	3.9	16.6	24.3	0.2	6.9	407	32	2.3
		Max	28.9	32.1	43.5	224.2	43.7	5.2	18.2	37.0	7.3	8.3	507	207	6.2
		Mean	24.2	11.2	34.5	170.6	33.8	4.6	17.5	33.2	2.5	7.7	453	104	3.9
		S.D.	5.4	14.7	9.0	55.5	9.2	0.6	0.7	6.0	3.3	0.6	44	81	1.7

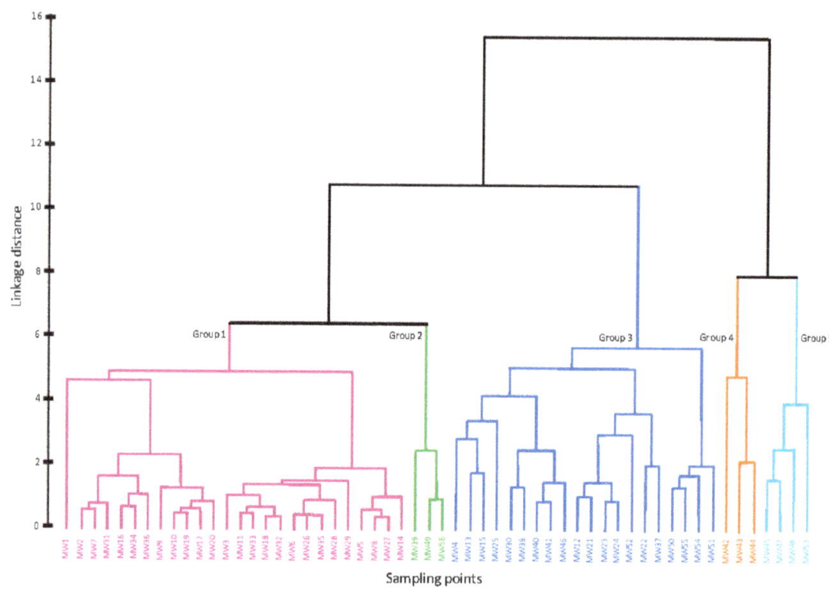

Figure 11. Dendrogram for groundwater samples divided into five groups.

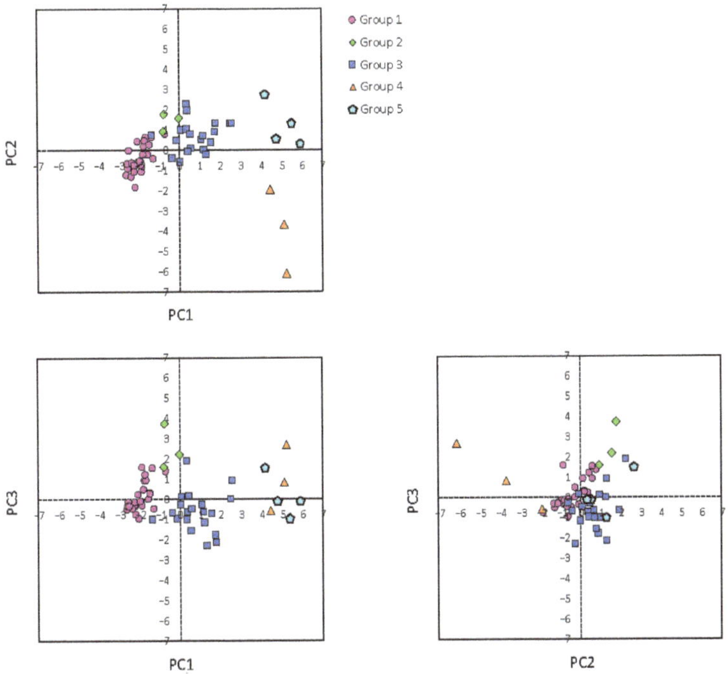

Figure 12. Scatter plots for PC1, PC2, and PC3.

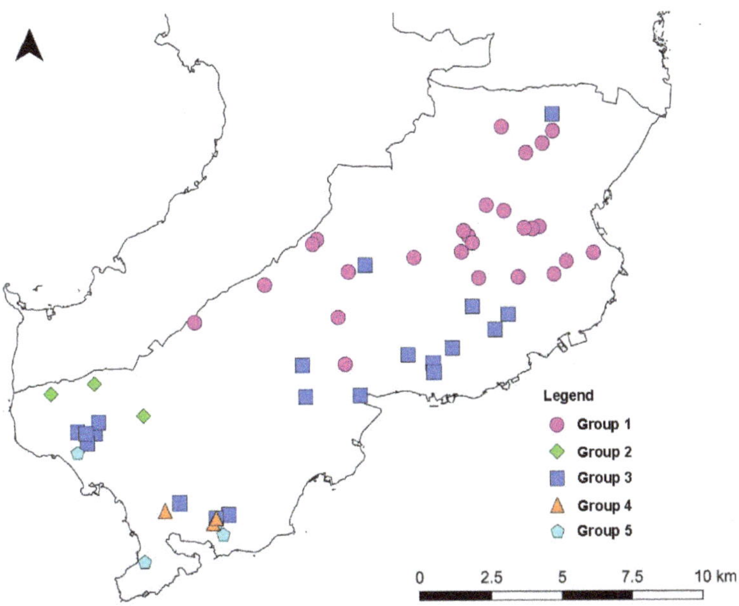

Figure 13. Spatial location of each group.

4. Conclusions

To improve the understanding of cause and effects of nitrate pollution and hydrochemical characteristics of groundwater in the Shimabara Peninsula, groundwater samples were collected from 56 municipal wells in Minami-Shimabara City. Major dissolved ions and pH, EC, ORP, and DO were analyzed. The ORP and DO values suggested that denitrification may be responsible for decreasing nitrate concentrations. The major groundwater composition was Ca-HCO_3 type. In addition, Na-HCO_3, Mg-HCO_3, and Ca-(SO_4+NO_3) types were also observed at a few locations. This water chemistry is formed by rock weathering, precipitation, and mixing with saltwater from seawater intrusion. NO_3-N concentrations exceeded Japanese drinking water standards (10 mg/L) at one location. The pollution is related to agricultural land use. The high Cl^- and low SO_4^{2+} concentrations at this point suggest that the pollution source is manure and/or livestock waste. PCA showed that processes controlling water chemistry are explained by three principal components: PC1 corresponds to dissolved constituents in groundwater and denitrification, PC2 represents ion exchange and low nitrate pollution, and PC3 represents severe nitrate pollution. HCA classified the 56 water samples 5 five groups. These can be broadly divided into 3 categories: the first characterized by high ion concentration especially in Na^+ and HCO_3^- (Group 4 and 5); the second representing the intermediate ion concentration group (Group 3), and the third with low ion concentration (Group 1 and 2).

The study revealed that the extent of nitrate pollution in Minami-Shimabara City is small. In other words, NO_3-N concentrations are lower than in Shimabara and Unzen Cities, and the percentage of NO_3-N exceeding the standard limits is also small. The sampling campaign was between July and August when rainfall amount was large. In the investigations conducted during this period, a decrease in nitrate concentration was observed in Shimabara City due to dilution caused by rainfall [8,10]. Therefore, future surveys are needed at different seasons. It is also necessary to introduce analysis of nitrogen oxygen stable isotopes of NO_3^- to clarify the nitrate source.

Since this study found high Na^+ concentrations in some wells, we additionally evaluated the suitability of groundwater for agricultural use. Groundwater was evaluated using the USSL diagram and FAO guideline. In some places, due to high EC, SAR, Na, NO_3-N, HCO_3, and pH, caution is necessary when using groundwater for irrigation. In addition, groundwater with high SAR should not be used for some types of crop species. In Minami-Shimabara City, agricultural wells have been installed in addition to the municipal wells that were investigated in this study. Since groundwater is affected by seawater in some areas [17], it is necessary to collect groundwater samples for agricultural wells to evaluate their suitability for agricultural production.

Author Contributions: Conceptualization, methodology, supervision, K.N.; investigation, K.N. and Z.-Q.Y.; formal analysis, data curation, H.A., K.N. and Z.-Q.Y.; writing—original draft preparation, K.N. and H.A.; writing—review and editing, K.N. and R.B.; funding acquisition, K.N. All authors have read and agreed to the published version of the manuscript.

Funding: This work was supported by JSPS KAKENHI Grant Number JP20K12209.

Institutional Review Board Statement: Not applicable.

Informed Consent Statement: Not applicable.

Data Availability Statement: The data presented in this study are available on request from the corresponding author.

Acknowledgments: The authors thank the Water Works Bureau of Minami-Shimabara City for support of the water sampling.

Conflicts of Interest: The authors declare no conflict of interest.

References

1. Current Status of Groundwater Utilization. Available online: https://www.mlit.go.jp/mizukokudo/mizsei/mizukokudo_mizsei_tk1_000062.html (accessed on 12 October 2022).
2. Groundwater Quality Measurement Results for FY2022. Available online: https://www.env.go.jp/water/report/r03-03/post_3.html (accessed on 12 October 2022).

3. Kuroda, K.; Fukushi, T. Groundwater Contamination in Urban Areas. In *Groundwater Management in Asian Cities*; Takizawa, S., Ed.; Springer: Tokyo, Japan, 2008; pp. 125–149.
4. Fuoco, I.; Marini, L.; De Rosa, R.; Figoli, A.; Gabriele, B.; Apollaro, C. Use of reaction path modelling to investigate the evolution of water chemistry in shallow to deep crystalline aquifers with a special focus on fluoride. *Sci. Total Environ.* **2022**, *830*, 154566. [CrossRef] [PubMed]
5. Fuoco, I.; De Rosa, R.; Barca, D.; Figoli, A.; Gabriele, B.; Apollaro, C. Arsenic polluted waters: Application of geochemical modelling as a tool to understand the release and fate of the pollutant in crystalline aquifers. *J. Environ. Manag.* **2022**, *301*, 113796. [CrossRef] [PubMed]
6. Shimabara Area Peninsula Development Plan. Available online: https://www.mlit.go.jp/common/001124224.pdf (accessed on 15 October 2022).
7. The Second Term of Shimabara Peninsula Nitrate Load Reduction Project, Revised Edition in FY2020. Available online: https://www.pref.nagasaki.jp/shared/uploads/2021/04/1617338157.pdf (accessed on 12 October 2022).
8. Nakagawa, K.; Amano, H.; Asakura, H.; Berndtsson, R. Spatial trends of nitrate pollution and groundwater chemistry in Shimabara, Nagasaki, Japan. *Environ. Earth. Sci.* **2016**, *75*, 234. [CrossRef]
9. Amano, H.; Nakagawa, K.; Berndtsson, R. Groundwater geochemistry and nitrate-contaminated agricultural site. *Environ. Earth. Sci.* **2016**, *75*, 1145. [CrossRef]
10. Nakagawa, K.; Amano, H.; Persson, M.; Berndtsson, R. Spatiotemporal variation of nitrate concentrations in soi and groundwater of an intensely polluted agricultural area. *Sci. Rep.* **2021**, *11*, 2598. [CrossRef] [PubMed]
11. Nakagawa, K.; Amano, H.; Berndtsson, R. Spatiotemporal characteristics of groundwater chemistry in Unzen, Nagasaki, Japan. *Water* **2021**, *13*, 426. [CrossRef]
12. Minami-Shimabara City Waterworks Business Management Strategy Development Outsourcing. Available online: https://www.city.minamishimabara.lg.jp/common/UploadFileOutput.ashx?c_id=3&id=6250&sub_id=6&flid=32496 (accessed on 15 October 2022).
13. Vegetation Survey 1/25,000. Available online: http://gis.biodic.go.jp/webgis/sc-025.html?kind=vg67 (accessed on 17 October 2022).
14. Agriculture, Forestry, and Fisheries in Graphs and Statistics. Available online: http://www.machimura.maff.go.jp/machi/contents/42/214/details.html (accessed on 15 October 2022).
15. Seamless Digital Geological Map of Japan V2 1: 200,000. Available online: https://gbank.gsj.jp/seamless/ (accessed on 19 October 2022).
16. Sugimoto, T. Geology and petrology at Shimabara Peninsula, Kyushu, Japan -From recent results-. *J. Geotherm. Res. Soc. Jpn.* **2006**, *28*, 347–360.
17. Murakami, T. *1:50,000 Hydrogeological Maps of Japan Shimabara Peninsula, No.25*; Geological Survey of Japan: Kawasaki, Japan, 1975. Available online: https://www.gsj.jp/Map/EN/environment.html (accessed on 31 October 2022). (In Japanese)
18. Past Weather Data. Available online: https://www.data.jma.go.jp/obd/stats/etrn/index.php?prec_no=84&block_no=0922&year=&month=&day=&view= (accessed on 17 October 2022).
19. Rahman, A.; Mondal, N.C.; Tiwari, K.K. Anthropogenic nitrate in groundwater and its health risks in the view of background concentration in a semi arid area of Rajasthan, India. *Sci. Rep.* **2021**, *11*, 9279. [CrossRef] [PubMed]
20. Chotpantarat, S.; Parkchai, T.; Wisitthammasri, W. Multivariate statistical analysis of hydrochemical data and stable isotopes of groundwater contaminated with nitrate at Huay Sai royal development study center and adjacent areas in Phetchaburi province, Thailand. *Water* **2020**, *12*, 1127. [CrossRef]
21. Zainol, N.F.M.; Zainuddin, A.H.; Looi, L.J.; Aris, A.Z.; Isa, N.M.; Sefie, A.; Yusof, K.M.K.K. Spatial Analysis of Groundwater Hydrochemistry through Integrated Multivariate Analysis: A Case Study in the Urbanized Langat Basin, Malaysia. *Int. J. Environ. Public Health* **2021**, *18*, 5733. [CrossRef] [PubMed]
22. El-Kholy, R.A.; Zaghlool, E.; Isawi, H.; Soliman, E.A.; Khalil, M.M.H.; El-Aassar, A.-H.M.; Said, M.M. Groundwater quality assessment using water quality index and multivariate statistical analysis case study: East Matrouh, Northwestern coast, Egypt. *Environ. Sci. Poll. Res.* **2022**, *29*, 65699–65722. [CrossRef] [PubMed]
23. Kanellopoulos, C.; Argyraki, A. Multivariate statistical assessment of groundwater in cases with ultramafic rocks and anthropogenic activities influence. *Appl. Geochem.* **2022**, *141*, 105292. [CrossRef]
24. Wali, S.U.; Alias, N.B.; Harun, S.B.; Umar, K.J.; Gada, M.A.; Dankani, I.M.; Kaoje, I.U.; Usman, A.A. Water quality indices and multivariate statistical analysis of urban groundwater in semi-arid Sokoto Basin, Northwestern Nigeria. *Groundw. Sustain. Dev.* **2022**, *18*, 100779. [CrossRef]
25. Oxidant/Reduction (Redox). Available online: https://www.usgs.gov/mission-areas/water-resources/science/oxidationreduction-redox (accessed on 28 October 2022).
26. Gillham, R.W.; Cherry, J.A. Field Evidence of Denitrification in Shallow Groundwater Flow System. *Water Qual. Res. J.* **1978**, *13*, 53–72. [CrossRef]
27. Evangelou, V.P. *Environmental Soil and Water Chemistry: Principles and Applications*; John Wiley and Sons: New York, NY, USA, 1998; pp. 80–82.

28. Apollaro, C.; Tripodi, V.; Vespasiano, G.; De Rosa, R.; Dotsika, E.; Fuoco, I.; Critelli, S.; Muto, F. Chemical, isotopic and geotectonic relations of the warm and cold waters of the Galatro and Antonimina thermal areas, southern Calabria, Italy. *Mar. Pet. Geol.* **2019**, *109*, 469–483. [CrossRef]
29. Gibbs, R.J. Mechanisms Controlling World Water Chemistry. *Science* **1970**, *170*, 1088–1090. [CrossRef] [PubMed]
30. Elango, L.; Kannan, R. Rock–water interaction and its control on chemical composition of groundwater. *Dev. Environ. Sci.* **2007**, *5*, 229–243.
31. Mohamed, A.; Asmoay, A.; Alshehri, F.; Abdelrady, A.; Othman, A. Hydro-Geochemical Applications and Multivariate Analysis to Assess the Water–Rock Interaction in Arid Environments. *Water* **2022**, *12*, 6340. [CrossRef]
32. Giménez-Forcada, E. Dynamic of Sea Water Interface using Hydrochemical Facies Evolution Diagram. *Groundwater* **2010**, *48*, 212–216. [CrossRef] [PubMed]
33. Giménez-Forcada, E.; Sánchez San Román, F.J. An Excel Macro to Plot the HFE-Diagram to Identify Sea Water Intrusion Phases. *Groundwater* **2015**, *53*, 819–824. [CrossRef] [PubMed]
34. U.S. Salinity Laboratory Staff. *Diagnosis and Improvement of Saline and Alkali Soils*; Agricultural Handbook No.60.; U.S. Government Printing Office: Washington, DC, USA, 1954; 160p.
35. Ayers, R.S.; Westcot, D.W. *Water Quality for Agriculture*; Food and Agriculture Organization of the United Nations: Rome, Italy, 1985; pp. 6–11.
36. World Health Organization. *Guidelines for Drinking-Water Quality*; Fourth Edition Incorporating the First and Second Addenda; World Health Organization: Geneva, Switzerland, 2022; pp. 438–444.
37. Dubrovsky, N.M.; Burow, K.R.; Clark, G.M.; Gronberg, J.M.; Hamilton, P.A.; Hitt, K.J.; Mueller, D.K.; Munn, M.D.; Nolan, B.T.; Puckett, L.J.; et al. *The Quality of Our Nation's Waters—Nutrients in the Nation's Streams and Groundwater, 1992–2004*; U.S. Geological Survey Circular 1350.; U.S. Geological Survey: Reston, VA, USA, 2010; 174p.

water

MDPI

Article

PFAS in the Drinking Water Source: Analysis of the Contamination Levels, Origin and Emission Rates

Dauren Mussabek [1,2,*], Anna Söderman [1], Tomomi Imura [2], Kenneth M. Persson [1,3,4], Kei Nakagawa [2], Lutz Ahrens [5] and Ronny Berndtsson [1,6]

1 Division of Water Resources Engineering, LTH Lund University, 22100 Lund, Sweden
2 Institute of Integrated Science and Technology, Nagasaki University, Nagasaki 852-8521, Japan
3 Sweden Water Research AB, 22370 Lund, Sweden
4 South Sweden Water Supply AB (Sydvatten AB), 21532 Malmö, Sweden
5 Department of Aquatic Sciences and Assessment, Swedish University of Agricultural Sciences (SLU), 75007 Uppsala, Sweden
6 Centre for Advanced Middle Eastern Studies (CMES), Lund University, 22100 Lund, Sweden
* Correspondence: dauren.mussabek@tvrl.lth.se; Tel.: +46-46-2229470

Abstract: Groundwater contamination caused by the use of the aqueous film-forming foam (AFFF) containing per- and polyfluoroalkyl substances (PFAS) was investigated in southern Sweden. ΣPFAS concentrations in groundwater ranged between 20 and 20,000 ng L^{-1}; PFAS composition was primarily represented by PFOS and PFHxS. The PFAS chain length was suggested to have an impact on the contaminant distribution and transport in the groundwater. PFAS profiling showed that the use of PFSAs- and PFCAs/FTSAs-based PFAS-AFFF can be a contributor to PFAS contamination of the drinking water source (groundwater). PFAS emission was connected to PFAS-AFFF use during the fire-training and fire-fighting equipment tests at the studied location. PFAS emission per individual fire training was (semi-quantitatively) estimated as [$1.4 < 11.5 \pm 5.7 < 43.7$ kg] (n = 20,000). The annual emission estimates varied as [$11 < 401 \pm 233 < 1125$ kg yr^{-1}] (n = 1005) considering possible [$2 < 35 \pm 20 < 96$] individual fire-training sessions per year.

Keywords: AFFF; PFAS; groundwater

Citation: Mussabek, D.; Söderman, A.; Imura, T.; Persson, K.M.; Nakagawa, K.; Ahrens, L.; Berndtsson, R. PFAS in the Drinking Water Source: Analysis of the Contamination Levels, Origin and Emission Rates. *Water* **2023**, *15*, 137. https://doi.org/10.3390/w15010137

Academic Editor: Andrea G. Capodaglio

Received: 28 November 2022
Revised: 22 December 2022
Accepted: 27 December 2022
Published: 30 December 2022

1. Introduction

PFAS-containing aqueous film-forming foam (PFAS-AFFF) has been used by firefighters for several decades [1]. PFAS-AFFF has a wide application in extinguishing hydrocarbon-fuel fires (class B fires); which is primarily due to the thermal stability of the active surfactant [2] and its ability to lower the surface tension at low concentrations [3]. From its early deployment in aircraft rescue in 1978 (Los Angeles International airport (3M Light Water), PFAS-AFFF has become an effective and perhaps a universal solution in fire safety (in fuel fires) at airfields worldwide [1,4].

However, PFAS-AFFF use during fire-training activities, firefighting equipment tests, and emergency events can lead to contamination of the surrounding aquatic environment [5,6]. The ubiquitous occurrence of PFAS in surface water and groundwater in proximity to firefighting training locations has been reported in various studies worldwide [7–10]; and is the major historical and current source of PFAS in Sweden [11]. The source water contamination with PFAS can lead to human exposure via drinking water [12–14]. The evidence of negative health effects and cases of human exposure stresses the importance of further investigation on PFAS distribution in the environment [12,14–17]. Nearly 50% of Sweden's drinking water originates from groundwater reservoirs situated in highly permeable glaciofluvial deposits that are susceptible to contamination [11].

PFAS spread and distribution in the aquatic environment are often connected to the water solubility and persistence of the compounds [18,19]. Far-field transport conditions have been studied in surface water [20,21] and groundwater [22,23]. Furthermore, the PFAS circulation in the urban (and/or industrial) water and waste handling cycles has been connected to the occurrence in water treatment plants [24,25] and landfills [26]. However, due to the surfactant nature of the contaminant and complex interaction mechanisms in the carrier-phase vs. media interface, the interpretation of the PFAS distribution in the corresponding transport domain requires further investigation [27–30]. Moreover, the PFAS distribution can be affected by various macroscopic conditions related to the environmental features of the area and the emission source. This can be connected to a large variation in the previously reported field-derived distribution predictors [22,31,32]. PFAS sorption and transport mechanisms in natural systems are yet subject to further investigation.

Assessment and interpretation of the water contamination by PFAS require an understanding of a cluster of interconnected processes, including analysis of the contamination levels, spatial distribution, as well as emission history. Furthermore, in contamination linked to PFAS-AFFF use, the application scenarios, equipment and purpose of the application can play a significant role in PFAS emission, as well as contamination levels and composition. PFAS-AFFF application can be continuous (historical) or related to an emergency event, thus affecting the consequent contaminant distribution [8,33]. It is often not known how and under which conditions or with which chemical speciation PFAS-AFFF has been practiced. Therefore, detailed analysis and understanding of the processes behind water contamination by PFAS and related distribution mechanisms are important.

Another important aspect in the assessment of PFAS contamination is the analytical limitations. Most of the conventional analysis methods are, in large measure, designed for quantification of the target substances, thus, despite the sufficient resolution, the detected PFAS composition is often restricted to the method [9,34]. In the contamination cases related to PFAS-AFFF, it is important to consider the presence of various structural isomers, as well as the structure of the functional group (e.g., a fraction of the branched isomers (PFOA and PFOS) can constitute 20–30% of the mass) [3,35,36].

The general purpose of the present study is to improve the knowledge in the analysis of the historical water source contamination by PFAS and address the related aspects of PFAS-AFFF application in fire-training activities and equipment tests. The conducted investigation included: (i) analysis of contamination levels and PFAS composition, (ii) assessment of the PFAS and PFAS-AFFF emission scenarios and (iii) assessment of the contaminant transport conditions.

2. Materials and Methods

2.1. Study Site

2.1.1. Area Description

The studied area is located in Ronneby Municipality (Blekinge County) in Sweden (Figure S1 in Supplementary Materials). The landscape of the area is mainly represented by hilly terrain with woods surrounding the urbanized areas. There were two main objects of interest considered in the present study. These included the F17 airfield (Blekinge Air Force Wing) and Brantafors waterworks (Ronneby Municipality). The F17 (active since 1944) has been used by both military and civil air traffic. The airfield area is restricted and surrounded by wood in the north and west, and lake Sänksjön in the north. The F17 hosts both the flight exercise area and an operational wing [37]. The Brantafors waterworks has been in operation since the 1970s and at later development stages it supplied the Ronneby Municipality with drinking water. At Brantafors, source water for municipal water supply was extracted from the groundwater reservoir at four main extraction locations: north (GW1), south (GW3 and GW4), and east (GW2) (Figure S1 in Supplementary Materials). The waterworks have been reconstructed and modified during their operation. There was limited information available for the period prior to the 1990s. Prior to 2010, due to the water quality (according to former standards), the water treatment process included aeration, pH adjustment, and UV disinfection. Brantafors was later reconstructed, and the treatment was extended with aeration, chemical precipitation, rapid sand filtration and UV disinfection.

2.1.2. Surface Water and Hydrogeology of the Area

There are several surface water bodies in the studied area, including the Hasselstadsbäcken creek, the Klintabäcken creek, the Ronnebyån river, and lake Sänksjön (Figure S1 in Supplementary Materials). The study area is located within the larger main catchment area of "Ronnebyån". Furthermore, the area is divided into two sub-catchment areas which are divided straight through the F17 airfield. Hasselstadsbäcken originates in the wetland area south-west of Lake Sänksjön and eventually discharges into Sörbybäcken creek, a tributary to Ronnebyån. Klintabäcken originates from the wetland areas in north and east of Sänksjön and discharges into Ronnebyån. Ronnebyån flows in a south direction and discharges into the Ronnebyfjärden Bay of the Baltic Sea. Lake Sänksjön is a kettle lake (formed by the retrieving glacier) and is primarily fed by groundwater and surface runoff (Figures S1 and S4 in Supplementary Materials).

The study area is located in one of the largest delta formations in South-eastern Sweden (Bredåkradeltat) which was formed during the withdrawal of the last ice age. The area is mainly covered by glaciofluvial materials consisting of sand and silt. However, an esker formation is also present (Bredåkraåsen) and it runs parallel to the Klintabäcken (Figure S1 in Supplementary Materials). The esker mainly consists of sand, gravel, and rocks. It is located directly on the bedrock and is partially covered by glaciofluvial material.

The bedrock is primarily represented by Karshamngranit (a granite) consisting of quarts, feldspar and mica; the bedrock surface is characterized by deep and long fractures in the direction from north to south (Figures S4 and S5 in Supplementary Materials).

The area contains a groundwater reservoir which is represented by the esker and coarse-grained glaciofluvial material. It is mainly an unconfined aquifer; however local confined conditions can occur due to fine glaciofluvial material. The aquifer contains local subsystems and a transient groundwater divide is formed following the runoff area of Klintabäcken, which runs along the airfield to the east. To the west, the reservoir runs towards Ronnebyån where it is confined by the bedrock which is close to the surface at this area (Figure S4 in Supplementary Materials). The coarse-grained glaciofluvial material contains a lower groundwater extraction potential of 1–5 L s^{-1} whereas the esker formation contains a higher potential of 5–25 L s^{-1}. Close to Ronnebyån the potential is even higher due to the added infiltration from the river (Figure S4 in Supplementary Materials). The primary groundwater flow direction is north to south following the flow of Klintabäcken (Figures S3 and S4 in Supplementary Materials).

2.1.3. PFAS Emission Source

The PFAS emission was primarily related to the use of PFAS-AFFF during fire-training activities and firefighting equipment tests at F17. The emission sources were connected to the designated fire-training and equipment test sites on the F17 airfield territory. The main emission source was identified at the fire-training facility (FTF), located in the east of the F17 airfield (Figure S1 in Supplementary Materials). However, there are several potential sources (training and test locations) suggested by former F17 personnel, including the north-end and south-end of the airstrip and the fire station in the south of the airstrip (Figure S1 in Supplementary Materials).

At FTF, the AFFF release was primarily connected to the aircraft rescue operation training (including open fire suppression). The training was mainly performed on a designated concrete platform (Figure S2 in Supplementary Materials). Prior to 2004, the fire training was conducted by military personnel. Since 2004, the operation was taken over by the civil fire rescue services. Due to several reasons, not least confidentiality rules, the exact training protocol prior to 2004, as well as possible changes in exercise routine and location at later stages, have not fully been clarified. Therefore, the PFAS-AFFF application in fire-training and equipment test activities was considered to continue until 2014, when the PFAS contamination was detected. Later, PFAS-AFFF applications were consequently ceased, and training are further conducted with an alternative instead.

According to former military personnel, during the period until 2004, airfield fire safety was provided by the use of a specialized emergency vehicle (heavy terrain vehicle type 4112). It was designed for fire safety and rescue missions and is commonly used in operations across the country. The main training objective included a simulation of the rapid-fire suppression and a crew rescue in the aircraft crash. The rescue training (including various activities) was carried out during one week (including several sessions per day) of the military service period and some occasional training activities throughout the period. Training with AFFF was conducted on 2 to 4 occasions a year (according to former personnel).

2.2. PFAS Measurements in Groundwater

PFAS concentration was measured in the groundwater samples collected at the extraction points GW1, GW3, and GW4. Duplicate water samples were collected directly in the well (from the water surface) in 1 L PP bottles. Samples were stored in dark at 4 °C. Prior to analysis, water samples were placed into the sonication bath for 20 min. Samples were consequently transferred into a 10 mL glass injection vial and spiked with an internal standard mix prior to analysis.

Additionally, PFAS measurements in groundwater were taken using Polar Organic Chemical Integrative Sampler (POCIS) with solid phase sorbent (Oasis HLB). The individual POCIS preparation was performed as described by Gobelius et al. [24]. Individual samplers (without cage) were set in a column configuration (Figure S6 in Supplementary Materials) and placed into the groundwater well. There were two sampler columns deployed with eight individual samplers (4 duplicates) and three individual samplers (1 triplicate), respectively. Sampler columns were deployed in groundwater extraction wells upstream (POCIS ×8) and downstream (POCIS ×3) of the groundwater aquifer at approximately 15 and 20 m depth at extraction points GW1 and GW3, respectively (Figure S4 in Supplementary Materials). For the well at GW3, a single deployment was set for 4 weeks, and at well GW1, four deployments were set for 1-to-4-week intervals. Individual samples were stored frozen prior to extraction and analysis. There were 13 samples extracted and analyzed (including 2 procedural blank samples). Sample extraction and analysis were performed according to the methods described by Gobelius et al. [24].

In total, 29 PFAS compounds were analyzed in the present study, including five perfluoroalkane sulfonates ($C_{4, 6, 7, 8, 10}$ PFSAs) (PFBS, PFHxS, PFHpS, PFOS and PFDS), thirteen perfluoroalkyl carboxylates ($C_{3\text{–}13, 15, 17}$ PFCAs) (PFBA, PFPeA, PFHxA, PFHpA, PFOA, PFNA, PFDA, PFUnDA, PFDoDA, PFTriDA, PFTeDA, PFHxDA and PFOcDA), three perfluorooctane sulfonamides (FOSAs) (FOSA, MeFOSA and EtFOSA), two perfluorooctane sulfonamidoethanols (FOSEs) (MeFOSE and EtFOSE), three perfluorooctane sulfonamidoacetic acids (FOSAAs) (FOSAA, MeFOSAA and EtFOSAA), and three fluorotelomer sulfonates FTSAs (6:2 FTSA, 8:2 FTSA and 10:2 FTSA). A mix of 16 mass-labeled internal standards (IS) (i.e., $^{13}C_8$-FOSA, d_3-MeFOSAA, d_5-EtFOSAA, d_3-MeFOSA, d_5-EtFOSA, d_7-MeFOSE, d_9-EtFOSE, $^{13}C_4$-PFBA, $^{13}C_2$-PFHxA, $^{13}C_4$-PFOA, $^{13}C_5$-PFNA, $^{13}C_2$-PFDA, $^{13}C_2$-PFUnDA, $^{13}C_2$-PFDoDA, $^{18}O_2$-PFHxS and $^{13}C_4$-PFOS) was applied for internal calibration (<98% purity, Wellington Laboratories, Guelph, ON, Canada).

Duplicate samples and procedural blanks were analyzed using LC/UPLC system (Thermo Fisher Scientific, Waltham, MA, USA). A Hypersil GOLD aQ column (20 mm × 2.1 mm i.d., 12 μm particles, Thermo Fisher Scientific, Waltham, MA, USA) was used as an extraction column for online solid phase extraction. ACQUITY UPLC BEH-C18 column (100 mm × 2.1 i.d., 1.7 μm particles, Waters Corporation, Wilmslow, UK) was used as the analytical column. Injection volumes were 1.0 mL for all samples. A triple-stage quadrupole MS/MS TSQ Quantiva (Thermo Fisher Scientific, Waltham, MA, USA) was used as the detection sensor. Analysis data evaluation was performed using TraceFinder™ 3.3 software (Thermo Fisher Scientific, Waltham, MA, USA). The limit of quantification (LOQ) for the PFAS analysis was determined as the lowest calibration point in the linear range (from 0.1 ng L^{-1} to 2000 ng L^{-1}) if the S/N ratio was higher than 3.

2.3. PFAS Emission Estimates

There was limited information available on the AFFF use at the study site, as well as AFFF type or/and composition utilized in the past. PFAS emissions were estimated using Monte Carlo simulations and accessible historical records and accounts. The contaminant release was assessed in connection to suggested fire-training activities, equipment utilization, and AFFF release scenarios. PFAS was considered an active AFFF surfactant and its content in AFFF was estimated based on reported AFFF compositions [38]. The surfactant (PFAS) and water release per individual training session (fire-training and/or equipment test activity) were estimated according to the suggested equipment specifications and AFFF utilization scenarios. The emission scenarios were estimated as a cumulative release by individual training session scenarios per simulated period. An individual annual emission scenario was estimated from subsampled individual release scenarios considering a possible variation of the training sessions per day (*k*), number of exercises per period (*j*), and number of events per year (*i*):

$$PFAS_{annual} = \sum_{1}^{i}\sum_{1}^{j}\sum_{2}^{k} m_{\ per\ session} \tag{1}$$

Data evaluation, calculations, and analyses were performed using Microsoft Excel (Microsoft, proprietary), Argo (Booz Allen Argo, open source), and MATLAB (MathWorks, proprietary) software. There was limited information available (by the date) on an exact PFAS composition in related PFAS-AFFF formulations (Table S3 in Supplementary Materials) [34,39,40]. Although approximation on suspected PFAS composition was possible, due to the historical nature of the contamination and possible variation in AFFF types, the PFAS emission estimates were restricted to "blind" PFAS. Thus, simulation estimates are subject to uncertainty.

3. Results and Discussion

3.1. Groundwater Contamination

In the analyzed groundwater samples corresponding to extraction points GW1, GW3, and GW4, there were 12 out of 29 analyzed PFAS (PFBS, PFHxS, PFOS, PFDS, PFHpA, PFHxA, PFOA, PFNA, PFOcDA, 6:2 FTSA, 8:2 FTSA and FOSA) detected (Table S1 in Supplementary Materials). PFAS analysis indicated high groundwater contamination levels at GW1 and GW3 with ∑PFAS concentrations of 4200 ± 40 ng L^{-1} and 20,000 ± 1900 ng L^{-1}, respectively, followed by relatively low levels at GW4 with ∑PFAS of 18 ± 3 ng L^{-1}.

In the groundwater corresponding to GW4, PFAS composition was primarily represented by PFOS (35%), PFHxS (32%) and FOSA (33%). At GW3, PFAS composition was dominated by PFOS (48%) and PFHxS (33%); followed by PFHxA (7%), PFOA (6%), PFBS (5%), and remaining compounds. Similarly, PFAS composition at GW1 was primarily represented by PFOS (72%) and PFHxS (19%), followed by remaining substances (<9%).

Based on the PFAS contamination levels, the concentration gradient is suggested to have propagated from GW1 (north) to GW3 (south), which agrees with primary groundwater flow direction from north to south (Figures S3 and S4 in Supplementary Materials). PFAS concentrations at GW1 and GW3 are suggested to indicate a unified emission source affecting the groundwater from the north.

A shift in relative composition was observed for PFSAs and PFCAs (for GW1 vs. GW3) and suggestively connected to contaminant mobility. For PFCAs (PFHxA-PFHpA-PFOA), the observed composition shifted from 58-12-30% at GW1 to 46-13-41% at GW3 (downstream), whereas composition of PFSAs (PFBS-PFHxS-PFOS) changed from 2-21-77% at GW1 to 5-39-56% at GW3. Furthermore, the difference in detected levels (GW1 vs. GW3) for PFDA, PFNA, 6:2 FTSA, 8:2 FTSA and FOSA is attributed to a possible effect of the molecular chain length on PFAS transport. However, PFAS distribution in the groundwater can be affected by different transport processes (including groundwater extraction and groundwater table variation) and sorption processes related to the characteristics of the soil [22,27,41]. Therefore, until the contaminant transport conditions are sufficiently established, the suggested effect of the PFAS chain length on transport should be considered with some precaution.

Analysis of the POCIS deployed in groundwater wells at GW1 and GW3 showed a slightly better method sensitivity (Table S2 in Supplementary Materials). For PFCAs, detected in the samples corresponding to both GW1 and GW3, the observed PFAS composition was extended to PFPeA, PFHxA, PFHpA, PFOA, PFNA and PFDA. Furthermore, it was possible to conduct semi-quantitative measurements for some branched PFAS, thus extending the inventory to L-PFHxS, B-PFHxS, L-PFOS, B-PFOS, L-FOSA and B-FOSA. However, since the on-site calibration of POCIS was not possible and PFAS concentrations were estimated using the previously reported sampling rates [24], the analysis results were mainly considered for a qualitative assessment.

Overall, PFAS composition in POCIS was identical to previously measured in groundwater samples and primarily represented by PFSAs (81% (GW1) and 65% (GW3)) and PFCAs (18% (GW1) and 35% (GW3)). Based on detected PFAS composition (Figure 1) and cross-evaluation with reported PFAS-AFFF formulations (Table S3 in Supplementary Materials), there are a few possible PFAS-AFFF types in connection to the PFAS emission.

		6:2 FTSA <1%			8:2 FTSA <1%
PFPeA 6–16%	PFHxA 8–11%	PFHpA 1–2%	PFOA 2–5%	PFNA <1%	PFDA <1%
PFBS 6–11%		PFHxS 37%		PFOS 17–37%	
				FOSA <1%	
				MeFOSA <1%	

Figure 1. PFAS profile detected in the POCIS deployed at GW1 and GW3 (sorted left to right according to fluorinated chain length).

The PFSAs based PFAS-AFFF (often referred as Legacy foam) is suggested to be the major contributor to PFAS emission. This is in connection to the PFSAs (PFBS, PFHxS and PFOS) profile observed in groundwater samples. The PFCAs composition, on the other hand, is attributed to PFCAs/FTSAs based PFAS-AFFF and alternatively to FTSAs based PFAS-AFFF (assuming possible FTSAs to PFCAs transformation) [3,36,39]. Furthermore, traces of perfluoroalkane sulfonamide and sulfonamido substance such as FOSA and MeFOSA were considered as an indicator of the fluorination method used for the production of the corresponding PFAS. FOSA and MeFOSA can be linked to the PFAS synthesis involving electrochemical fluorination [36]. Thus, the detected FOSA and MeFOSA are associated with the material used in the production of detected PFSAs (PFBS, PFHxS and PFOS) and indirectly confirm the PFSAs-based formulation of PFAS-AFFF. PFCAs (PFPeA,

PFHxA, PFHpA, PFOA, PFNA and PFDA) detected in the groundwater samples can be attributed to electrochemical fluorination, as well as to fluorotelomer (i.e., telomerization) processes [36]. FTSAs, on the other hand, can be associated with the fluorotelomer process-based origin. Assessment of the PFSAs vs. PFCAs and FTSAs traces, as well as analysis of the perfluoroalkane sulfonamides and sulfonamido substances (PFAS precursor compounds), are useful in the identification of the synthesis processes and related PFAS-AFFF origins [9,40,42].

PFAS-AFFF identification based on measured PFAS composition in groundwater, however, neglects possible effects of the transport and retention. Thus, the detected PFAS composition may not fully represent the actual PFAS composition in PFAS-AFFF. The assumptions on the PFCAs/FTSAs based PFAS-AFFF should be considered with precaution. Further investigation is necessary to establish the transport conditions and secure the connection between PFAS in groundwater and at the emission source.

Furthermore, in the present study, PFAS analysis was conducted considering a group of target substances (29 PFAS) and, despite the quantification of some branched isomers (PFHxS, PFOS and FOSA), the detected substances do not fully represent the PFAS composition. The analytical representation (of the composition) can be fortified by use in combination with methods such as total organic fluoride (TOF) and total oxidable precursors (TOP) assay; furthermore, with the use of non-targeted analysis/screening.

3.2. Assessment of the PFAS Emission

The individual fire-training session scenarios were simulated considering possible variations in the AFFF stock solution composition, surfactant composition (PFAS content), AFFF solution (used for foam aggregation), and equipment utilization. For the AFFF stock solution composition, the prior population boundaries were assigned as $[0.5 < 0.065 \pm 0.21 < 0.9]$ for water content, $[0.1 < 0.17 \pm 0.054 < 0.24]$ for surfactant content (not restricted to PFAS), and [0.5–1] as an additional parameter for the surfactant composition (or PFAS content) [34,38]. AFFF (stock solution) dilution scenarios were estimated based on reported and suggested by F17 former personnel (with a prior population set as $[0.01 < 0.03 < 0.04]$) [43]. With suggested equipment utilization scenarios, the PFAS emission per individual training session [kg] was estimated as $[1.4 < 11.5 \pm 5.7 < 43.7]$ (n = 20,000, average deviation (adev) = 4.5). Similarly, the release of water [L] corresponding to content in AFFF, was estimated as $[945 < 1985 \pm 594 < 3032]$ (n = 20,000, adev = 515). Furthermore, with suggested exercise routine, the annual PFAS emission rate [kg yr^{-1}] was estimated as $[6 < 414 \pm 436 < 5462]$ (note: values above correspond to one individual simulation (n = 20,000)).

The annual release scenarios were evaluated based on samples from the simulated individual emission scenarios (per fire-training session). Subsampling and calculations were conducted using suggested fire-training exercise routines (including 2–6 individual sessions per day, 1–5 training days, with 1–5 occurrences per year). Consequently, 1005 annual release scenarios were evaluated (Figure 2).

The estimated annual emission scenarios varied in the range of $[11 < 401 \pm 233 < 1125$ kg yr$^{-1}]$, corresponding to $[2 < 35 \pm 20 < 96]$ individual fire-training sessions per year (n = 1005) (Figure 2).

With further approximation, the annual PFAS emission can be estimated for the suggested most-likely exercise routine scenario as 3 consequent training days per occurrence and 2–4 occurrences per year. Thus, for the scenario with 6–60 individual sessions carried out per year (with 1–5 sessions, 3 training days, and 1–4 occurrences per year), the annual emission range is approximately 100–700 kg per year. However, since the annual emission scenarios are based on a rather limited sample, given estimates are strictly relative to the evaluated population. Further investigation is necessary to validate the estimate's accuracy.

Due to limited data available on both possible PFAS-AFFF composition and historical records of AFFF use at the studied site, it was not possible to estimate the individual PFAS emission. The estimates of PFAS release were conducted with no restriction to the actual PFAS composition in the PFAS-AFFF stock solution. Furthermore, the overall surfactant composition in the PFAS-AFFF was considered to include fluorinated (PFAS) and non-fluorinated surfactants.

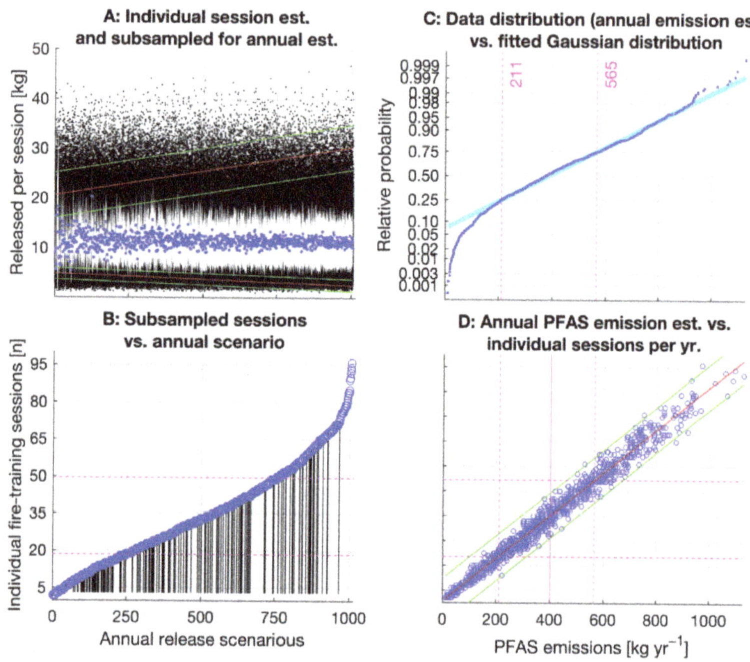

Figure 2. Evaluation the PFAS emission scenarios. (**A**) black dots represent simulated individual PFAS emission scenarios (corresponding to an individual fire-training session); blue dots represent the mean of the subsampled n-scenarios with white area corresponding to ± one standard deviation; red line at top and bottom represent the maximum and minimum of the subsampled n-scenarios, respectively, with green lines corresponding to ± one standard error. Vertical axe corresponds the variation in individual emission scenarios and horizontal axe corresponds to set of individual samples (n = 20,000 × 1005). (**B**) blue circles indicate the number of subsampled session scenarios in A; vertical black lines correspond to suggested most-likely exercise routine with 3 consequent training days per annual exercise occurrence. (**C**) comparison of the distribution of the simulated annual release scenarios (blue dots) vs. fitted Gaussian distribution (cyan); vertical dashed magenta lines correspond to first and third quantiles of the data. (**D**) Estimates on annual PFAS emission (blue line) vs. number of individual training sessions, shown as a linear fit with green line corresponding to ± three standard deviations; vertical magenta line corresponds to estimated median, vertical black line corresponds estimated mean, and dashed magenta lines correspond to first and third quantiles of the data.

In the present study, provided PFAS emissions were based on available records/suggestions and conducted considering a range of both PFAS-AFFF contents and application rates. However, an accurate assessment of emissions, requires a thorough clarification of the AFFF application history. This includes the PFAS-AFFF stock solution formulations, PFAS/surfactant composition in AFFF, and AFFF application routine (frequency and duration). To the authors' knowledge, this is the first study attempting to reconstruct the historical PFAS emission as in connection to PFAS-AFFF release scenarios and fire-training activities.

3.3. PFAS Transport Considerations

The PFAS emission was connected to the use of PFAS-AFFF in fire-training and equipment tests at FTF. The spatial distribution of PFAS and contamination of the aquatic environment was primarily associated with transport in the dissolved state, including transport with the surface runoff, transport with groundwater, and further distribution with raw water extraction (Figure 3).

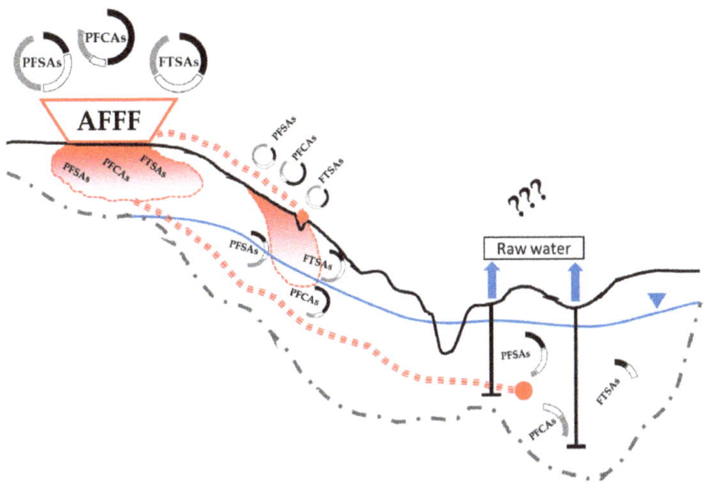

Figure 3. Conceptualization of the PFAS release, transport, and distribution at the studied site.

The initial stage of emission was associated with fire-training activities at FTF (including the simulation of the aircraft rescue mission). Primary AFFF application was connected to the dispatch of the rescue team/vehicle and putdown of an open fuel fire. According to former personnel, the exercise included an initial blast (on-vehicle) of the rescued aircraft with AFFF. Consequently, when the simulated aircraft fire is partially contained, a secondary suppression was conducted (off-vehicle) with a targeted application of AFFF. It was also suggested, that fire training sessions were possibly followed by clean-up stages where the aircraft crash simulation area was flushed with water. It is therefore assumed, that an individual training session could contribute to the generation of a substantial amount of liquid. Thus, including AFFF release, AFFF dissociation and possible use of water (in area flushes), the mass generated during the exercise session is considered sufficient for possible PFAS transport with surface runoff. The transport and distribution with runoff, however, can be affected by topographical features of the area surrounding the FTF and soil saturation. Furthermore, the PFAS transport with surface runoff is suggested to have been affected by the re-direction of surface flows with the drainage system at FTF. However, limited information was available on the drainage operation period as well as its transport capacity. It is presumed that the initial PFAS transport with surface runoff is followed by distribution processes associated with hydrological processes (distribution and retention) of much longer response time. Further PFAS distribution, associated with infiltration processes and transport in unsaturated soil, is considered as a primary contributor to PFAS accumulation and retention in the porous media, as well as to the shift in PFAS composition, prior to further advective transport. However, due to complex interaction mechanisms in the porous media, the PFAS transport in unsaturated soil is difficult to decipher with certainty, not least due to limited data available on the contaminant emission. The advective stage of the PFAS transport is primarily subject to transport in a dissolved state (with groundwater). Although the PFAS transport is assumed to primarily occur in soil of high permeability and mostly connected to free-water mobility, the retardation/sorption process is relevant to certain soil fractions and compositions. Ultimately, the PFAS sorption and transport in aqueous-solid, solid-air, and aqueous-air interfaces are complex and not well-studied processes. The PFAS sorption/interaction can be affected by the mineral composition of the heterogeneous media as well as porosity and pore-size distribution [22,28,30,41]. Further investigation is required to better understand the PFAS distribution in soil and groundwater. Consequent PFAS transport was primarily subjected to groundwater extraction, processing, and distribution with drinking water. Since water treatment, prior to plant modification, was restricted to treatment barriers insensitive to PFAS, the drinking water is assumed to have been contaminated for an extended period.

In the present study, several corresponding transport estimates were introduced, including PFAS sorption and transport in soil and far-field transport with groundwater. However, an accurate estimate was not possible due to data limitations on PFAS sorption parameters, emission rates and durations, and inconsistency in observed PFAS concentration and composition. Further investigations are necessary for the validation of the transport estimates.

In cases of historical contamination, as considered in the present study, the observed contamination levels can be represented by a series of inconsistent AFFF release events. Thus, representing a series of short-lasting contaminant spikes of different magnitude (levels) and related PFAS composition. The far-field transport processes, however, are strongly connected to the hydrogeological features of the area and related response time variability. It is therefore suggested important to consider the scale-related features of both PFAS emission and distribution processes.

3.4. Remarks on Regulations

There are currently several regulations on PFAS; these are, however, implemented on different institutional/industrial levels. The initial attempts on regulating the substances came from the manufacturers, possibly, as a response to the rising awareness of the environmental and health impacts of PFAS (e.g., a gradual phase-out of PFOS and PFOA) [44–46]. The UN Stockholm Convention on Persistent Organic Pollutants and EU registration, evaluation and authorization of chemicals (REACH) was established in the 2000s and remains as a primary regulatory tool (for listed individual substances) [47]. In the latest amendment by REACH (EU 2021/1297) the regulations were expanded to include long-chain (C9–C14) PFCAs, their salts and related substances [48]. To date, there are, however, no regulations on PFAS as a group/class of chemicals [47].

In the context of PFAS-AFFF, the European Chemicals Agency (ECHA) is aiming for an EU-wide restriction on PFAS in firefighting foam [49]. However, identification and disposal of the PFAS-AFFF, as well as a safe transition and replacement, are yet be addressed [50]. On the other hand, as in relation to the public health and risks of human exposure to PFAS, a revision of the acceptable level in water can be a strong factor, e.g., the EU drinking water directive (EU 2020/2184) limiting the drinking water levels to 100 ng L^{-1} (PFAS 10) and 500 ng L^{-1} (PFAS total) [51]. In Sweden, an additional limit has been set to 4 ng L^{-1} for PFOS, PFOA, PFHxS, and PFNA [51,52]. Ultimately, work regulating PFAS is an ongoing process that evolves with an increase in knowledge, as well as in public awareness. Both regulatory actions and an actual practical implementation are challenging and demanding tasks that require international efforts.

4. Conclusions

PFAS emission at the studied site was connected to PFAS-AFFF application during fire-training and fire-fighting equipment tests. The groundwater contamination was studied and PFAS contamination levels were established. Measured ΣPFAS concentrations in groundwater ranged between 20 and 20,000 ng L^{-1}. The PFAS composition in analyzed samples was dominated by PFOS and PFHxS. The PFAS chain length is suggested to have an impact on the distribution and transport of groundwater. Based on PFAS profiling, PFAS contamination of groundwater was primarily connected to the application of PFSA-based and PFCA-/FTSA-based PFAS-AFFF. AFFF utilization and PFAS release scenarios were studied, and PFAS emissions per individual fire-training session were estimated as [$1.4 < 11.5 \pm 5.7 < 43.7$ kg] (n = 20,000). The PFAS annual emissions were estimated as [$11 < 401 \pm 233 < 1125$ kg yr^{-1}] (n = 1005, corresponding to [$2 < 35 \pm 20 < 96$] individual fire-training sessions per year). Further investigation is necessary regarding the PFAS-AFFF formulations and application routine. Distribution of PFAS in the aquatic environment is associated with transport in surface runoff, vertical advective-diffusive transport in unsaturated soil, advective transport with groundwater, and further distribution with raw water extraction. Further investigation is necessary for the assessment of the PFAS transport conditions.

Supplementary Materials: The following supporting information (in supplementary materials) can be downloaded at: https://www.mdpi.com/article/10.3390/w15010137/s1, Figure S1: study area description, including: airfield territory, PFAS emission sources, surface water bodies and groundwater extraction locations; Figure S2: topography of the studied area, including: airfield territory, surface water bodies, PFAS emission sources and spread area; Figure S3: description of the soil layer depth at the studied area, including: airfield territory, PFAS emission sources and spread area, surface water bodies and groundwater extraction locations; Figure S4: hydrogeology of the studied area, including: airfield territory, PFAS emission sources and spread area, surface water bodies and groundwater extraction locations, groundwater confinement and capacities; Figure S5: bedrock lithology at the studied area, including: airfield territory, PFAS emission sources and surface water bodies; Figure S6: photo of the POCIS prior to deployment into the groundwater well; Table S1: PFAS concentrations measured in triplicate groundwater samples collected from groundwater extraction wells corresponding to GW1, GW3 and GW4; Table S2: PFAS detected in POCIS and corresponding PFAS concentrations estimated for groundwater measured at GW3 and GW1; Table S3: example of analyzed PFAS compositions corresponding to PFAS-AFFF reported for US market and measured PFAS concentrations and relative composition in PFAS-AFFF reported for on Swedish market.

Author Contributions: Conceptualization, D.M. and K.M.P.; methodology, D.M. and. L.A.; formal analysis, D.M. and L.A.; investigation, D.M.; data curation, D.M.; writing—original draft preparation, D.M. and A.S.; writing—review and editing, K.M.P., K.N. and R.B.; visualization, D.M. and T.I.; supervision, K.M.P., K.N., L.A. and R.B.; funding acquisition, D.M. All authors have read and agreed to the published version of the manuscript.

Funding: Present study was partially funded by the Åke och Greta Lissheds Foundation (2017-00069) and Japan Society for the Promotion of Science (SP19502). The APC was funded by Lund University.

Data Availability Statement: The data that support the findings of this study are available on request from the corresponding author [D.M.]; some information is restricted due to confidentiality and privacy considerations.

Acknowledgments: We would like to thank Mattias Andersson (Ronneby Miljö and Teknik AB) for extensive help with the investigation and field studies.

Conflicts of Interest: The authors declare no conflict of interest.

References

1. Cousins, I.T.; Vestergren, R.; Wang, Z.Y.; Scheringer, M.; McLachlan, M.S. The precautionary principle and chemicals management: The example of perfluoroalkyl acids in groundwater. *Environ. Int.* **2016**, *94*, 331–340. [CrossRef] [PubMed]
2. Gramstad, T.; Haszeldine, R.N. 512. Perfluoroalkyl derivatives of sulphur. Part VI. Perfluoroalkanesulphonic acids CF3·[CF2]·SO3H (n = 1–7). *J. Chem. Soc.* **1957**, *6*, 2640–2645. [CrossRef]
3. Kissa, E. *Fluorinated Surfactants and Repellents*; Marcel Dekker: New York, NY, USA, 2001; Volume 97, p. 640.
4. Kleiner, E.; Jho, C. Recent developments in 6: 2 fluorotelomer surfactants and fosam stabilizers. In Proceedings of the 4th Reebok Foam Seminar, Bolton, UK, 6–7 July 2009; pp. 6–7.
5. Levine, A.D.; Libelo, E.; Bugna, G.; Shelley, T.; Mayfield, H.; Stauffer, T. Biogeochemical assessment of natural attenuation of JP-4 contaminated ground water in the presence of fluorinated surfactants. *Sci. Total Environ.* **1998**, *208*, 179–195. [CrossRef]
6. Moody, C.; Field, J. Determination of Perfluorocarboxylates in Groundwater Impacted by Fire-Fighting Activity. *Environ. Sci. Technol.* **1999**, *33*, 2800–2806. [CrossRef]

7. Moody, C.A.; Martin, J.W.; Kwan, W.C.; Muir, D.C.G.; Mabury, S.C. Monitoring perfluorinated surfactants in biota and surface water samples following an accidental release of fire-fighting foam into Etobicoke Creek. *Environ. Sci. Technol.* **2002**, *36*, 545–551. [CrossRef] [PubMed]
8. Awad, E.; Zhang, X.M.; Bhavsar, S.P.; Petro, S.; Crozier, P.W.; Reiner, E.J.; Fletcher, R.; Tittemier, S.A.; Braekevelt, E. Long-Term Environmental Fate of Perfluorinated Compounds after Accidental Release at Toronto Airport. *Environ. Sci. Technol.* **2011**, *45*, 8081–8089. [CrossRef] [PubMed]
9. Karrman, A.; Elgh-Dalgren, K.; Lafossas, C.; Moskeland, T. Environmental levels and distribution of structural isomers of perfluoroalkyl acids after aqueous fire-fighting foam (AFFF) contamination. *Environ. Chem.* **2011**, *8*, 372–380. [CrossRef]
10. Kwadijk, C.J.A.F.; Kotterman, M.; Koelmans, A.A. Partitioning of Perfluorooctanesulfonate and Perfluorohexanesulfonate in the Aquatic Environment after an Accidental Release of Aqueous Film Forming Foam at Schiphol Amsterdam Airport. *Environ. Toxicol. Chem.* **2014**, *33*, 1761–1765. [CrossRef]
11. Banzhaf, S.; Filipovic, M.; Lewis, J.; Sparrenbom, C.J.; Barthel, R. A review of contamination of surface-, ground-, and drinking water in Sweden by perfluoroalkyl and polyfluoroalkyl substances (PFASs). *Ambio* **2017**, *46*, 335–346. [CrossRef]
12. Andersson, E.M.; Scott, K.; Xu, Y.; Li, Y.; Olsson, D.S.; Fletcher, T.; Jakobsson, K. High exposure to perfluorinated compounds in drinking water and thyroid disease. A cohort study from Ronneby, Sweden. *Environ. Res.* **2019**, *176*, 108540. [CrossRef]
13. Ryota, S. Kyoto University Professors Detect PFOS Concentration of 4 Times National Average in Ginowan Residents' Blood. *Ryukyu Shimpo*, 17 May 2019. Available online: http://english.ryukyushimpo.jp/2019/05/21/30452/(accessed on 20 December 2022).
14. Shin, H.M.; Vieira, V.M.; Ryan, P.B.; Steenland, K.; Bartell, S.M. Retrospective Exposure Estimation and Predicted versus Observed Serum Perfluorooctanoic Acid Concentrations for Participants in the C8 Health Project. *Environ. Health Perspect.* **2013**, *121*, A113. [CrossRef]
15. Kerger, B.D.; Copeland, T.L.; DeCaprio, A.P. Tenuous dose-response correlations for common disease states: Case study of cholesterol and perfluorooctanoate/sulfonate (PFOA/PFOS) in the C8 Health Project. *Drug Chem. Toxicol.* **2011**, *34*, 396–404. [CrossRef]
16. Li, Y.; Fletcher, T.; Mucs, D.; Scott, K.; Lindh, C.H.; Tallving, P.; Jakobsson, K. Half-lives of PFOS, PFHxS and PFOA after end of exposure to contaminated drinking water. *Occup. Environ. Med.* **2018**, *75*, 46–51. [CrossRef]
17. Oberg, M.; Da Silva, A.; Ringblom, J.; Scott, K.; Lindh, C.; Jakobsson, K. Probabilistic risk of decreased levels of triiodothyronine following chronic exposure to PFOS and PFHxS. *Toxicol. Lett.* **2018**, *295*, S224. [CrossRef]
18. Ahrens, L. Polyfluoroalkyl compounds in the aquatic environment: A review of their occurrence and fate. *J. Environ. Monit.* **2011**, *13*, 20–31. [CrossRef] [PubMed]
19. Moller, A.; Ahrens, L.; Surm, R.; Westerveld, J.; van der Wielen, F.; Ebinghaus, R.; de Voogt, P. Distribution and sources of polyfluoroalkyl substances (PFAS) in the River Rhine watershed. *Environ. Pollut.* **2010**, *158*, 3243–3250. [CrossRef] [PubMed]
20. Filipovic, M.; Berger, U.; McLachlan, M. Mass Balance of Perfluoroalkyl Acids in the Baltic Sea. *Environ. Sci. Technol.* **2013**, *47*, 4088–4095. [CrossRef] [PubMed]
21. Filipovic, M.; Laudon, H.; McLachlan, M.; Berger, U. Mass Balance of Perfluorinated Alkyl Acids in a Pristine Boreal Catchment. *Environ. Sci. Technol.* **2015**, *49*, 12127–12135. [CrossRef]
22. Weber, A.K.; Barber, L.B.; LeBlanc, D.R.; Sunderland, E.M.; Vecitis, C.D. Geochemical and Hydrologic Factors Controlling Subsurface Transport of Poly- and Perfluoroalkyl Substances, Cape Cod, Massachusetts. *Environ. Sci. Technol.* **2017**, *51*, 4269–4279. [CrossRef]
23. Shin, H.M.; Steenland, K.; Ryan, P.B.; Vieira, V.M.; Bartell, S.M. Biomarker-Based Calibration of Retrospective Exposure Predictions of Perfluorooctanoic Acid. *Environ. Sci. Technol.* **2014**, *48*, 5636–5642. [CrossRef]
24. Gobelius, L.; Persson, C.; Wiberg, K.; Ahrens, L. Calibration and application of passive sampling for per- and polyfluoroalkyl substances in a drinking water treatment plant. *J. Hazard. Mater.* **2019**, *362*, 230–237. [CrossRef]
25. Sinclair, E.; Kannan, K. Mass loading and fate of perfluoroalkyl surfactants in wastewater treatment plants. *Environ. Sci. Technol.* **2006**, *40*, 1408–1414. [CrossRef]
26. Busch, J.; Ahrens, L.; Sturm, R.; Ebinghaus, R. Polyfluoroalkyl compounds in landfill leachates. *Environ. Pollut.* **2010**, *158*, 1467–1471. [CrossRef]
27. Hellsing, M.S.; Josefsson, S.; Hughes, A.V.; Ahrens, L. Sorption of perfluoroalkyl substances to two types of minerals. *Chemosphere* **2016**, *159*, 385–391. [CrossRef]
28. Jeon, J.; Kannan, K.; Lim, B.J.; An, K.G.; Kim, S.D. Effects of salinity and organic matter on the partitioning of perfluoroalkyl acid (PFAs) to clay particles. *J. Environ. Monitor.* **2011**, *13*, 1803–1810. [CrossRef]
29. Nouhi, S.; Ahrens, L.; Pereira, H.C.; Hughes, A.V.; Campana, M.; Gutfreund, P.; Palsson, G.K.; Vorobiev, A.; Hellsing, M.S. Interactions of perfluoroalkyl substances with a phospholipid bilayer studied by neutron reflectometry. *J. Colloid Interface Sci.* **2018**, *511*, 474–481. [CrossRef]
30. Pereira, H.C.; Ullberg, M.; Kleja, D.B.; Gustafsson, J.P.; Ahrens, L. Sorption of perfluoroalkyl substances (PFASs) to an organic soil horizon—Effect of cation composition and pH. *Chemosphere* **2018**, *207*, 183–191. [CrossRef]
31. McCarthy, C.; Kappleman, W.; DiGuiseppi, W. Ecological Considerations of Per- and Polyfluoroalkyl Substances (PFAS). *Curr. Pollut. Rep.* **2017**, *3*, 289–301. [CrossRef]

32. Zareitalabad, P.; Siemens, J.; Hamer, M.; Amelung, W. Perfluorooctanoic acid (PFOA) and perfluorooctanesulfonic acid (PFOS) in surface waters, sediments, soils and wastewater—A review on concentrations and distribution coefficients. *Chemosphere* **2013**, *91*, 725–732. [CrossRef] [PubMed]
33. Filipovic, M.; Woldegiorgis, A.; Norstrom, K.; Bibi, M.; Lindberg, M.; Osteras, A.H. Historical usage of aqueous film forming foam: A case study of the widespread distribution of perfluoroalkyl acids from a military airport to groundwater, lakes, soils and fish. *Chemosphere* **2015**, *129*, 39–45. [CrossRef] [PubMed]
34. Place, B.J.; Field, J.A. Identification of Novel Fluorochemicals in Aqueous Film-Forming Foams Used by the US Military. *Environ. Sci. Technol.* **2012**, *46*, 7120–7127. [CrossRef]
35. Benskin, J.P.; De Silva, A.O.; Martin, J.W. Isomer Profiling of Perfluorinated Substances as a Tool for Source Tracking: A Review of Early Findings and Future Applications. *Rev. Environ. Contam. Toxicol.* **2010**, *208*, 111–160. [CrossRef]
36. Buck, R.C.; Franklin, J.; Berger, U.; Conder, J.M.; Cousins, I.T.; de Voogt, P.; Jensen, A.A.; Kannan, K.; Mabury, S.A.; van Leeuwen, S.P. Perfluoroalkyl and polyfluoroalkyl substances in the environment: Terminology, classification, and origins. *Integr. Environ. Assess. Manag.* **2011**, *7*, 513–541. [CrossRef]
37. F17. Available online: https://www.forsvarsmakten.se/sv/organisation/blekinge-flygflottilj-f-17/ (accessed on 20 December 2022).
38. Rupert, W.; Verdonik, D.; Hanauska, C. *Environmental Impacts of Fire Fighting Foams*; Hughes Associates Inc.: Baltimore, MD, USA, 2005.
39. Kemikalieinspektionen. *Chemical Analysis of Selected Fire-Fighting Foams on the Swedish Market 2014*; Swedish Chemicals Agency: Sundbyberg, Sweden, 2015.
40. Kärrman, A.; Bjurlid, F.; Hagberg, J.; Ricklund, N.; Larsson, M.; Stubleski, J.; Hollert, H. *Study of Environmental and Human Health Impacts of Firefighting Agents*; Technical Report; MTM Research Centre, Örebro University: Örebro, Sweden, 2016.
41. Johnson, R.L.; Anschutz, A.J.; Smolen, J.M.; Simcik, M.F.; Penn, R.L. The adsorption of perfluorooctane sulfonate onto sand, clay, and iron oxide surfaces. *J. Chem. Eng. Data* **2007**, *52*, 1165–1170. [CrossRef]
42. Moody, C.A.; Field, J.A. Perfluorinated surfactants and the environmental implications of their use in fire-fighting foams. *Environ. Sci. Technol.* **2000**, *34*, 3864–3870. [CrossRef]
43. Turekova, I.; Karol, B. The Environmental Impacts of Fire-Fighting Foams. *Res. Pap. Fac. Mater. Sci. Technol. Slovak Univ. Technol.* **2010**, *18*, 111–120. [CrossRef]
44. Bock, A.R.; Laird, B.E. CHAPTER 1 PFAS Regulations: Past and Present and Their Impact on Fluoropolymers. In *Perfluoroalkyl Substances: Synthesis, Applications, Challenges and Regulations*; The Royal Society of Chemistry: London, UK, 2022; pp. 1–21.
45. Brennan, N.M.; Evans, A.T.; Fritz, M.K.; Peak, S.A.; von Holst, H.E. Trends in the Regulation of Per- and Polyfluoroalkyl Substances (PFAS): A Scoping Review. *Int. J. Environ. Res. Public Health* **2021**, *18*, 10900. [CrossRef]
46. Reisch, M. 3M supports PFAS regulation. *Chem. Eng. News* **2019**, *97*, 12.
47. Kemikalieinspektionen. Per- and Polyfluoroalkyl Substances (PFAS). Available online: https://www.kemi.se/en/chemical-substances-and-materials/highly-fluorinated-substances (accessed on 19 December 2022).
48. EU. *Commission Regulation (EU) 2021/1297 of 4 August 2021 Amending Annex XVII to Regulation (EC) No 1907/2006 of the European Parliament and of the Council as Regards Perfluorocarboxylic Acids Containing 9 to 14 Carbon Atoms in the Chain (C9-C14 PFCAs), Their Salts and C9-C14 PFCA-Related Substances*; EU: Maastricht, The Netherlands, 2021.
49. ECHA. *Proposal to Ban 'Forever Chemicals' in Firefighting Foams Throughout the EU (ECHA/NR/22/05)*; ECHA: Helsinki, Finland, 2022.
50. Swedish Civil Contingencies Agency (MSB). Släckmedel för Räddningstjänst. Available online: https://www.msb.se/sv/amnesomraden/skydd-mot-olyckor-och-farliga-amnen/raddningstjanst-och-raddningsinsatser/slackmedel-for-raddningstjanst/ (accessed on 20 December 2022).
51. Livsmedelsverket. PFAS i Dricksvatten och Livsmedel—Kontroll. Available online: https://www.livsmedelsverket.se/foretagande-regler-kontroll/dricksvattenproduktion/riskhantering-pfas-i-dricksvatten-egenfangad-fisk/ (accessed on 20 December 2022).
52. Livsmedelsverket; Lindfeldt, E.; Gyllenhammar, I.; Strandh, S.; Halldin Ankarberg, E. *Kartläggning av Per- och Polyfluorerade Alkylsubstanser (PFAS) i Sveriges Kommunala rå- Och Dricksvatten*; Livsmedelsverket: Uppsala, Sweden, 2021.

water

MDPI

Article

Decision Support for Lake Restoration: A Case Study in Swedish Freshwater Bodies

Maja Sellergren [1,2], Jing Li [3], Stina Drakare [1] and Sebastian Thöns [4,*]

1 Department of Aquatic Sciences and Assessment, Swedish University of Agricultural Sciences, 750 07 Uppsala, Sweden
2 Department of Earth Sciences, Uppsala University, 752 36 Uppsala, Sweden
3 Water Resources Engineering, Department of Building and Environmental Technology, LTH, Lund University, 221 00 Lund, Sweden
4 Department of Building and Environmental Technology, LTH, Lund University, 221 00 Lund, Sweden
* Correspondence: sebastian.thons@kstr.lth.se

Abstract: A considerable number of lakes in Sweden have high phosphorus internal loading from the sediments which cause cyanobacterial blooms every summer. Due to potential risks with such blooms for human health, drinking water supply, and ecosystem services, measures need to be taken to control the phosphorus content. Measures to control the phosphorus input from the surrounding land has been in focus. However, the measures have not been sufficient. This is because phosphorus deposited at the bottom of the lakes for many years are finally starting to leak to the water phase when the decomposition of sediments leads to anoxic conditions. In order to determine effective and efficient lake restoration measures, methods for lake restoration decision support by a multi-criteria analysis and the application of a decision analysis are developed. The multi-criteria analysis includes the determination of costs, longevity, and efficacy of six common lake restoration measures to reduce internal phosphorous loads in two lakes selected as a case study. The results show that aluminum treatment combines a highest efficacy with a high-cost efficiency being thus the optimal identified measure. The method involves adding an aluminum solution to the lakes' sediment, which binds phosphorus, preventing it to be released to the water column. The multi-criteria model is integrated to a decision analytical model. The decision analytical model is used to identify the monetary socio-economic and environmental boundaries for the implementation of the optimal lake restoration measure.

Keywords: Bayesian decision analytical model; lake restoration; aluminum treatment; internal phosphorus load

Citation: Sellergren, M.; Li, J.; Drakare, S.; Thöns, S. Decision Support for Lake Restoration: A Case Study in Swedish Freshwater Bodies. *Water* **2023**, *15*, 668. https://doi.org/10.3390/w15040668

Academic Editors: Akira Kawamura and Kei Nakagawa

Received: 19 December 2022
Revised: 13 January 2023
Accepted: 30 January 2023
Published: 8 February 2023

1. Introduction

In European water policy, the Water Framework Directive (WFD) aims to improve the chemical and ecological quality of European surface waters to achieve good ecological status in all rivers, lakes, coastal, and transitional waters [1]. Currently, more than half of European water bodies are in a degraded condition, and nutrient enrichment [2] is one of the main problems [3]. Eutrophication in lakes often leads to a negative loop, where high levels of phosphorus (P) in the water lead to excessive phytoplankton growth. When the phytoplankton die and fall to the bottom, they are decomposed which requires oxygen which leads to low oxygen levels of the bottom water of the lake. Low oxygen levels in turn releases P from the bottom sediment and makes it available to the phytoplankton. In this way, P added to a lake years ago can cause eutrophication today. This phenomenon is called internal loading [4], and it is a common problem in lake restoration projects when preventing nutrients from land from entering a lake will not be enough to prevent blooms of phytoplankton.

Phosphorus (P) can be released from sediments via several processes. If there is oxygen in the bottom water, P is strongly bound to metals in the sediment, such as iron, aluminum, and calcium. These complexes are difficult to dissolve and make the P unavailable to phytoplankton and other plants, and it does not contribute to eutrophication [5]. However, changes in oxygen and pH can dissolve complexes, making the P bioavailable by diffusion from the sediment to the overlying water column [6]. For example, the binding of P to iron (Fe), is sensitive to low oxygen conditions which lead to a decrease in the reduction-oxidation (redox) potential at the sediment-water interface, causing release of Fe bound P (PFe) from the sediment to the water column via the reduction in ferric Fe (Fe III) to water soluble ferrous Fe (Fe II) [7]. P release from sediment is also affected by pH because the solubility of Fe and Al increases at both low and high pH, whereas the solubility of calcium (Ca) bound P (PCa) decreases at low pH [8]. Elevated P release from Fe and Al complexes during high pH conditions has been observed in several studies [8,9], which is generally caused by excessive photosynthetic activity/phytoplankton growth, which increases the pH by fixation of carbon dioxide during photosynthesis [9].

Many efforts have been done for lake management, such as assessment of external and internal nutrient loading [10], identifying and analyzing existing lake management strategies [11], and lake restoration approaches [12].

There are mainly two ways of reducing P loading, either through removing the P from the lake or lake sediment or through stabilizing P in the sediments. The most studied methods for such lake restoration are aluminum treatment, Phoslock, oxygenation, mixing of the water, dredging, and reduction fishing. Each method with different success rate, longevity, and costs [13,14]. Aluminum treatment, Phoslock, oxygenation, and mixing aim to bind P in the sediments. Dredging and reduction fishing physically remove P [14].

Stora and Lilla Ullfjärden are two lakes situated close upstream to Lake Mälaren, the third largest freshwater lake in Sweden. High P concentrations have been a problem for a long time in this lake system, and it causes phytoplankton blooms every summer. Due to large volumes of toxin-producing cyanobacteria in Lakes Stora and Lilla Ullfjärden, they pose great potential threats downstream in Lake Mälaren, which is used for drinking water for approximately 2 million people [15].

Until now, the external phosphorus load has been reduced by minimizing known sources of P leakage from land, e.g., by better manure management and fields fallowed only during growth season in summer [16,17]. Despite these measures, the P concentration in the lakes remains high due to internal loading of P. The costs for removing P from the lakes are high and local authorities have been hesitating for many years to use such methods to improve status of the lakes more drastically. So far, no standards are available for local decision makers to determine which lake restoration measures to implement in lakes with internal loading of P.

Based on the outlined challenges, this paper aims to develop methods, and to advance scientific understanding for lake restoration decision support by the performance of a multi-criteria analysis and the application of a decision analysis. Both is exemplified through a case study. The multi-criteria analysis encompasses costs, longevity, and efficacy of six lake restoration measures determined by a meta-analysis. Additionally, it is examined whether the cost of the selected treatment can be justified by highlighting the economic values of the lakes. The ecosystem services of the lakes are identified, and each service is assigned an estimated economic value. A Bayesian decision analytical model is used to determine the optimal lake restoration measure and the monetary socio-economic and environmental boundaries for implementation.

This is, to the knowledge of the authors, the first application of a using multi-criteria model integrated to a decision analytical model for water resources management. The study gives answers to the questions: (i) What is the most suitable lake restoration measure to reduce the internal loading in a specific case regarding costs, longevity, and efficacy? (ii) Can the cost of the selected treatment be justified by highlighting the economic values of the lakes?

2. Materials and Method

2.1. Case Study

2.1.1. Lakes Stora and Lilla Ullfjärden

The two lakes belong to the innermost part of the large, multi-basin Lake Mälaren and are separated by narrow passages, partly from each other, partly from the next basin which eventually empties into the Baltic Sea, with the outlet in central Stockholm, Sweden. Stora and Lilla Ullfjärden can be considered independent lakes, instead of coves to Lake Mälaren, Figure 1. Lilla Ullfjärden is one of the deepest lakes in the multi basin Lake Mälaren system with a maximum depth of 53 m and mean depth of 22 m. The lake area is 1.88 km^2 and the drainage area 8.51 km^2, dominated by forest (69%), and only a small area of farmland and urban area. Stora Ullfjärden has a maximum depth of 27 m and a mean depth of 15.2 m. The lake area is 2.8 km^2 and the drainage area 48 km^2 including the upstream Lilla Ullfjärden. Farmland are dominating (39%) followed by forest (36%), farmland (10%) and urban areas (5%) [16,17]. Both lakes are dimictic and usually covered by ice and snow during winter. The lakes are interesting in many respects. First, the lakes and their surroundings have a high nature value with a nature reserve at the lake shore, the Uppland hiking trail passing by, a public bathing area, and possibilities of fishing. Second, Stora Ullfjärden is home to a highly threatened underwater plant, and the deep Lilla Ullfjärden harbors glacial relicts, i.e., cold-water species that have been trapped and remained here since the ice age [18].

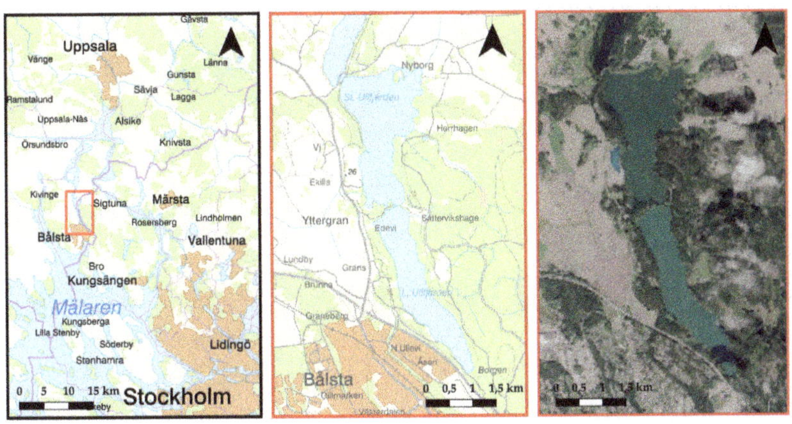

(a) (b) (c)

Figure 1. (**a**) Overview map showing the location of the lakes, shared by Stockholm County and Uppsala County, Sweden © Lantmäteriet; (**b**) Lilla Ullfjärden is an innermost cove of Lake Mälaren entering Stora Ullfjärden which empties into the next basin of Lake Mälaren © Lantmäteriet; (**c**) Satellite image from 31 August 2021 showing dark waters north of Stora Ullfjärden, Stora Ullfjärden with some cyanobacterial bloom, and Lilla Ullfjärden with a strong bloom. Original images: ESA Copernicus Sentinel Data, Syke [19].

2.1.2. Need for Lake Restauration

According to the EU Water Framework Directive (WFD), good ecological status must apply in all lakes [20]. The ecological status based on total P of Stora Ullfjärden is bad (51 µg P/L), and in Lilla Ullfjärden poor (59 µg P/L) which means that measures need to be done to reduce the concentration to around 9 µg P/L, which is a P-level that corresponds to good ecological status based on phytoplankton [5,6,21] and will reduce the problems with cyanobacterial blooms.

2.2. *Multi-Criteria Analysis of Lake Restoration Measures*

2.2.1. Cost, Longevity and Efficacy

A meta-analysis comprising costs, longevity, and efficacy was performed for the six well-studied measures to reduce internal phosphorous loading. Input data for the analysis can be found in Appendix A, which is based on a literature review. Many references are retrieved from the report Internal load of phosphorus in Swedish lakes [13], but several references were added or removed if they were not considered relevant, such as all lakes with a lake surface area of less than 10 hectares. Included in the analysis are both results from completed studies and estimated results from planned measures.

To compare costs based on studies conducted in different years, inflation information was used to recalculate the cost to a corresponding monetary value in 2021. In cases where cost-information came from a study abroad, the cost was first recalculated to Swedish currency (SEK) using the exchange rate that prevailed in the year the measurement was conducted.

2.2.2. Cost in Long Term

Lake restauration measures have different longevity and thus costs over a considered period accumulate according to inflation and discounting rates. To illustrate this, the total cost was calculated for three different time horizons, 5, 10, and 50 years. A measure that is implemented far in the future will have a lower cost due to discounting. The discount rate used was 1.0225 determined with a discount rate for a public decision maker and the current inflation developments [22].

Since the aluminum treatment has an average longevity of 12 years (Table A1, Appendix A) the cost after 5 and 10 years will be the same as after 1 year, Equation (1). For reduction fishing, with an average longevity of 3 years, the start-up cost will have to be paid twice in 5 years (start year and third year), Equation (2), and 4 times in 10 years (start year, third year, sixth year, and ninth year), Equation (3). For oxygenation and mixing an operating cost is also added each year. For oxygenation with a longevity of 20 years, the start-up cost will have to be paid three times in 50 years (start year, 20th year and 40th year), and the operating cost is paid every year during the 50 years, Equation (4).

Total cost 5 years, Aluminum treatment (longevity 12 years):

$$\text{Total cost} = \text{startup cost} \tag{1}$$

Total cost 5 years, Reduction fishing (longevity 3 years):

$$\text{Total cost} = \text{startup cost} + (\text{startup cost}/1.02253) \tag{2}$$

Total cost 10 years, Reduction fishing (longevity 3 years):

$$\text{Total cost} = \text{startup cost} + (\text{startup cost}/1.02253) + (\text{startup cost}/1.02256) + (\text{startup cost}/1.02259) \tag{3}$$

Total cost 50 years, Oxygenation (longevity 20 years):

$$\text{Total cost} = \text{startup cost} + (\text{startup cost}/1.022520) + (\text{startup cost}/1.022540) + (\text{operating cost} \times 50/1.022550) \tag{4}$$

2.3. *Decision Analysis of Lake Restoration Measures*

2.3.1. Bayesian Decision Analysis

Bayesian decision analysis constitutes a comprehensive methodology for decision support when the outcomes cannot be predicted with certainty [23]. A recent research effort constitutes connecting the decision theory to environmental impact quantification [24]. Rational decisions are about choosing the option that provides the maximum expected utility or the minimal risk according to the expected utility theorem [25]. In this study, Bayesian decision analysis is based on a risk analysis. Each system state is assigned a probability of occurrence and consequences. Risks are calculated as the product of the probability and the consequences:

$$R(A) = P(A) \times c(A) \tag{5}$$

R is the total risk of system state A, P is the probability that system state A occurs, and c is the expected consequences of system state A. The system state can be modified, in our case with lake restoration measures, r_i. The measures will influence the probability of the system state, i.e., $P(A|r_i)$, and will have costs $c(r_i)$. The total risks and cost for the measure r_i are then calculated with Equation (6):

$$R(A|r_i) = P(A|r_i) \times c(A) + c(r_i) \tag{6}$$

The objective function to determine a decision about a restorative measure is then the minimization of the expected value (operator E) of the costs and consequences overall system states A_j according to the expected utility theorem, Equation (7). The type of decision analysis is called a prior decision analysis.

$$R_{Prior} = \min_{r_i} E_{A_j}[c(A_j, r_i)] \tag{7}$$

2.3.2. Probability of Good Ecological Water Status

The probability of good water status in a lake after aluminum treatment is based on a previous project in Lake Växjö, in the south of Sweden [26]. Lake Växjö was treated with aluminum in May to August 2018. Before the lake was treated, reduction fishing was conducted in the lake to stimulate the establishment of underwater vegetation and improve the possibilities of achieving the intended effect with the aluminum treatment [26]. The target value for good water status in Lake Växjö regarding total phosphorus was 15.2 μg/L [27]. The probabilities of achieving good water status in Stora and Lilla Ullfjärden were calculated based on the outcome of the aluminum treatment of Lake Växjö assuring similar success rate (Figure 2). The probability of good ecological water status before measure is 4%, and 92% after aluminum treatment.

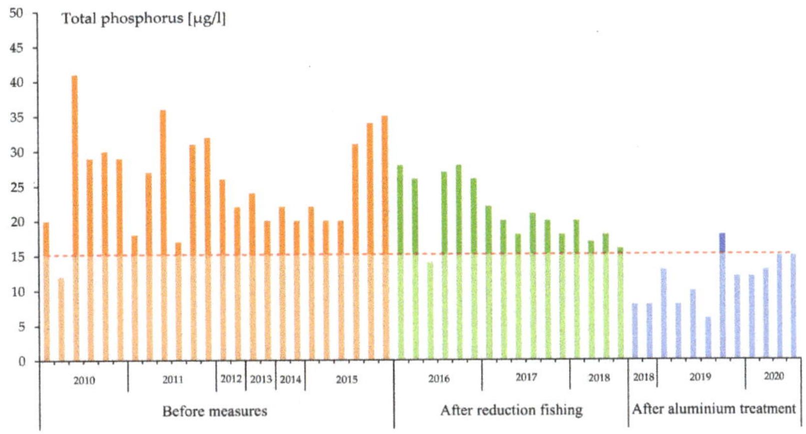

Figure 2. Total phosphorus concentration in Lake Växjö before measures (year 2010–2015), after reduction fishing (year 2015–May 2018) and after aluminum treatment (May 2018–2020) [28]. The red dashed line indicates the limit of good status (15.2 μg/L) in the lake with respect to total phosphorus.

2.3.3. Economic Value of the Lakes

Ecosystem services may be valued economically, not least for activation of resources in society [29]. Listing the services of the specific lakes gives an overview of many aspects that contribute to the value of lakes [30], which is performed based on studies from the same region of Sweden. The various ecosystem services contributions by lakes have an economic value that is either comprehensive (has a market value), can be valued (by using estimates and assumptions), or is not comprehensive (has no market value). In this way, all aspects and attributes surrounding the lakes can be highlighted [30].

2.3.4. Socio-Economic Profitability Assessment of Aluminum Treatment

Restoration of lakes is costly. However, a lake that does not attain good water status also imply consequences and costs. Examples of a bad status are, e.g., increased costs for an adjacent drinking water treatment plant, there may be increased costs for bathing sites located at the lake or reduced value for nearby houses, etc. Since it is difficult to calculate the cost of not having a good ecological status, this study examines what economic consequences need to be surpassed to justify a measure. If the cost of the measure cannot justify the benefit that the measure implies, it is not socio-economic profitable for the measure to be conducted. These consequences are determined with the Bayesian decision analysis. Since there is no value for the cost if good status is not achieved (cost not good status), the Bayesian decision analysis can be used to determine the minimum consequences of a not good lake status to compensate for the costs of the measure.

3. Results

3.1. Multi-Criteria Analysis of Lake Restoration Measures

3.1.1. Cost, Longevity and Efficacy

A summary of the scientific literature-based data in Appendix A is presented in Table 1 and shows the average value for start-up cost, operating cost, longevity, and efficacy for six lake restauration measures focusing on reducing internal phosphorus loading. For more details and the literature sources, please see Appendix A.

Table 1. Average value for start-up cost, operating cost, longevity, and efficacy for the six lake restauration measures to decrease internal phosphorous loading. For the start-up and operating cost, the cost is recalculated to a cost valid in 2021 using an inflation calculator [31].

Measures:	Start-Up Cost	Operating Cost	Longevity	Efficacy
	[SEK/hectare]	[SEK/year/hectare]	[years]	[%]
Al. treatment:	43,300	-	12	90
Phoslock:	139,200	-	30 [1]	60
Oxygenation:	58,900	3700	20 [2]	40
Mixing:	13,600	700	20 [3]	0
Dredging:	185,800	-	2.5	25
Reduction fishing:	36,700	-	3	5

Notes: [1] As the binding capacity does not decrease with time [32] a service life of 30 years is estimated. [2] This is the expected longevity of the pump aggregate to provide oxygen (air), which is expected to be 20 years [13]. [3] Just as for oxygenation, the longevity is assumed to be the expected longevity of the aggregate, which is estimated to 20 years.

3.1.2. Cost in Long Term

Lake restauration measures have different longevity and thus costs accumulate according to inflation and discounting rates over the periods of 5, 10, and 50 years. A measure that is implemented far in the future will have a lower cost due to discounting. In Table 2 the cost of each lake restauration measure of the considered periods are documented.

From Table 2 it is understood that dredging is by far the most expensive option regardless of time horizon. Mixing remains the least expensive option regardless of time horizon followed by aluminum treatment as the second least expensive option.

Table 2. Cost of each lake restauration measure depending on time, based on start-up cost, operating cost, and longevity. A discounting rate of 2.25% have been used.

Measures:	Cost 5 Years	Cost 10 Years	Cost 50 Years
	[SEK/hectare]	[SEK/hectare]	[SEK/hectare]
Al. treatment:	43,300	43,300	136,200
Phoslock:	139,200	139,200	210,600
Oxygenation:	149,100	162,100	332,600
Mixing:	16,700	19,200	39,400
Dredging:	361,500	690,000	2,305,000
Reduction fishing:	71,000	133,200	385,600

3.1.3. Cost-Effectiveness Comparison

To make a cost-effective comparison of the lake restauration measures, a simplified multi-criteria analysis is performed. In Table 3 a ranking (0 to 5) for the efficacy and the costs of a period of 50 years is introduced and the ranks are added.

Table 3. Results of the multi-criteria analysis for possible lake restoration measures to reduce internal phosphorous loading. By efficacy is meant reduction of internal load, making the option with the highest efficacy to receive the rank 5 and the lowest the rank 0, based on Table 1. For the category cost 50 years, a low cost is preferred, making the option with the lowest cost to receive a 5 and the highest cost a 0, based on Table 3. A high total score is desirable.

Measures:	Efficacy	Cost 50 Years	Total
Aluminum treatment:	5	4	**9**
Phoslock:	4	3	**7**
Mixing:	0	5	**5**
Oxygenation:	3	1	**4**
Reduction fishing:	1	2	**3**
Dredging:	2	0	**2**

Aluminum treatment scores the highest in the multi-criteria analysis. This is because it is the method that most effectively reduces the internal phosphorus load and is also the second least expensive method in the long term.

3.2. Decision Analysis of Lake Restoration Measures

3.2.1. Economic Value of the Lakes

To estimate the value of the lakes Stora and Lilla Ullfjärden, all services that contribute to the value are examined and listed in Table 4.

Table 4. Summary of all services contributing to the value of Stora and Lilla Ullfjärden.

Services:	Relevance
Surface water of good quality:	Drinking water production downstream.
Capture nutrients and pollutants:	Bottom sediments act as settling basins for nutrients, metals, and Polycyclic Aromatic Hydrocarbons (PAHs).
Tourism:	Shop at the bathing site Ekillabadet.
Bathing site:	Ekillabadet, public bathing site.
Fishing:	Extensive recreational fishing in the lakes.
Boat life:	Boat club in Lilla Ullfjärden.
Lake view for nearby houses:	Increased property value.

Table 4. *Cont.*

Services:	Relevance
National nature value:	Particularly valuable species occur in the lakes, such as the underwater plant Baltic water-plantain (Alisma wahlenbergii), several species of glacial relicts (Crustacea), and red-listed fish species.
Recreation values:	Lake shore with two Natura 2000 areas next to the lakes. The Uppsala trail passes through the area.
Aesthetic values:	The lakes contribute to emotional well-being.
Scientific values:	The lakes have been the subject of limnological and hydrological research for more than a century (but are more sparingly investigated since the 70s).
Establishment of the WFD:	It states that lakes, watercourses, and coastal waters must reach good ecological and good chemical surface water status. The current water status must not deteriorate in any respect.

In Table 5, each service is qualitatively assigned an expected value. In this way, all aspects and attributes surrounding the lakes can be highlighted. Table 5 shows that the majority of the services do not comprehensibly influence the financial value of Stora and Lilla Ullfjärden.

Table 5. Summary of qualitative expected values for the services of Stora and Lilla Ullfjärden.

Services:	Value [1]	Reduced Value if Good Status Is Not Achieved [2]	Valuing Possibility
Surface water of good quality:	High	Minor	Comprehensible
Capture nutrients and pollutants:	High	High	
Tourism:	Minor	Middle	
Bathing site:	Middle	Middle	Can be valued
Fishing:	Minor	Middle	
Boat life:	Minor	Minor	
Lake view for nearby houses:	High	Minor	
National nature value:	High	Middle	Not comprehensible
Recreation values:	High	Minor	
Aesthetic values:	High	High	
Scientific values:	Middle	Middle	
Establishment of the WFD:	High	High	

Notes: [1] The expected Value is presented on the qualitative scale: "High" (millions of SEK), "Middle" (hundred thousand SEK), "Minor" (less than one hundred thousand SEK). [2] The expected Reduced value if good status is not achieved is presented on the qualitative scale: "High" (75–100% reduced value), "Middle" (25–75% reduced value), "Minor" (less than 25% reduced value).

3.2.2. Socio-Economic Profitability Assessment of Aluminum Treatment

With Bayesian decision analysis, it is possible to determine the economic consequences of a not good lake status by weighting the total risks and expected costs against a good lake status with restorative measures. Figure 3 illustrates finding the tipping point, when the expected not good status consequences with and without the measure are balanced. The balancing is achieved by reducing the probability of a not good status with the measure according to its efficacy.

The inputs needed for the Bayesian decision model to calculate the cost of not achieving good status are the probability of good status and measure cost (start-up cost + operating cost). The lake restauration measure with the lowest total risk is chosen (Equation (7)), which constitutes aluminum treatment, see also Section 3.1.3. The input to the analysis is presented in Table 6.

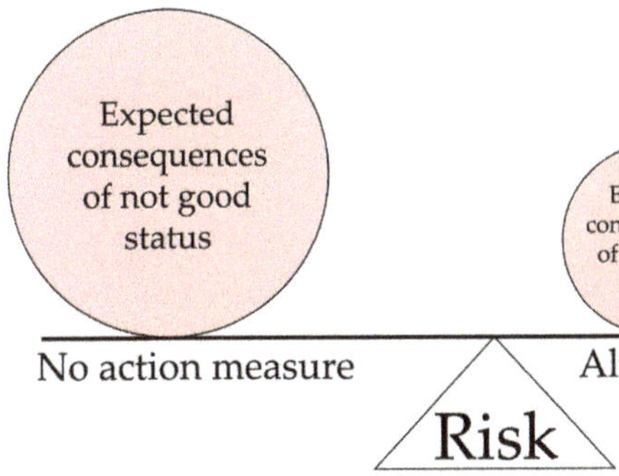

Figure 3. Illustration of the tipping point for when total risks and expected costs is the same for the options "no action measure" and "aluminum treatment".

Table 6. Input data in the Bayesian decision analysis. The cost for aluminum treatment has been multiplied by 500 because the combined lake surface area of Lilla and Stora Ullfjärden is 500 hectares.

Probabilities and Costs:	Probability of Good Ecological Water Status	Areal Cost 12 Years	Total Cost 12 Years
Unit:	[%]	[SEK/hectares]	[SEK]
No measure:	4	0	0
Aluminum treatment:	92	43,300	21,650,000

Figure 4 shows how the input data (Table 6) is used in the Bayesian decision analysis to determine the cost for a not good status, which result in the alternatives "no measure" and "aluminum treatment" to have an equal risk. If the cost of a not good status increases further than shown, aluminum treatment will be the option with the lowest risk.

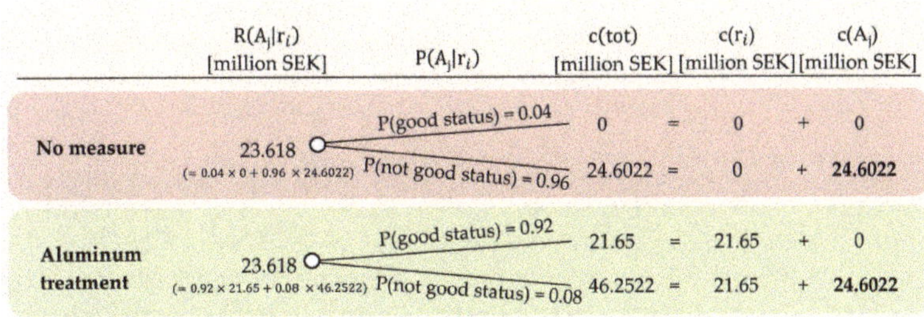

Figure 4. Bayesian decision analysis where the alternative aluminum treatment is compared to the alternative no measure. $R(A_j|r_i)$ is the total risk of system state A_j (good/not good status), for the lake restoration measure r_i (no measure or aluminum treatment), P is the probability and c is the consequences (measure cost and cost good/not good status).

From Figure 4, it can be stated that if the cost not good status is higher than SEK 24.6 million over a 12-year period, the lakes should be treated with aluminum.

4. Discussion and Limitations

The measures presented in this study represent six well-studied measures to reduce internal phosphorus loading. These measures will only have a long-term effect together with continuous efforts to reduce the external phosphorus supply [14].

The six measures to reduce the internal phosphorus loading have different longevities. The costs have been calculated, discounted and accumulated to total costs representative for different time spans. For 10 years, the aluminum method (with a longevity 12 years) only needs to be conducted once, while reduction fishing (longevity 3 years) needs to be conducted four times. Apart from mixing, which according to previous studies does not show a reduction in internal load, the aluminum method is the least costly method, on a 1, 5, 10, and 50 year horizon. It is also the measure that reduces the internal load most effectively.

The probability of good ecological status after a measure is conducted, has been determined based on a previous lake restoration project in Lake Växjösjön, Sweden [26]. From this project there are measurement data for the lake's phosphorus levels before measures, after reduction fishing, and after aluminum treatment. The fact that reduction fishing was conducted before the aluminum treatment may affect the good status probability of the aluminum treatment. Since reduction fishing did not have a high effect and a longevity of only 3 years, it was concluded that the effect of reduction fishing before aluminum treatment can be neglected.

The ecologic societal services of Stora and Lilla Ullfjärden are presented in Tables 4 and 5. The expected value of the services and the expected reduced value if good status is not achieved are based on similar studies. Listing the lake's services in this way is done to give an overview of the various services a lake contributes, to give examples of how the value of different services can be estimated and, above all, to highlight and emphasize the importance of all the lake's services, even services that do not have a market value.

With the help of the Bayesian decision analysis, it can be shown that the lakes should be treated with aluminum if the consequences of poor water status is higher than SEK 24.6 million over a 12-year period. If this cost is compared with Table 5, which shows expected reduced values for all of Stora and Lilla Ullfjärdens' services if good status is not achieved, the decision whether the lakes should be treated with aluminum treatment or not can be supported. It should further be noted that the value of the lakes lies mainly in services that do not have a direct monetary value, such as national nature value including valuable species occurring in the lakes, recreational values, and aesthetic values. However, if the status does not meet a good condition, then this intangible value reduction can be very significant.

Regardless of which measure is chosen for the case study lakes, more site-specific studies need to be conducted to dimension the measure and to confirm the behavior similarities of Lake Växjösjön and the Stora and Lilla Ullfjärden lakes. If aluminum treatment is chosen as a measure, sediment samples need to be taken to investigate which phosphorus fractions are dominating in the sediments, as this affects what dose of aluminum is needed. A dose that is too high can cause aluminum remaining in inorganic form, which is toxic to plants and animals. However, a dose that is too low instead gives a too short-lived result causing continuous summers of toxin-producing cyanobacteria, which it is also toxic to plants and animals. Therefore, a long-term, adequate monitoring program, including proper determination of external loading, is crucial to document the effect of aluminum treatment on sediment phosphorus release and water quality in the lake [33].

We have some further limitations of the study. First, our reference data for the determination of the status probabilities are limited to one lake restoration project in a Swedish lake. Different type of lakes might have other chemical and biological characteristics that might influence the effect of treatment methods. Future work may include different types of lakes. Second, we did not include the external loads effect and how they impact the total phosphorus load and the water quality dynamics.

5. Summary and Conclusions

According to the WFD, good ecological surface water status must apply in all lakes [20]. However, this is not the case in Stora and Lilla Ullfjärden, two lakes close to Lake Mälaren, which is of high value for the capital of Sweden and the surrounding areas. Several measures have already been conducted in the lakes to reduce eutrophication. These measures had the aim of reducing the external phosphorus load, i.e., the nutrient released from surrounding soil. Despite these measures, Stora and Lilla Ullfjärden still suffer from phytoplankton blooms every year. Because the lakes have a long history of external phosphorus loading, phosphorus has accumulated in the bottom sediments and causes internal phosphorus loading. To reach the desired total phosphorus concentration in the lakes, the internal phosphorus load must be decreased by further measures.

In order to determine effective and efficient lake restoration measures, methods for lake restoration decision support by a multi-criteria analysis and the application of a decision analysis are developed and exemplified with a case study. The multi-criteria analysis is developed to determine costs, longevity, and efficacy of six common lake restoration measures to reduce internal phosphorous loads. The multi-criteria model is integrated to a Bayesian decision analytical model. The decision analytical model is used to determine the optimal lake restoration measure and boundaries for measure implementation.

From the study it can be concluded that aluminum treatment is the most reasonable choice to reduce the internal load in the case study lakes Lilla and Stora Ullfjärden. This well-studied method effectively reduces the internal load of phosphorus, and the method is also cost-effective both in the short and long term compared to other measures. However, more site-specific studies are required for dose determination and transferability of measure characteristic underlying this study.

Aluminum treating both lakes, which have a total lake surface area of 500 hectares, is estimated to cost around SEK 21.7 million. However, not achieving good ecological status in the lakes may have monetary consequences and may touch intangible nature values (as discussed). The lakes have a high nature value and contribute to many cultural ecosystem services, such as bathing, fishing, and recreation. With the help of Bayesian decision analysis, a lower boundary value for these socio-economic consequences in the Stora and Lilla Ullfjärden lakes has been established, namely 24.6 million SEK aggregated over a 12-year period.

Future research efforts can be directed towards accounting for measurements to support the restoration measure implementation, which would specifically address the dose determination for aluminum treatment. The performed multi criteria and the decision analyses may also form the basis for the development of a decision support tool in the DiCyano project (funded by VINNOVA).

Author Contributions: Conceptualization, M.S., J.L., S.D. and S.T.; methodology, M.S., J.L., S.D. and S.T.; validation, J.L., S.D. and S.T.; formal analysis, M.S.; investigation, M.S.; data curation, M.S.; writing—original draft preparation, M.S.; writing—review and editing, M.S., J.L., S.D. and S.T.; visualization, M.S.; supervision, J.L. and S.D.; project administration, J.L.; funding acquisition, J.L. All authors have read and agreed to the published version of the manuscript.

Funding: This research was supported by VINNOVA (the Swedish Innovation Agency), grant number "2020-03365" in the project named A smart digital platform for managing Cyanotoxin risk in drinking water supply and beyond (DiCyano).

Data Availability Statement: Not applicable.

Acknowledgments: The authors wish to express their sincere thanks to Stephan Köhler at the Department of Aquatic Sciences and Assessment, Swedish University of Agricultural Sciences, Uppsala, Sweden, for his guidance; to Zanna Zielfeldt Nikas, water action coordinator at Upplands-Bro municipality, for her well-written comments; to Walter Cassel, doing research in the lakes Stora and Lilla Ullfjärden, for the enjoyable collaboration.

Conflicts of Interest: The authors declare no conflict of interest.

Appendix A

Table A1. Aluminum treatment: background data for the multi-criteria analysis.

Source	Lake	Lake Type	Method	Treatment Year	Treated Area	Start-Up Cost	Start-Up Cost	Reduced Internal Loading	Longevity
				[Year]	[Hectare]	[SEK]	[SEK/Hectare]	[%]	[Year]
[34]	Långsjön	Shallow	Sediment injection	2006	29	3,530,756	121,750	90	>9
[34]	Flaten	Deep	Sediment injection	2000	40	3,212,260	80,307	95	>15
[35]	Calhoun	Deep	Hypolimnetic water application	2001	130	1,833,000	14,057	100	>14
[36]	Harriet	Deep	Hypolimnetic water application	2001	47	1,034,687	22,015	85	5
[35]	Cedar	Deep	Hypolimnetic water application	1996	60	1,657,000	27,849	95	13
[13]	Spring	Deep	Hypolimnetic water application	2014	39	3,907,000	23,679		>1
[13]	Long	Deep	Hypolimnetic water application	2009	102	1,360,000	61,818		>6
[13]	Medical Lake	Deep	Hypolimnetic water application	1977	122	934,000	14,594		>40
[13]	McCarron	Deep	Hypolimnetic water application	2005	77	680,000	30,909		>10
[13]	Bryant	Deep	Hypolimnetic water application	2008	74	2,120,000	29,444		>7
[37]	Hjälmaren	Deep	Hypolimnetic water application		20,000	615,000,000	30,750	100	
[26]	Växjösjön	Deep	Hypolimnetic water application and sediment injection	2018	65	5,400,000	69,948	60	
[38]	S.Bergundasjön	Shallow	Hypolimnetic water application and sediment injection	2019	310	15,500,000	35,880		
Average value aluminum treatment:							**43,300 (±31,500)**	**90 (±14)**	**>12**

Table A2. Phoslock: background data for the multi-criteria analysis.

Source	Lake	Lake Type	Treatment Year	Treated Area	Start-Up Cost	Start-Up Cost	Reduced Internal Loading	Longevity
			[Year]	[Hectare]	[SEK]	[SEK/Hectare]	[%]	[Year]
[39]	Lake Rauwbraken	Shallow	2008	4	517,170	129,292.5		
[39]	Lake De Kuil	Shallow	2009	7	1,461,999	208,857		
[13]	The average cost of some small lakes		2013			79,565		
[13]	Summary, Table 10B						30–90	
Average value Phoslock:						**139,200 (±6500)**	**60 (±30)**	**30 ***

Note: * Because the binding capacity does not decrease with time [32] a longevity of 30 years is estimated.

Table A3. Oxygenation: background data for the multi-criteria analysis.

Source	Lake	Treatment Year	Treated Area	Start-Up Cost	Start-Up Cost	Operating Cost	Operating Cost	Reduced Internal Loading	Longevity
		[Year]	[Hectare]	[SEK]	[SEK/Hectare]	[SEK/Year]	[SEK/Year/Hectare]	[%]	[Year]
[13]	Pine Lake	1981	36	765,278	21,258	297,608	8267		
[40]	San Vincent Reservoir	1975	405	7,834,985	19,346	607,211	1499		
[13]	Average 15 lakes	2001	113	20,156,230	178,374	1,004,491	8889		
[41]	Stubbs Bay (Lake Minnetonka)	2004	81	4,336,573	53,538	340,123	4199		
[13]	Tegel	2002	400	32,916,685	82,292	565,759	1414		
[13]	JC Boyle Reservoir	2009	170	3,188,657	18,757	398,582	2345		
[13]	Marston Reservoir	2009	251	19,929,104	79,399	117,343	468		
[13]	Bear Creek Lake	2002	45	2,994,680	66,548	261,925	5821		
[42]	Cherry Creek Reservoir	2002	342	3,492,338	10,212	88,182	258		
[13]	Summary, Table 10B							30–50	20 *
Average value oxygenation:					**58,900 (±52,800)**		**3684 (±3300)**	**40 (±10)**	**20 ***

Note: * The expected lifetime of the aggregate [13].

Table A4. Mixing: background data for the multi-criteria analysis.

Source	Lake	Lake Type	Treatment Year	Treated Area	Start-Up Cost	Start-Up Cost	Operating Cost	Operating Cost	Reduced Internal Loading	Longevity
			[Year]	[Hectare]	[SEK]	[SEK/Hectare]	[SEK/Year]	[SEK/Year/Hectare]	[%]	[Year]
[43]	Iskmosunden	Shallow	2014	30	332,857	11,095	11,095	370		
[44,45]	Jordan Lake	Water tank	2014	4636	10,856,617	2342	6,263,433	1351	0	
[44]	Lake Houston	Water tank	2006	243	6,680,995	27,494	82,054	338	0	
Average value mixing:						**13,644 (±12,800)**		**686 (±580)**	**0**	**20 ***

Note: * The expected lifetime of the aggregate [13].

Table A5. Dredging: background data for the multi-criteria analysis.

Source	Lake	Lake Type	Treatment Year	Treated Area	Start-Up Cost	Start-Up Cost	Reduced Internal Loading	Longevity
			[Year]	[Hectare]	[SEK]	[SEK/Hectare]	[%]	[Year]
[46]	Finjasjön	Shallow	1991	111	95,985,075	864,730	0	
[26]	Trummen	Shallow	1970	99	21,346,647	215,623	0	
[26]	Växjösjön	Deep	1990	87	45,037,065	517,667	0	
[26]	South Bergundasjön	Shallow	1992	430	50,514,179	117,475	0	
[47]	Lillsesjön alt. 1		2014	55	5,422,886	98,598		
[13]	Clear Lake	Shallow	2009	1468	79,716,418	54,303		
[13]	Half Moon Lake		1991	53,4	256,159	4797		
[13]	Lilly Lake		1991	35,6	366,374	10,291		
[13]	Lenox Lake		1991	13,4	400,959	29,922		

Table A5. *Cont.*

Source	Lake	Lake Type	Treatment Year	Treated Area	Start-Up Cost	Start-Up Cost	Reduced Internal Loading	Longevity
[13]	Nutting Lake		1991	31,6	1,337,916	42,339		
[13]	Collins Park Lake		1991	24,3	2,137,369	87,958		
[13]	Summary, Table 10B						0–50	
[48]	Dongqian Lake, China							2
[49]	Lake Taihu, China							3
Average value dredging:						**185,791 (±268,000)**	**25**	**2.5**

Table A6. Reduction fishing: background data for the multi-criteria analysis.

Source	Lake	Treatment Year	Treated Area	Start-Up Cost	Start-Up Cost	Reduced Internal Loading	Longevity
		[Year]	[Hectare]	[SEK]	[SEK/Hectare]	[%]	[Year]
[50]	Östhammarfjärden	2011	121.6	1,915,876	15,756		
[51]	Borringesjön		295	1,409,950	4779		
[51]	Yddingesjön		213	1,030,348	4837		
[51]	Havgårdsjön		57	292,836	5137		
[46]	Finjasjön	1993	111	11,192,836	100,836		
[52]	Ringsjön	2005	39,5	6,766,026	171,292		
[53]	Växjösjön	2013	79	5,979,626	75,691		
[53]	Trummen	2013	76	2,184,338	28,741		
[13]	Ryssbysjön	2010			6152		
[13]	Average of four lakes	1989	263	2,375,224	9018		
[13]	Nokomis	2014	81	964,189	11,904		
[13]	Nokomis	2010	81	477,008	5889		
[13]	Summary, Table 10B					0–10	2–5
Average value reduction fishing:					**36,669 (±52,700)**	**5**	**3**

References

1. European Commission. Implementation Report Water Framework Directive and Floods Directive—Questions and Answers. Available online: https://shortest.link/bqgl (accessed on 18 December 2022).
2. ScienceDirect. Available online: https://www.sciencedirect.com/topics/earth-and-planetary-sciences/nutrient-enrichment (accessed on 24 November 2022).
3. Poikane, S.; Kelly, M.G.; Salas Herrero, F.; Pitt, J.A.; Jarvie, H.P.; Claussen, U.; Leujak, W.; Lyche Solheim, A.; Teixeira, H.; Phillips, G. Nutrient criteria for surface waters under the European Water Framework Directive: Current state-of-the-art, challenges and future outlook. *Sci. Total Environ.* **2019**, *695*, 133888. [CrossRef] [PubMed]
4. Pettersson, K. Mechanisms for internal loading of phosphorus in lakes. *Hydrobiologia* **1998**, *373*, 21–25. [CrossRef]
5. James, W.F.; Sorge, P.W.; Garrison, P.J. Managing internal phosphorus loading and vertical entrainment in a weakly stratified eutrophic lake. *Lake Reserv. Manag.* **2015**, *31*, 292–305. [CrossRef]
6. Agstam-Norlin, O. Restoration of Nutrient Rich Lakes Restoration of Nutrient Rich Lakes. Ph.D. Thesis, Swedish University of Agricultural Sciences, Uppsala, Sweden, 2022.
7. Ramm, K.; Scheps, V. Phosphorus balance of a polytrophic shallow lake with the consideration of phosphorus release. *Hydrobiologia* **1997**, *342*, 43–53. [CrossRef]
8. Gomez, E.; Durillon, C.; Rofes, G.; Picot, B. Phosphate adsorption and release from sediments of brackish lagoons: pH, O2 and loading influence. *Water Res.* **1999**, *33*, 2437–2447. [CrossRef]
9. Boers, P.C.M. The influence of pH on phosphate release from lake sediments. *Water Res.* **1991**, *25*, 309–311. [CrossRef]
10. Kowalczewska-Madura, K.; Dondajewska-Pielka, R.; Gołdyn, R. The Assessment of External and Internal Nutrient Loading as a Basis for Lake Management. *Water* **2022**, *14*, 2844. [CrossRef]

11. Mardiatno, D.; Faridah, F.; Listyaningrum, N.; Hastari, N.R.F.; Rhosadi, I.; da Costa, A.D.S.; Rahmadana, A.D.W.; Lisan, A.R.K.; Sunarno, S.; Setiawan, M.A. A Holistic Review of Lake Rawapening Management Practices, Indonesia: Pillar-Based and Object-Based Management. *Water* **2023**, *15*, 39. [CrossRef]
12. Alhamarna, M.Z.; Tandyrak, R. Lakes restoration approaches. *Limnol. Rev.* **2021**, *21*, 105–118. [CrossRef]
13. Huser, B.; Löfgren, S.; Markensten, H. *Internbelastning Av Fosfor I Svenska Sjöar Och Kustområden*; Report 2016: 6; Swedish University of Agricultural Sciences, Water and Environment: Uppsala, Sweden, 2016; Available online: https://pub.epsilon.slu.se/13724/7/huser_b_et_al_161026.pdf (accessed on 5 May 2022).
14. Erlandsson Lampa, M.; Witter, E. *Metod För Påverkanstypen Historisk Förorening—Internbelastning*; Luleå, Härnösand, Västerås, Kalmar, Göteborg: Vattenmyndigheterna i Samverkan, Sweden, 2018. Available online: https://viss.lansstyrelsen.se/ReferenceLibrary/55065/Metod%20Internbelastning.pdf (accessed on 6 May 2022).
15. Drakare, S.; Wallman, K.; Almlöf, K.; Segersten, J.; Sveriges, R. *Fokus På Mälaren 2019-Sammanfattande Resultat Från Miljöövervakning Och Forskningsprojekt Knutna Till Samarbetet Mellan SLU Och Mälarens Vattenvårdsförbund*; Report 2020: 3; Swedish University of Agricultural Sciences, Department of Aquatic Sciences and Assessment: Uppsala, Sweden, 2020. Available online: https://pub.epsilon.slu.se/16975/7/drakare_s_et_al_200514.pdf (accessed on 27 May 2022).
16. VISS Lilla Ullfjärden. Available online: https://viss.lansstyrelsen.se/Waters.aspx?waterMSCD=WA90764558 (accessed on 7 March 2022).
17. VISS Mälaren-Stora Ullfjärden. Available online: https://viss.lansstyrelsen.se/Waters.aspx?waterMSCD=WA47437003 (accessed on 7 March 2022).
18. Gustafsson, A. *Lilla Ullfjärden (Object Data Sheet)*; Naturvatten AB: Upplands-Bro Municipality, Sweden, 2015. Available online: https://www.upplands-bro.se/download/18.56ea77fc1686d89c2f852fa/1548081133208/Sjöar-LillaUllfjärden.pdf (accessed on 10 April 2022).
19. Syke, Tarkka. Available online: http://wwwi4.ymparisto.fi/i4/eng/tarkka/index.html?type=RGB&date=2021-08-31&datespan=1&name=DEFAULT&lang=en&zoom=11.118612663152941&lat=59.64168&lon=17.45038 (accessed on 12 May 2022).
20. The European Parliament and the Council of Union European. Directive 2000/60/EC of the European Parliament and of the Council Establishing a Framework for Community Action in the Field of Water Policy. *Off. J Eur. Commun.* **2000**. Available online: https://eur-lex.europa.eu/resource.html?uri=cellar:5c835afb-2ec6-4577-bdf8-756d3d694eeb.0004.02/DOC_1&format=PDF (accessed on 27 April 2022).
21. Swedish Agency for Marine and Water Management. Available online: https://www.havochvatten.se/vagledning-foreskrifter-och-lagar/foreskrifter/register-vattenforvaltning/klassificering-och-miljokvalitetsnormer-avseende-ytvatten-hvmfs-201925.html (accessed on 26 November 2022).
22. SCB. Inflation i Sverige 1830–2021. Available online: https://www.scb.se/hitta-statistik/statistik-efter-amne/priser-och-konsumtion/konsumentprisindex/konsumentprisindex-kpi/pong/tabell-och-diagram/konsumentprisindex-kpi/inflation-i-sverige/ (accessed on 24 May 2022).
23. Raiffa, H.; Schlaifer, R. *Applied Statistical Decision Theory*; Wiley: New York, NY, USA, 1961; ISBN 047138349X.
24. Dong, Y.; Miraglia, S.; Manzo, S.; Georgiadis, S.; Sørup, H.J.D.; Boriani, E.; Hald, T.; Thöns, S.; Hauschild, M.Z. Environmental sustainable decision making—The need and obstacles for integration of LCA into decision analysis. *Environ. Sci. Policy* **2018**, *87*, 33–44. [CrossRef]
25. Von Neumann, J.; Morgenstern, O. *Theory of Games and Economical Behavior*; Princeton University Press: Princeton, NJ, USA, 1947.
26. Olofsson Madestam, H. *Fosforfastläggning i Växjösjön*; Synlab: Halmstad, Sweden, 2020.
27. VISS Växjösjön. Available online: https://viss.lansstyrelsen.se/Waters.aspx?waterMSCD=WA26248924 (accessed on 26 October 2022).
28. Mörrumsåns Vattenråds Samlade Recipientkontroll (SRK). Available online: http://webstar.vatten.slu.se/db.html (accessed on 17 May 2022).
29. Kinell, G.; Söderqvist, T. *Ekonomisk Värdering Med Scenariometoder—En Vägledning Som Stöd För Genomförande Och Upphandling*; Report 6469; The Swedish Environmental Protection Agency: Stockholm, Sweden, 2011.
30. Morrison, G. *Mälarens Värde*; Chalmers University: Gothenburg, Sweden, 2009. Available online: https://miljobarometern.stockholm.se/content/docs/vp/mal/Malarens_varde_Chalmers_2009.pdf (accessed on 10 May 2022).
31. Ekonomifakta, Räkna på Inflationen. Available online: https://www.ekonomifakta.se/Fakta/Ekonomi/Finansiell-utveckling/Rakna-pa-inflationen/ (accessed on 18 May 2022).
32. Lürling, M.; Faassen, E.J. Controlling toxic cyanobacteria: Effects of dredging and phosphorus-binding clay on cyanobacteria and microcystins. *Water Res.* **2012**, *46*, 1447–1459. [CrossRef] [PubMed]
33. Agstam-Norlin, O.; Lannergård, E.E.; Rydin, E.; Futter, M.N.; Huser, B.J. A 25-year retrospective analysis of factors influencing success of aluminum treatment for lake restoration. *Water Res.* **2021**, *200*, 117267. [CrossRef] [PubMed]
34. Schütz, J.; Rydin, E.; Huser, B.J. A newly developed injection method for aluminum treatment in eutrophic lakes: Effects on water quality and phosphorus binding efficiency. *Lake Reserv. Manag.* **2017**, *33*, 152–162. [CrossRef]
35. Huser, B.J.; Futter, M.; Lee, J.T.; Perniel, M. In-lake measures for phosphorus control: The most feasible and cost-effective solution for long-term management of water quality in urban lakes. *Water Res.* **2016**, *97*, 142–152. [CrossRef] [PubMed]
36. Huser, B. Aluminum application to restore water quality in eutrophic lakes: Maximizing binding efficiency between aluminum and phosphorus. *Lake Reserv. Manag.* **2017**, *33*, 143–151. [CrossRef]

37. Karlsson, M.; Malmaeus, M.; Rydin, E. *Åtgärder Mot Internbelastning Av Fosfor I Hjälmaren*; Report Nr C 381; IVL Swedish Environmental Research Institute: Stockholm, Sweden, 2019. Available online: https://www.ivl.se/download/18.34244ba71728fcb3f3fab6/1591706073502/C381.pdf (accessed on 27 April 2022).
38. Hedrén, A. *Statusrapport Åtgärder Växjösjöarna*; Planning Department Technical Administration: Växsjö Municipality, Sweden, 2019.
39. Spears, B.M.; Mackay, E.B.; Yasseri, S.; Gunn, I.D.M.; Waters, K.E.; Andrews, C.; Cole, S.; de Ville, M.; Kelly, A.; Meis, S.; et al. A meta-analysis of water quality and aquatic macrophyte responses in 18 lakes treated with lanthanum modified bentonite (Phoslock®). *Water Res.* **2016**, *97*, 111–121. [CrossRef] [PubMed]
40. Lorenzen, M.W. *A Guide to Aeration/Circulation Techniques for Lake Management*; Environmental Protection Agency: Portland, OR, USA, 1977. Available online: http://hdl.handle.net/2027/mdp.39015041330062 (accessed on 26 May 2022).
41. Barr Engineering. *Stubbs Bay Feasibility Study*; Barr Engineering: Minneapolis, MN, USA, 2004. Available online: https://minnehahacreek.org/sites/minnehahacreek.org/files/Stubbs%20Bay%20Feasibility%20Study--Final.pdf (accessed on 23 April 2022).
42. Ruzzo, W.P. *Cherry Creek Reservoir Destratification System*; PE, LLC: Denver, CO, USA, 2008. Available online: https://www.cherrycreekbasin.org/wp-content/uploads/2014/04/Cherry-Creek-Reservoir-Destratification-System.pdf (accessed on 26 May 2022).
43. Leander, B. Iskmosundens Syresättningsprojekt. Iskmo, Finland. 2015. Available online: https://iskmosunden.fi/iskmosundens-syresattningsprojekt/ (accessed on 23 April 2022).
44. Town of Cary—Cary Proposes Jordan Lake Aeration System for Improved Water Quality. Available online: https://www.townofcary.org/Home/Components/News/News/10836/715?arch=1&npage=95 (accessed on 16 May 2022).
45. US Army Corps of Engineers. *Finding of No Significant Impact Jordan Lake Aeration System*; US Army Corps of Engineers: Wimlington, NC, USA, 2013. Available online: https://www.saw.usace.army.mil/portals/59/docs/recreation/jordanlake/signed_jordan_demonstraton_project_fonsi_july2014.pdf (accessed on 23 April 2022).
46. Restaurering av Finjasjön. Available online: https://docplayer.se/19147081-Hassleholms-kommun-gatukontoret-restaureringen-av-finjasjon.html (accessed on 19 May 2022).
47. Lillesjön Åtgärdsförslag Och Fortsatta Arbeten. Available online: https://docplayer.se/7346234-Informationsmote-2014-01-28-grimstorp-lillesjon-atgardsforslag-och-fortsatta-arbeten.html (accessed on 19 May 2022).
48. Jing, L.; Bai, S.; Li, Y.; Peng, Y.; Wu, C.; Liu, J.; Liu, G.; Xie, Z.; Yu, G. Dredging project caused short-term positive effects on lake ecosystem health: A five-year follow-up study at the integrated lake ecosystem level. *Sci. Total Environ.* **2019**, *686*, 753–763. [CrossRef] [PubMed]
49. Liu, C.; Zhong, J.; Wang, J.; Zhang, L.; Fan, C. Fifteen-year study of environmental dredging effect on variation of nitrogen and phosphorus exchange across the sediment-water interface of an urban lake. *Environ. Pollut.* **2016**, *219*, 639–648. [CrossRef] [PubMed]
50. Sandström, O. Reduktionsfiske Som Metod För Att Minska Övergödningen I Östhammarsfjärdarna. The Uppland Foundation Report 2011/2, Sweden. 2011. Available online: https://www.upplandsstiftelsen.se/UserFiles/Archive/5417/Rapporter/2011_2_Reduktionsfiske.pdf (accessed on 25 April 2022).
51. Davidsson, T. Börringesjön, Yddingesjön Och Havgårdssjön—Vattenkvalitet Och Åtgärdsförslag. The Ecology Group on Behalf of the Segeån Watercourse Association, Landskrona, Sweden. 2003. Available online: https://segea.se/wp-content/uploads/2020/12/E1_Yddinge_Borringe_Havgard_2003.pdf (accessed on 25 April 2022).
52. Nilsson, R. Projekt Ringsjön—Årsrapport 2006. Höörs Kommun, Sweden. 2006. Available online: https://docplayer.se/23733176-Projekt-ringsjon-arsrapport-2006.html (accessed on 25 April 2022).
53. Förstudie Avseende Förutsättningar Att Med Reduktionsfiske Biomanipulera Sjösystemet Trummen, Växjösjön, S. & N. Bergundasjöarna I Växjö Kommun. Available online: https://docplayer.se/12847907-Forstudie-avseende-forutsattningar-att-med-reduktionsfiske-biomanipulera-sjosystemet-trummen-vaxjosjon-s-n-bergundasjoarna-i-vaxjo-kommun.html (accessed on 17 May 2022).

water

MDPI

Article

A Detailed Analysis on Hydrodynamic Response of a Highly Stratified Lake to Spatio-Temporally Varying Wind Field

Hieu Ngoc Le [1], Tetsuya Shintani [2,*] and Keisuke Nakayama [2]

1 Department of Civil and Environmental Engineering, Tokyo Metropolitan University, Tokyo 192-0397, Japan
2 Department of Civil Engineering, Kobe University, Kobe 657-8501, Japan
* Correspondence: shintani@tmu.ac.jp

Abstract: Wind is generally considered an important factor driving the transport and mixing processes in stratified enclosed systems such as lakes and reservoirs. Lake Abashiri is one of the instances of such a system. For these systems, typically, the temporally unsteady but spatially uniform nature of wind has been assumed for simplicity. However, the spatial non-uniformity of wind could significantly alter compound hydrodynamic responses. In this study, such responses were investigated under the continuous imposition of different inhomogeneous wind conditions by applying numerical models and integrated analysis. The resultant tracer transport in both uniform and non-uniform wind cases was insignificant for the total study period of 9 days. However, under the short interval of T_i, where T_i is the internal fundamental period, different behaviors of both surface particle transport and the internal wave field were identified. Particularly, surface mass transport responses to higher spatial wind variance were obviously different from those in the uniform case. In addition, internal wave spectra under strong wind magnitude, which has low spatial variances, became identical to that of uniform wind; however, there were some discrepancies in the non-uniform case in the wave spectra under the influence of weak-to-moderate wind of high spatial variances. The results could provide an in-depth understanding of the lake's hydrodynamic response to inhomogeneous wind which could improve water management in lakes and reservoirs.

Keywords: stratified lake hydrodynamics; inhomogeneous wind quantification; numerical simulation

Citation: Le, H.N.; Shintani, T.; Nakayama, K. A Detailed Analysis on Hydrodynamic Response of a Highly Stratified Lake to Spatio-Temporally Varying Wind Field. *Water* **2023**, *15*, 565. https://doi.org/10.3390/w15030565

Academic Editors: Luca Martinelli and Giuseppe Pezzinga

Received: 28 November 2022
Revised: 23 January 2023
Accepted: 25 January 2023
Published: 1 February 2023

1. Introduction

Water quality in lakes and reservoirs has implications for its beneficial applications, such as fisheries' management, water supply quality, phytoplankton population, and control of the manganese level [1–3]. Such conditions are controlled by both nutrient loadings and physical factors such as mixing, and transport processes driven by winds [4]. As a major source of energy in lakes, coastal waters, and some estuaries, wind forcing plays an important role in the formulation of lake mixing, circulation, and internal seiche excitation [5–7]. Wind blowing over a lake rapidly distributes its momentum to water and consequently induces surface current and transport of floating objects such as phytoplankton along the wind direction. In addition, turbulent mixing induced by wind can affect the structure of the stratification, leading to changes in hydrodynamics and, in turn, species composition, proliferation of toxic algae, and dwindling of drinking water supplies [8,9].

Information on wind-induced transport, mixing, and internal waves is typically explained by numerical approaches, whereby the source of wind input is frequently retrieved from a point-based stationary wind field to spatially represent the phenomenon [5,7,10,11]. However, such an assumption is often oversimplified, although its dependency on the surrounding orography is crucial, resulting in a lack of spatial representation of the inhomogeneity of actual meteorological conditions [12]. Moreover, spatial wind fields often vary across large lakes due to pressure gradients over the water surface [13], and across small lakes due to sheltering effects [14,15]. Consequently, wind-induced lake hydrodynamics are insufficiently explained with the inaccurate surface boundary condition. Therefore, the coupling effects of the horizontally non-uniform wind and the real bathymetry of the concerned study sites are needed to realistically simulate the phenomena [16,17]. This could be achieved through a stand-alone uncoupled hydrodynamic model interacting with spatially interpolated wind [18,19], climate products [20], and numerical downscaling methods [12,21].

Generating the spatial variability of wind either by observation or by numerical simulation illustrated improvements in hydrodynamic simulations that were underpinned in a number of studies. For example, the simulation in Lake Kinneret in Israel improved the wave amplitudes and reduced the overall phase error after the spatially inhomogeneous wind stress was appropriately estimated [22]. The flow structure in Lake Belau changed drastically under the induction of spatial variation in the wind [23]. In Lake Ledro, TKE and its dissipation rate were reduced because of the representation of spatially inhomogeneous wind, as they are sensitive to the wind intensity [14]. The assumption of wind uniformity may often fail to reproduce the natural circulation condition and its characteristics in Spermberg Reservoir [24]. The spatial variations in the wind field separated the natural internal seiche excitation into different modes, as well as acting as a controller for higher horizontal modes in Lake Iseo [15]. Furthermore, the spatial variations in wind-generated wind curl, formulating seasonal regimes of Ekman layers resulting from the accumulation of particulate organic matter (POM) at the Tokyo Bay head, are driven by the pumping velocity [25]. Normally, comparisons of the response of lake hydrodynamics to spatial uniform and non-uniform wind fields are done with visualization, without being quantified. Quantification of these phenomena based on wind inhomogeneity levels is therefore necessary to elucidate their in-depth features.

To understand the complicated response of hydrodynamics to spatial and temporal variations in wind, meteorological wind-induced lake hydrodynamics were numerically simulated in Lake Abashiri from 21 November to 29 November 2011. To the best of our knowledge, this study is the first to quantify the interactions between spatial variations in wind forcing and lake hydrodynamics in Lake Abashiri. Therefore, under different spatial and temporal wind conditions, there are three main objectives in this study: (1) to quantify and understand the effects of wind forcing on surface mass transport and water current; (2) to quantify the effects of wind forcing on tracer movement; and (3) to estimate the potential mixing and to quantify the effects of wind forcing on internal seiches. The overall aim is to elucidate and quantify the differences in lake hydrodynamics between spatially uniform and spatially non-uniform wind.

2. Materials and Methods

2.1. Site Background

Lake Abashiri in Hokkaido (Figure 1a) is a small lake with a surface area of 32.87 km^2, a shoreline of 42 km, a maximum depth of 16.1 m, a mean depth of 7.2 m, and an altitude of 0.4 m. Lake Abashiri connects Abashiri River at the southernmost part of the catchment area and estuary at the northeast end (Figure 1b). Seawater frequently flows backward from the sea to the Abashiri River, so that the lake is stratified mainly with salinity. Lake Abashiri is an example of a highly stratified system linked by upstream freshwater intrusion and downstream seawater intrusion. This complex system forms a strong halocline, distinguishing the lake's layers. The lower layer of saline water is deoxidized up to <0.1%Sat due to a long residence time [26]. The lake provides important recreational activities, such as fishing and sports, throughout the season. However, eutrophication has recently become a concern due to the nutrient loading [27]. Lake Abashiri can emit 11% of the total methane in Lake Hokkaido and produce dissolved methane (DM) concentrations five times higher than the estimated one-dimensional methane model (DM concentrations were estimated by surface area) [27]. Such an issue could be problematic if the lower-layer gas was released from the lake. As an important factor of such a system, continuous wind exceeding 10 m/s could upwell the lake, releasing CH4 to the environment [7].

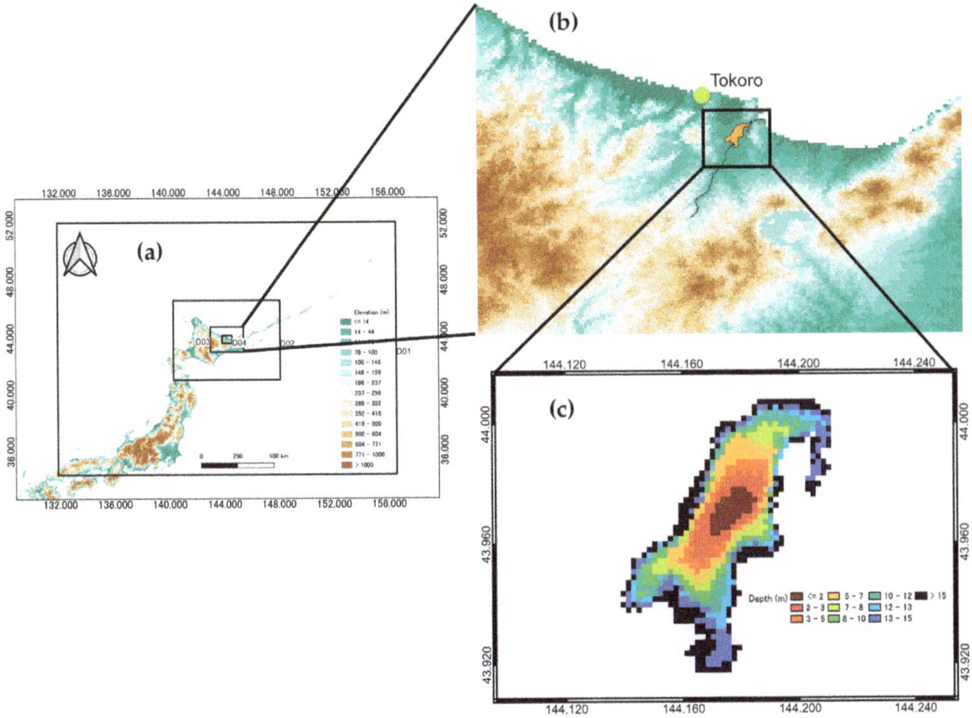

Figure 1. Maps of (**a**) Lake Abashiri with four nesting domains, (**b**) the station location, and (**c**) Lake Abashiri bathymetry.

2.2. Observation Data

Hourly values of wind speed and direction taken from Tokoro station (Figure 1b) from 21 November 2007, to 30 November 2007 were obtained from the Automated Meteorological Data Acquisition System (AMeDAS) which was then used to validate the WRF model. AMeDAS is a collection of Automatic Weather Stations (AWS) and is monitored by the Japan Meteorological Agency (JMA) [28].

2.3. Weather Research and Forecasting Model (WRF)

This study employed WRF for numerically downscaling and simulating wind inhomogeneities [29] with four nested domains (Figure 1a), with 200 × 200, 190 × 190, 178 × 178, and 166 × 166 cells and horizontal resolutions of 10 km, 3.3 km, 1.1 km, and 333 m, respectively. The temporal resolution was set for every 20 min. Table 1 illustrates WRF configurations and the domain setup for the simulation. In the resolution from the outermost to the innermost domain, the former covers almost the entire country of Japan to capture overall atmospheric phenomena, while the latter includes Lake Abashiri surrounded by a relatively flat area. In this context, hourly ERA5 [30] reanalysis data were dynamically downscaled from 0.25° spatial resolution. ERA5 is the most recent global reanalysis product that has benefited from a decade of advancing model forecasting and assimilation methods [31]. In addition, studies have reported that ERA5 is good at improving simulation in complex terrains and outperforms several other global reanalysis products in terms of surface wind [32–34]. We used the rapid radiative transfer model (RRTM) scheme for longwave radiation and the Dudhia scheme for shortwave radiation. Furthermore, we applied the Monin–Obukhov scheme [35], the Noah-MP land-surface model, and the Mellor–Yamada–Janjic TKE scheme for the surface layer, land surface, and planetary boundary layer, respectively. In addition, substitution of higher land-use resolution taken from Global Map Japan [36] and observation-based dynamical nudging to the initial conditions were implemented to further improve model performance. Each run began at 00:00 UTC and generated outputs every 20 min.

Table 1. Weather research and forecasting model configurations.

Domains	Domain 1	Domain 2	Domain 3	Domain 4
Resolution	10 km × 10 km	3.3 km × 3.3 km	1.1 km × 1.1 km	333 m × 333 m
No. of grid points	200 × 200	190 × 190	178 × 178	166 × 166
Microphysics	WSM3	WSM3	WSM3	WSM3
Longwave radiation	RRTM	RRTM	RRTM	RRTM
Shortwave radiation	Dudhia	Dudhia	Dudhia	Dudhia
Cumulus parameterization	Kain–Fritsch	Kain–Fritsch	Kain–Fritsch	None
Land-surface option	Noah-MP	Noah-MP	Noah-MP	Noah-MP
Surface layer option	Monin–Obukhov	Monin–Obukhov	Monin–Obukhov	Monin–Obukhov
Planetary boundary layer	Mellor–Yamada–Janjic	Mellor–Yamada–Janjic	Mellor–Yamada–Janjic	Mellor–Yamada–Janjic

2.4. Fantom-Refined Hydrodynamic Model

The Fantom-Refined [37,38] model was developed following equations of continuity and 3-D Navier–Stokes with incompressible and Boussinesq approximation together with thermal flux (details of the model can be found in Appendix A). The model is further improved with a local mesh refinement (LMR) technique based on a structured grid which uses rectangular grids similar to a tree-based grid. This method uses two different perspectives of a horizontal grid, macro, and micro views. The macro view determines the refining unit (hereafter the "container"), and each container is discretized in a structured manner with arbitrary horizontal resolution (hereafter the "micro view"). This method introduced the concept of a container that defines local groups with constant grid resolution in each direction. With this container, we may specify any computational cell in the domain with a simple, modifiable LMR and straightforward indexing. The study, however, applied only the macro view for simplification and for computational efficiency.

The horizontal resolution for the hydrodynamic model was set at 200 m in each direction (Figure 2a). In terms of vertical resolution, finer grids (0.1 m) were set at the middle of the lake's depth to represent the halocline fluctuations visibly and accurately, while coarser grids were used at the remaining depths (Figure 2b). Vertical temperatures were distributed evenly along the depth, and salinity was initialized with measured data, in which the halocline existed at a depth of approximately 5 m below the water surface.

Complexity exists when a clear understanding of lake hydrodynamics is challenged by both spatially varying drivers (wind and bathymetry), whereby lake hydrodynamics potentially react differently under the spatial distribution of the former, which is controlled by the surrounding orography. Meanwhile, variations in depth and geometry of the latter are of great concern in further alternating the mechanism. A deep comprehension of how a specific main driver affects hydrodynamics is the key to gaining a step-by-step understanding of the complex physical behavior of lake hydrodynamics. Therefore, flat bathymetry of 16 m was assumed to separate the complex effects of the real bathymetry.

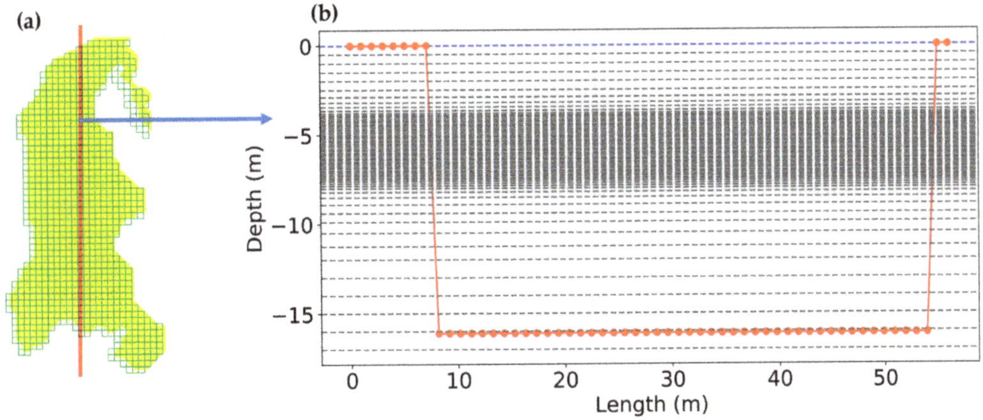

Figure 2. Fantom-Refined 3D configurations; (**a**) (square boxes) horizontal LMR mesh resolution, and (**b**) (black dashed line) vertical resolution along the fetch length. Blue dashed line is the water surface and the red connected dotted line is the side view bathymetry of the red line drawn in (**a**).

To investigate the effect of spatial and temporal variations in wind on hydrodynamics in the lake, a comparative study was conducted between unsteady but spatially uniform wind, and unsteady and spatially non-uniform wind (Table 2). By a simple interpolation technique, the horizontal and temporal resolutions of the wind field extracted from the fourth domain of the WRF model were refined to 200 m and 10 s, respectively, to match those of the hydrodynamic model. The uniform wind field was created by taking wind data at the center grid of the lake. Wind shear stress applied over the water surface was computed explicitly with the function of (U10) [38].

$$\tau_s = \rho_a C_D |U_w| U_w \quad (1)$$

where ρ_a is the air density, C_D is the drag coefficient, and U_w is the wind magnitude at 10 m above the water surface, i.e., U10. The model applied a bulk formula [39] to estimate sensible and latent heat fluxes, and short and long wave radiations from meteorological weather conditions.

Table 2. Fantom simulation durations setup. For every hour of time increment, the simulations of surface particle transports and internal seiches were reset every 8 h (T_i) for the entire 9 days of the simulation. Tracer transport and another case of internal seiche were simulated with no reset for the entire 9 days of the simulation.

	Duration of Simulation		**Time Steps**
	Spatial Uniformity	**Spatial Variations**	
Surface particle transports	T_i (8 h)	T_i (8 h)	10 s
Tracer transport	9 days (tracer was placed at the 21st)	9 days (tracer was placed at the 27th)	10 s
Internal seiche	9 days and T_i (8 h)	9 days and T_i (8 h)	10 s

2.5. Indicators for Spatial Variations in Wind Heterogeneity

Quantification of wind inhomogeneities is challenging due to the complex distinction between wind direction and wind speed. Recently, several studies have attempted to quantify the spatial variations in wind. For example, smaller root mean squared errors of vectors u and v indicated uniformly spatial variations and vice versa [14]. This technique, however, using the vectors u and v, cannot quantify the invariance of wind speed or wind direction, which could be directly addressed using 2D mean filter analysis [40].

The spatial consistency of the wind was determined based on the integrated coefficient of variance and a 2D mean filter [40].

$$f(x,y)_n = \frac{1}{(2n+1)^2} \sum_{(s,t)\in S_{xy}} g(s,z)_{n-1} \tag{2}$$

where $f(x,y)_n$ is the filtered image, n is an integer and can be selective with incremental size (e.g., 1, 2, 3, ... , m), m is arbitrarily chosen for grid size, $g(s,z)_{n-1}$ is the value in each pixel of the grid of the previous filtered image; and s and z are the corresponding coordinates of the grid itself. As noise reduction comes at the expense of blurring the image (a 2D visual representation of any pixel), if the resulting image ($f(x,y)_n$) has features in common with all the previous filtered image ($f(x,y)_0, \ldots, f(x,y)_{n-1}$), the wind field will be considered to be spatially homogeneous. Details of the computation procedures are described in Appendix B.

2.6. Indicators for Temporal Variations in Wind Heterogeneity

The metric Pearson correlation was applied to measure the linear relationship between two sets of data. It is the ratio of two variables' covariance to the sum of their standard deviations.

$$r_{xy} = \frac{\sum_{i=1}^{o}(x_i - \overline{x})(y_i - \overline{y})}{\sqrt{\sum_{i=1}^{o}(x_i - \overline{x})^2}\sqrt{\sum_{i=1}^{o}(y_i - \overline{y})^2}} \tag{3}$$

where overbar x and y are the average value of the sample population, and o is the sample size. To analyze the homogeneity of temporal wind field, a correlation map constructed between the center grid (the 2D time series data at the center of the chosen dataset) and the surrounding ones was calculated during the simulation period.

2.7. Wedderburn Number

The Wedderburn number, a non-dimensional and systematic relationship between the wind-induced shear stress and the baroclinic pressure gradient, was computed to identify the mixing condition in the reservoirs and lakes [41].

$$W = \frac{g' h_e^2}{u_*^2 L} \tag{4}$$

in which h_e is the epilimnion depth, L is the wind fetch length, u_* is the friction of wind velocity, and g' is the reduced gravity.

2.8. Surface Particle Transport, Tracer Simulation, and Internal Seiche

In principle, wind triggers two main important physical phenomena, lake circulation and surface and internal waves, both of which control the mass transport in lakes. To investigate the movement of the former, virtual particles are often placed and numerically simulated together with the atmospheric forcing interaction. This approach can be visually quantified by the Lagrangian tracking method, whereby the simulated particles often representing phytoplankton, fish eggs or larvae, are released, and the dynamics of the particle trajectories are recorded in time and space over the lake [12,19,21]. In this study, surface particles were placed uniformly on the water body to elucidate the effect of spatial and temporal variations of the wind field. In addition to the Lagrangian tracking method, tracer tracking is frequently adopted for its representation of either two- or three-dimensional transport of fluid containing pollutant [10,20]. To this aim, a normalized tracer concentration of 1.0 was set at the northern part of the lake.

Investigation on the development of the two-layer model of the temperature profile [42] was the key to further intensive empirical studies in an effort to classify the mixed-layer dynamics in lakes [43] and to readily determine the first vertical internal wave mode V1H1 [44]. To get a better insight into the internal waves that are generated by both homogeneous and inhomogeneous wind, a temporal variation of density interface for each simulation period was extracted. After that, spectral analysis was applied to evaluate and highlight the periodic evolution of internal seiches in both uniform and non-uniform cases. This technique can easily reveal hidden oscillatory motions in a time series structure or can detect the dominant frequencies of a periodic time series in terms of both spatial and temporal resolution [45]. Welch's method was used to estimate power spectral density (PSD), where the Hanning window function was adopted to weight the time series dataset.

Standing internal waves in confined basins can be categorized by nodal points on the vertical (V) and horizontal (H) components, V_aH_b, where a and b are the fundamental vertical and horizontal mode components, respectively [46]. Finally, we computed analytical internal waves based on the calculation described in [6] to theoretically identify the period of each mode. We also compared the performance of the numerical model forced by different simulation periods with the analytical model in a variety of wind conditions. Based on the procedure described in [6], the periods of internal seiches were estimated by a 1D multi-layer hydrostatic linear model as

$$T_{a,b} = \frac{2L}{b\sqrt{g\beta_a}} \quad (5)$$

where a and b are the fundamental vertical and horizontal modes, respectively, g is the gravitational acceleration, and β_a is the eigenvalues assuming a two-layer model system (a = 1). The fundamental period can also be calculated following the Merian formula:

$$T_i = \frac{2L}{\sqrt{g'\frac{h_1 h_2}{h_1+h_2}}} \quad (6)$$

where $g' = g\frac{\rho_2-\rho_1}{\rho_2}$ is the reduced gravity of the two-layer system, ρ_1 and ρ_2 are the densities of the two layers, respectively, h_1 is the thickness of the upper and h_2 is the thickness of the lower layer. Additionally, the fundamental period T_i can also be estimated as $T_{a=1,b=1}$.

3. Results

3.1. WRF Model Validation

Wind data taken from AMeDAS (Figure 3a) illustrated a strong north-easterly wind with a clear diurnal-cyclical alteration during the simulation period. Strong winds evolved during the early morning and weakened from the beginning of the afternoon until nocturnality, with a high frequency of maximum and minimum intensities of approximately 10 m/s and around 2 m/s, respectively. These periods with strong magnitudes and clear periodicity could potentially influence the synoptic-scale occurrences related to the lake's hydrodynamics. From the 24th to the 26th, low to moderate magnitudes were noted, with a tendency toward mild fluctuations. Validation between the model and observed data showed the general overestimation of model data, with RMSE and MBE values of 6.46 m/s and 0.17 m/s, respectively. There was also an overestimation of wind magnitudes on the first half of the 21st and the night before the 22nd. WRF showed minor phase differences from the 23rd, from the early morning to the late afternoon. A relatively consistent trend between the two datasets was seen from the 24th onward. In addition to wind magnitude, wind direction showed an overall consistent distribution between model and observation data (Figure 3b,c). Fortunately, the disagreement between numerical and observed data did not impose significant effects on the study, since the aim of this study was to distinguish uniform and non-uniform wind-induced lake hydrodynamics instead of perfectly achieved modelled data.

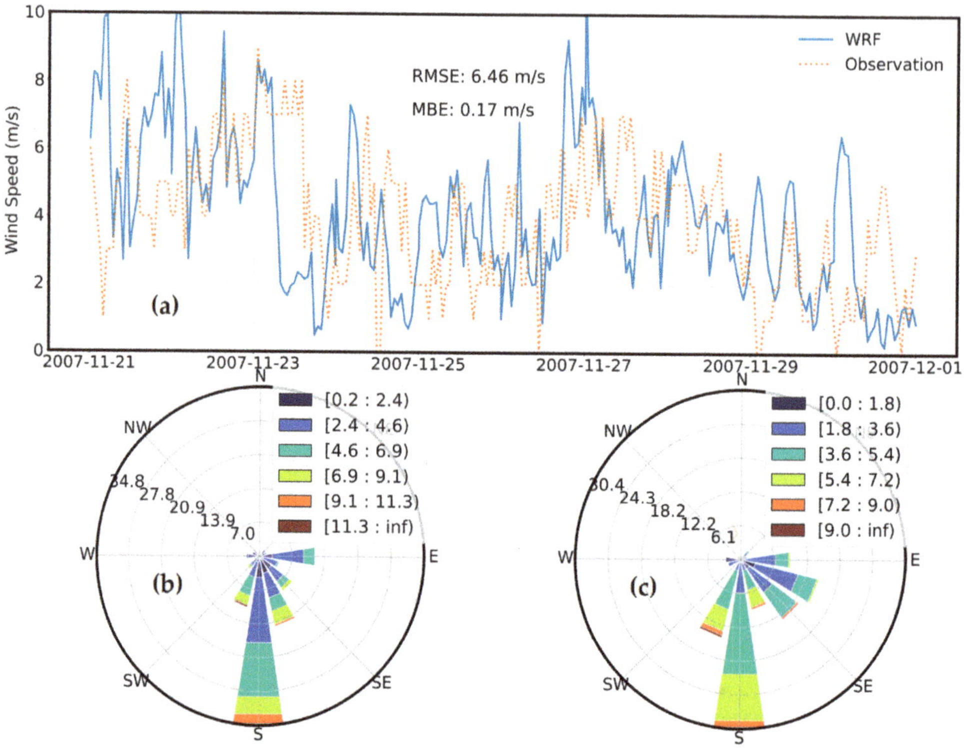

Figure 3. Validation in wind speed and wind direction between WRF and station data (Tokoro). (**a**) Validation in wind speed, (**b**) WRF model wind direction, and (**c**) Station wind direction.

3.2. Spatio-Temporal Wind Quantification

The spatial inhomogeneity levels of wind speed and direction were calculated with nine levels of the 2D mean filter (m = 9). The overall results are presented in (Figure 4a). The study period indicated general homogeneous wind direction in accordance with strong wind speed. Moreover, strong, and weak winds corresponded to low and high averaged coefficient of variance (Appendix B), $\overline{C}_v$ scores, revealing an inverse relationship between the wind magnitude and the quantification score. According to a rough wind inhomogeneity estimation [40], values of $\overline{C}_{v\_dir}$ less than 0.01 and $\overline{C}_{v\_spd}$ less than 0.66 are considered spatially homogeneous, while values of $\overline{C}_{v\_dir}$ higher than 0.01 and $\overline{C}_{v\_spd}$ higher than 0.76 demonstrate spatial inhomogeneity. In addition to the inverse relationship mentioned above, moderate wind (3–5 m/s) illustrated an ambiguous pattern in spatial distribution. As a result, the complexity of spatial distribution could potentially signify the difference in the lake's hydrodynamics. Based on the wind magnitude and the spatial inhomogeneity indicator, Figure 4b–d represent the spatial variations or spatial uniformity in wind. On the last day of the simulation (29 November 2007 23:50), very weak spatial wind (~1 m/s) produced insufficient momentum, disturbing the direction of the wind over the surface water body. Additionally, light air (~2 m/s) combined with moderate breeze (~4 m/s) (24 November 2007 15:00) formed a wind curl as a result of the shear stress differences in space. Figure 4c,d illustrated light (~4 m/s) to fresh breeze (~10 m/s), respectively. As we can see, near-spatial uniformity was noted in the wind over the lake's surface. Therefore, we can assume a spatially uniform wind field during the strong wind periods, while the moderate wind-induced periods can alter the lake's hydrodynamics due to the spatial variation of winds.

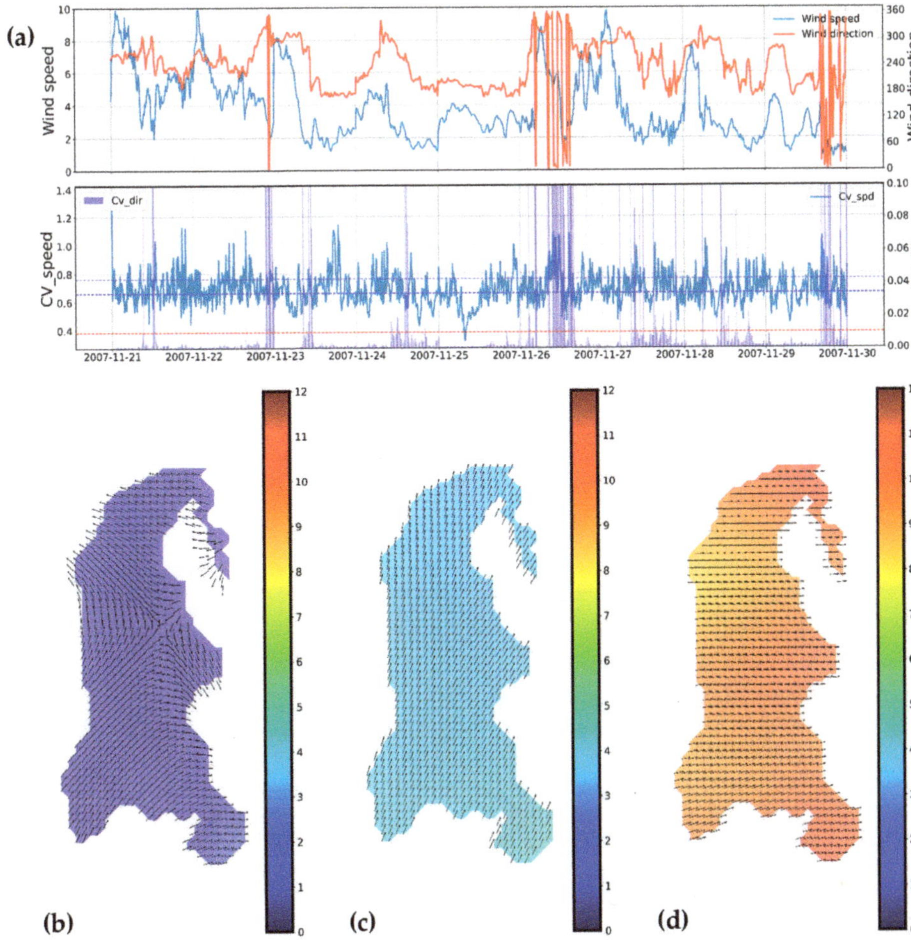

Figure 4. Maps of (**a**) Wind speed and direction inhomogeneity, (**b**) weak wind with high $\overline{C}_{v\_dir}$ = 0.1, $\overline{C}_{v\_spd}$ = 0.615 (29 November 2007 23:50), (**c**) moderate wind with high $\overline{C}_{v\_dir}$ = 0.0014, $\overline{C}_{v\_spd}$ = 0.863 (25 November 2007 04:00), (**d**) strong wind with low $\overline{C}_{v\_dir}$ = 0.001, $\overline{C}_{v\_spd}$ = 0.624 (21 November 2007 01:30).

Pearson correlation coefficients were estimated in the 2D time series wind data between the central grid and its surrounding grids over the simulation period (Figure 5). Correlation coefficients were calculated using Equation (3). For the wind direction, temporal inhomogeneity exhibited an overall positively moderate correlation from the middle point to its nearby grids, while weaker relationships were located farther from the center grid (Figure 5a). Therefore, insignificant spatial wind similarity covered the entirety of Lake Abashiri. This could be explained by the spatially disturbed wind direction during weak-to-moderate wind periods. On the other hand, surprisingly, wind speed illustrated relatively spatially homogeneous correlation coefficients all over the area, with moderate correlation coefficients. In conclusion, wind speed represented acceptable similarity (~0.5) but wind direction illustrated higher spatial disturbance over the simulation period.

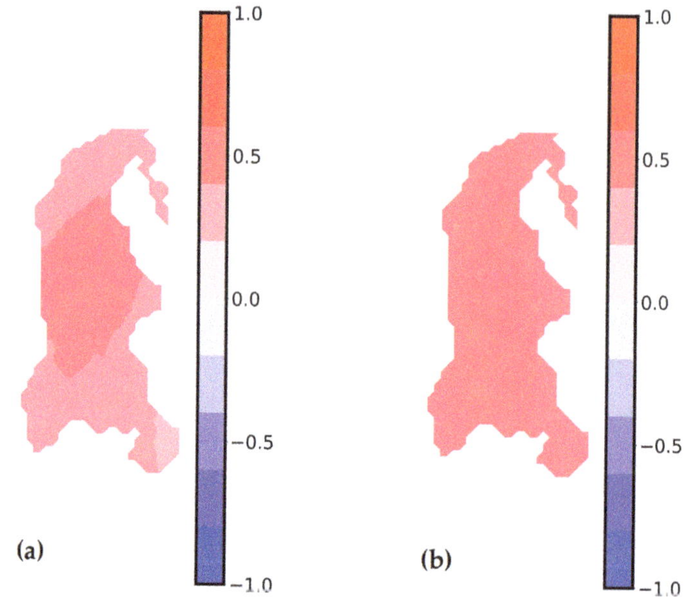

Figure 5. Maps of (**a**) correlation of wind direction between the center grid and the surrounding grids, (**b**) correlation of wind speed between the center grid and the surrounding grids.

3.3. Surface Particle Transport and Tracer Transport

Figure 6 shows the numerical simulations of wind-induced surface particle transport. Theoretically, particles traveled unidirectionally following the wind direction under spatially uniform wind. This is consistent under strong and moderate wind speeds, which inversely correlate to the low $\overline{C}_v$. Figure 6c,f display a constant movement of particles following the water currents driven by the strong and moderate winds, respectively. Therefore, low $\overline{C}_v$ generated translational motions in surface particles along the wind directions. In contrast to the strong wind, weak breezes generated the distinguishable motions of the water currents during a short time scale of the fundamental period, T_i (8 h), because of the large $\overline{C}_{v\_spd}$ and $\overline{C}_{v\_dir}$, showing the tendency of dispersing particles (Figure 6k). Compared to the spatial uniform wind, spatially distorted weak wind (~1 m/s) initiates a short distance of movement after the T_i period. However, particles started to travel omnidirectionally after providing sufficient drag force, causing the particles to scatter in different directions (Figure 6k). This means that after a long period ($>T_i$) of spatial variations in winds (large $\overline{C}_{v\_spd}$ and $\overline{C}_{v\_dir}$), the movement of surface particles would be far more different from that of spatially uniform wind.

Tracer mass transport was initially placed at the small estuary where the substances from the seaside were first provided in this lake. Figure 7 shows the concentration and distribution of the tracer under different wind conditions. Although the effects were relatively similar in both cases due to the prevailing wind direction, there were some discrepancies noted in the first few days of the simulation. For the first two days of strong prevailing NE wind, the tracer dispersion was mainly transferred to the estuary mount. As a result, the mass transport was blocked by the right-hand side boundary. The tracer was stirred up by the vertical circulation, which resulted in the tracer spreading over half of the water surface. Under uniform wind, the stable and homogeneous wind increased the concentration of tracer further south, while a lower concentration level was observed in the case of non-uniform wind (Figure 7b). The results of the inhomogeneous wind, which generated spatially disturbing water currents compared to that of spatially uniform wind, illustrated changes with less concentration in the southernmost part of the lake. By the end of the simulation, tracer mass transport was seen to be completely mixed up in the two wind cases, leaving no traits of a dominantly concentrated area.

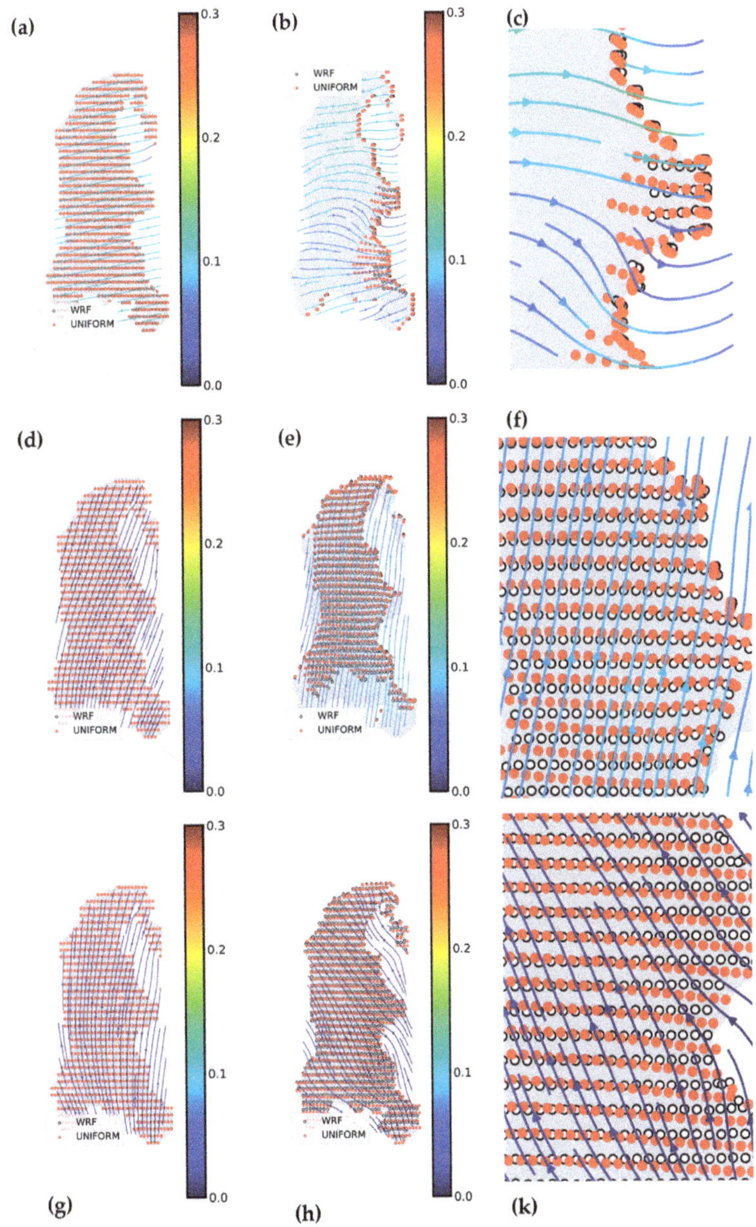

Figure 6. Every twenty minutes of surface particle movement and current vectors of eight hours' simulation, starting from 21 November 2007 00:00 (**a**); 25 November 2007 00:00 (**d**); and 29 November 2007 23:00 (**g**) to 21 November 2007 01:30 (**b**); 25 November 2007 00:00 (**e**); and 29 November 2007 23:50 (**h**), respectively. (**c**,**f**,**k**) are the zoom in of (**b**,**e**,**h**).

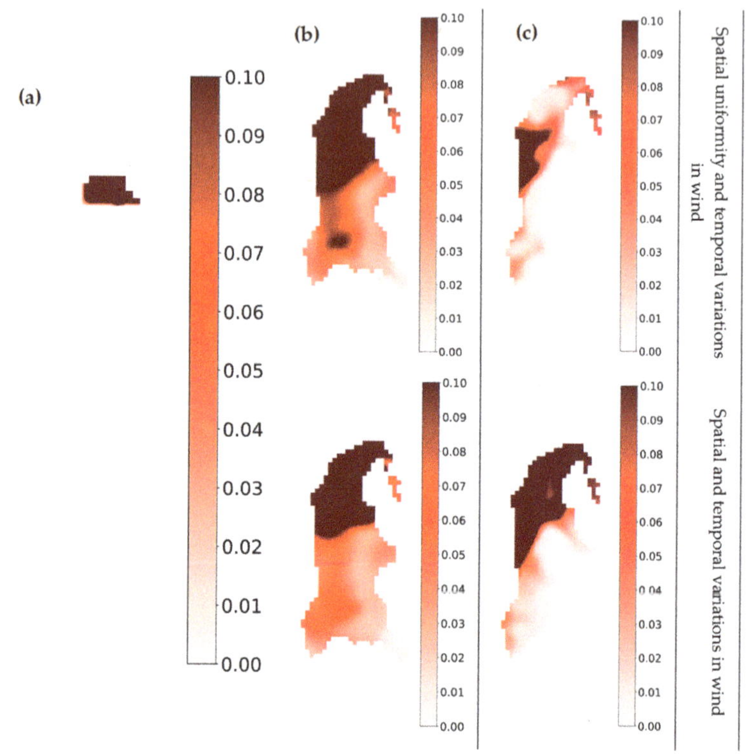

Figure 7. Simulations of (**a**) (21 November 2007 00:00) and (**b**) (23 November 2007 23:00) in which tracer mass transport was placed at the beginning of the simulation, (**c**) (30 November 2007 00:00) where tracer was placed after the 27th onward.

3.4. *Internal Seiche*

In general, the uniform and non-uniform wind shared similarities in inducing internal waves under the strongly stratified lake system (Figure 8). The depth of halocline computed under both wind cases did not present any complete upwelling under strong wind imposition. This was further confirmed by accessing the relationship between wind and the response of stratification (Figure 8c). There were roughly 96% of the total periods that are classified as "regime three" [43]. This indicated that the most prominent feature is internal seiching. Under this regime, strong wind durations are less than the entrainment time, illustrating the rare event of complete vertical mixing [43]. Regime four accounted for the remaining periods, showing the overall domination of buoyancy force. However, the complete dominance of regime three during the simulation period emphasized the importance of internal waves in this lake system.

A Wedderburn (W) (Figure 8a) number of 4 to 10 was often observed during the first period of strong prevailing NE wind force which gradually transferred energy and induced internal waves for approximately 5 days (Figure 8b). Strong oscillation was shown after the 26th and continued to highly fluctuate for roughly two days. This pattern was similar from the 28th onward. During this period, there was some time when $W < 1$ indicating potential mixing. This could explain why the resuspension events eventually happened under high wind speed [47]. However, the short periods of strong wind induced could not entirely mix up the strongly stratified lake system. After that, the imposition of weak-to-moderate wind did not generate enough energy, indicating the domination of buoyancy and damping effects.

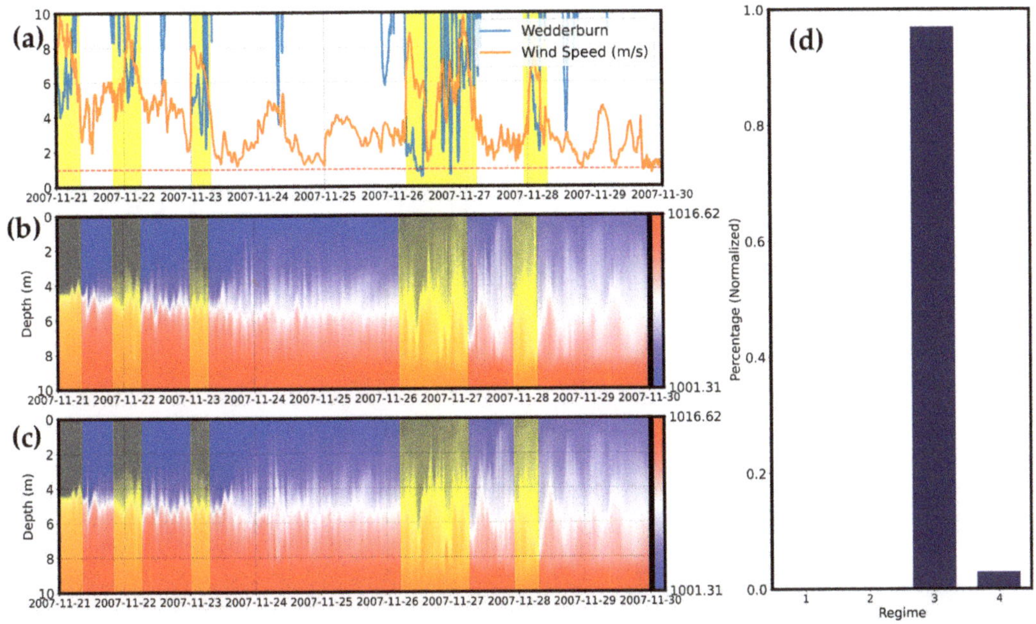

Figure 8. Mixing condition (**a**) wind speed and Wedderburn number, (**b**) depth of density difference excited by uniform wind, (**c**) depth of density difference excited by non-uniform wind, and (**d**) normalized percentage of the four regimes.

The depth of the halocline in both wind cases under strong wind illustrated noteworthy features. On the first few days of the simulation period, both spatially uniform and non-uniform wind exhibited their similar features because of the strong wind speed and low $\overline{C}_{v\_spd}$ and $\overline{C}_{v\_dir}$ (Figure 4a). The minor discrepancies started to exhibit on the next few days under the effects of weak-to-moderate wind speeds. Between the 26th and 28th, strong wind imposed on lake displayed similarities in the high fluctuations following the notable Wedderburn numbers. According to Figure 8b,c, after the 28th, the buoyancy and damping effects were much higher than those produced by low wind speeds. It is also noted that more mixings were seen under the imposition of spatially uniform wind (Figure 8b). Therefore, the effects of low-to-moderate wind speed accompanied by high fluctuation in wind direction must not be neglected.

4. Discussion

4.1. The Differences in Surface Mass Transport and Water Current between Spatially Uniform Wind and Spatially Non-Uniform Wind

The weak and spatially inhomogeneous wind forcing generated the uneven water current and could explain the significant differences in the particle distributions (Figure 6) among three cases during the short time scale of T_i, as shown in Figure 9.

Low $\overline{C}_{v\_spd}$ and $\overline{C}_{v\_dir}$ in the case of moderate to strong wind induced the water current to flow unidirectionally over the water surface. In addition to the high wind magnitudes, the wind direction did not change rapidly as it did in the weak wind condition (Figure 9a,b), which induced enough energy to push almost all the surface particles unidirectionally. This is consistent with the numerical simulation in [12]; the study illustrated that the horizontal transport of particles was stronger in winter Föhn due to the high and persistent magnitude of wind speed. As mentioned earlier, low values of $\overline{C}_{v\_spd}$ and $\overline{C}_{v\_dir}$ wind fields can indicate uniform wind fields. This can further reaffirm why wind speed is important and highly correlated with the wind direction as well as the water current.

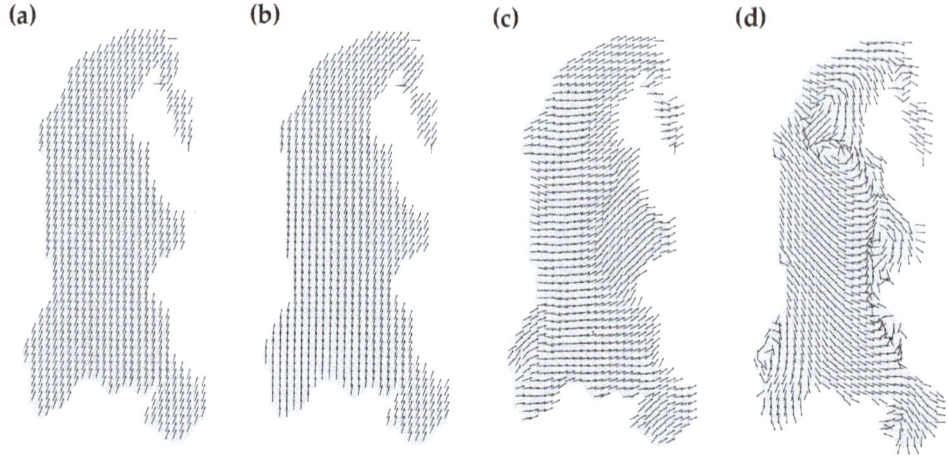

Figure 9. Vectors of water currents; (**a**) spatially uniform wind with strong wind magnitudes, (**b**) spatially non-uniform wind with strong wind magnitudes, (**c**) spatially uniform wind with low magnitudes, and (**d**) spatially non-uniform wind with low wind magnitudes. All pairs of simulation are at the same time.

On the other hand, in spatially non-uniform wind (Figure 9d), wind speeds were mostly below 2 m/s, and the low wind speed was often accompanied by omnidirectional wind direction [48], resulting in a disturbance of the movement of water current omnidirectionally across the lake. In contrast, spatially uniform wind (Figure 9c) produced a unidirectional water current in most parts of the lake, revealing the major differences in water current directions in both wind cases. Therefore, scattering effects would occur in the former wind case.

The assumptions of the flat bathymetry and the enclosed system, additionally, affirm that water-current curl (Figure 9d) can be formed under the imposition of spatial variations in wind due to the shear stress gradients where sheltering effects can be neglected in this lake [20]. A two-gyre circulation developed in Figure 9d, in which the clockwise gyre in the southern part coexisted with the counterclockwise gyre; meanwhile there was no such circulation under the strong wind condition. Depending on the lake size, this phenomenon is often observed when applying wind inhomogeneities as a governing factor [24]. Additionally, because Lake Abashiri is located in a flat and simple orography, the sheltering effects are minor, thus confirming the major effects of wind inhomogeneity. Based on the spatial distribution of water current (Figure 9), although the geometry of Lake Abashiri is simple, littoral zones in the south and at the middle generated clockwise flow, thus increasing the horizontal circulation (Figure 9d); meanwhile, this cannot be seen in spatially uniform wind conditions under the period of T_i (Figure 9c). As mentioned earlier, based on the calculated $\overline{C}_{v\_spd}$ and $\overline{C}_{v\_dir}$ scores, the high values when moderate winds were observed would create wind curls due to the gradient. The uneven distribution of wind shear stress can be sufficiently important for understanding the movements of water currents as well as particulate organic matter (POM).

4.2. *Correlation between Spatio-Temporal Indicators and Tracer Mass Transport*

A strong and prevailing north-easterly wind (NE) was the main factor controlling the similar movement of the tracer down to the south of the lake on the first few days; the tracer was finally mixed up entirely at the end of simulation period (Figure 7). As the tracer was completely mixed up after nine days of simulation, the complex alteration in hydrodynamics might be difficult to observe. To overcome that issue, a tracer placed on the 27th of November revealed the noteworthiness of spatially varied wind imposition (Figure 7c).

After one day, the tracer started reducing its concentration at the north boundary, while a higher concentration accumulated under non-uniform conditions and remained at the north of the lake. At the end of the simulation, concentrations remained higher in the case of spatial variations in wind, while the tracer concentration was fully mixed up with uniform wind.

Interestingly, the result obtained from the spatial variations in wind (Figure 7c) indicated that the movement of POM would transport it to the center of the lake, which is similar to the transportation of POM gathering at the center of the Tokyo Bay [25]. High moisture POM tends to accumulate at the center of the bay head [49,50]. Although the effects of wind curls and the assumptions of flat bathymetry in this study can be the factor producing the high concentrations at the lake center, real bathymetry can also increase the effect because it increases the water current velocity at the littoral zones. Further study is needed to understand the formation of positive and negative wind curl, and its effects on upwelling and downwelling under spatial variations in wind in combination with the real bathymetry.

4.3. *Effects of Temporal and Spatial Frequencies of Winds to Internal Seiche*

Overall, Figure 8 illustrates that the differences in internal waves of both wind cases were minor after nine days of the simulation. Noticeably, under strong wind during the first days of simulation (Figure 8), the height of the halocline responded comparably under both wind scenarios, wherein the differences as well as the major component causing the difference cannot be distinguished. To differentiate, the weak to strong wind-induced internal wave should be tested over a brief period. Simulations of internal waves within T_i were conducted to understand the differences under both wind conditions.

Figure 10 shows that the uneven distribution of wind direction was highlighted, with inverse correlation existing at southern part of the lake and a gradient from the center to the lake boundary in magnitude. Consequently, upwelling points were scattered, reducing the similar features of wind-induced internal wave. Considering the differences in spatial and temporal wind features, the results of spectral analysis generated a huge difference in PSD as well as frequency. While uniform wind coincides with V1H3 mode, non-uniform wind appeared to match to V1H2 mode. These alterations were obviously the main factors changing the behaviors of internal seiche in this system.

An eight-hour simulation of moderate wind (25 November 2007 00:00) indicated the overall stability of the lake's system. The temporal correlation of wind speed and direction showed a reasonable similarity of both spatial uniformity and non-uniformity (Figure 10). The moderate wind featured North-East (NE) as the general prevailing wind, resulting in most of the upwelling locations being located at the south end of the lake. This feature indicated a similar internal seiche in V1H2 mode, whereas both wind-induced internal wave cases disagreed in V1H3 mode, with a lower PSD in non-uniform wind compared to that of uniform wind. In addition, lower wind speed correlated to the PSD generated by the internal waves.

Finally, for the strong wind (~10 m/s), winds were predominantly easterly (E) during the period of (T_i). This could be seen from the upwelling locations accumulated at the western end of the lake boundary. Spectral analysis revealed that the prevailing wind (E) of spatially uniform and non-uniform wind excited internal waves with relatively similar power spectral densities at modes V1H1, V1H2, and V1H3. However, the numerical model did not match the mode V1H4. Moreover, both wind cases excited a similar PSD of internal waves, which reaffirms the similarity in both spatial uniformity and nonuniformity under strong wind fields.

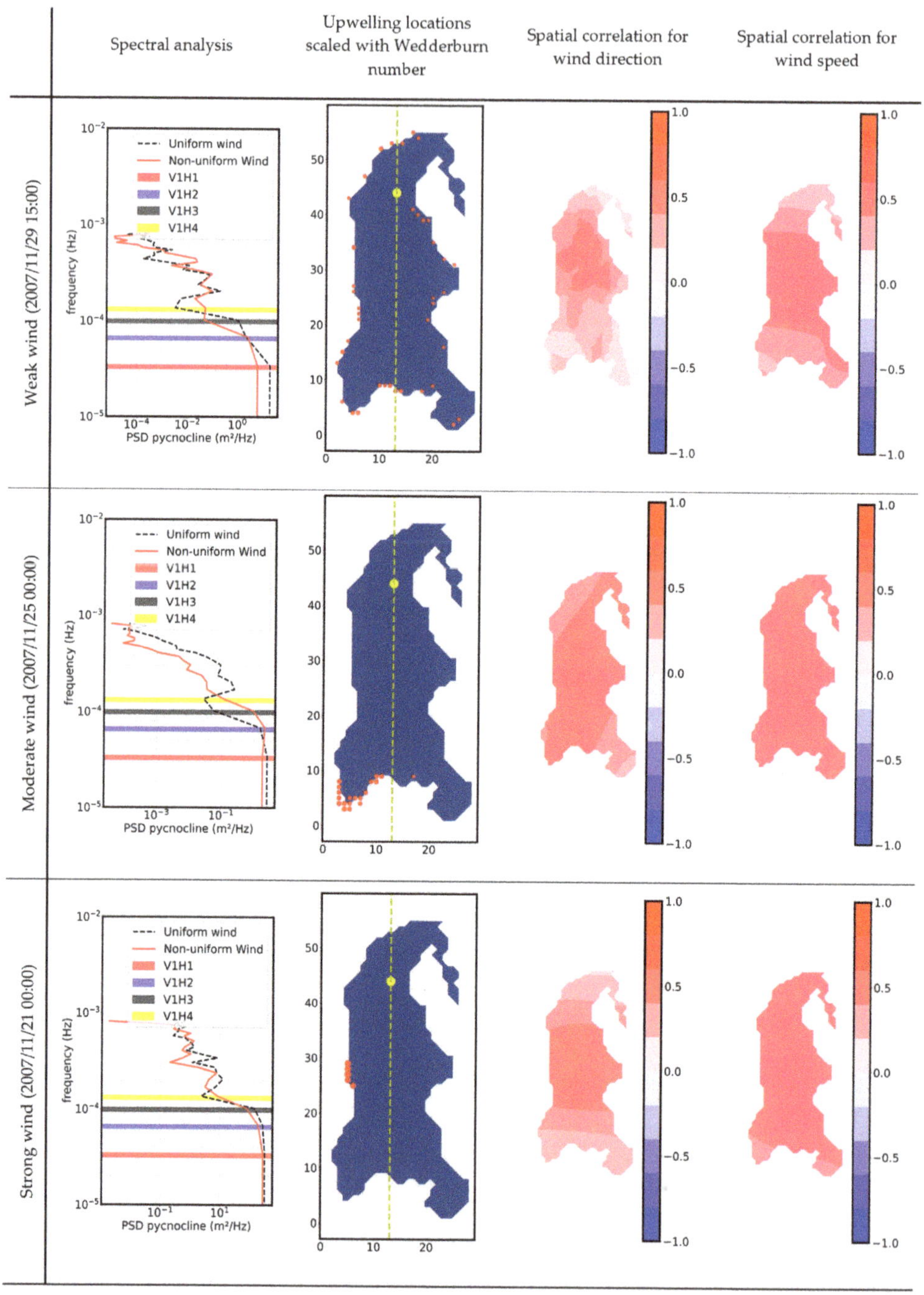

Figure 10. Statistical analysis of wind-induced internal seiche at a certain time of interest; yellow dash line is the fetch length, yellow dots on upwelling locations scaled with Wedderburn numbers are the point taken for calculating internal waves.

5. Conclusions

The study aimed to investigate the responses of the hydrodynamics of a lake under both spatially uniform and spatially non-uniform wind fields. Lake Abashiri was selected as the field, and the flat bottom of the lake was assumed for the analysis in order to focus on the effect of wind nonuniformity on the lake response. The realistic nonuniform weather was generated by the weather research and forecasting model with four level nesting grids over the area. To quantify the level of inhomogeneity of wind speed and wind direction, the wind products from WRF were then analyzed based on a 2D mean filter. Three of the lake's physical responses, including surface particle transport, tracer tracking, and internal waves, were examined under the different wind conditions. The analysis highlighted the key factors that effect changes when applying different wind scenarios.

Firstly, strong wind corresponding to a low level of variances both for wind speed and direction, $\overline{C}_{v\_spd}$ and $\overline{C}_{v\_dir}$, provided sufficient momentum to unidirectionally push the surface particles horizontally to the shoreline within the fundamental period of this stratified lake, T_i. On the other hand, wind curl was formed under the imposition of a weak-to-moderate wind field of high values of variance both in speed and direction. The surface particles moved only short distances but not unidirectionally, so they would be transported differently for the succeeding phase in the case of the uniform wind.

Secondly, when appropriately placing the tracer mass transport under wind of high variances, the possible effects of wind non-uniformity could be observed in the transport process in the POM concentration at the bay head of Lake Abashiri. Particularly, in the present study, a much higher concentration was seen in the more realistic wind condition, while a smaller concentration was seen in the case of spatially uniform wind.

Finally, the internal wave spectral analysis in both wind cases showed comparably similar patterns when considering the whole simulated period. Furthermore, the potential mixing, as defined by the Wedderburn number, showed that strong wind provided enough energy to fluctuate and potentially upwell the downwind along the fetch length. However, the system could not be mixed; this was probably due to the short duration of the strong wind and the assumption of uniform bathymetry. The analysis under the shorter time scale (T_i), revealed that the weaker the wind, the larger difference in power spectral density. This further elucidated the importance of weak-to-moderate wind effects on lake hydrodynamics.

This study emphasized the significant differences between spatially uniform and spatially non-uniform winds when comparing their interactions with surface mass transport, tracer tracking, and internal waves, excluding the effects of the bathymetry. Although it is important to consider the effects of bathymetry, they can, together with spatially non-uniform wind, alter and contribute to the complexity of lake hydrodynamics. As a result, future studies will include bathymetry effects to further analyze the sensitivity of lake hydrodynamics, coupled with spatial and temporal variations in wind.

Author Contributions: Conceptualization, H.N.L. and T.S.; Data curation, H.N.L.; Formal analysis, H.N.L.; Methodology, H.N.L.; Resources, T.S. and K.N.; Supervision, T.S.; Validation, T.S.; Visualization, H.N.L.; Writing—original draft, H.N.L.; Writing—review and editing, T.S. and K.N. All authors have read and agreed to the published version of the manuscript.

Funding: This research received no external funding.

Institutional Review Board Statement: Not applicable.

Informed Consent Statement: Not applicable.

Data Availability Statement: Not applicable.

Acknowledgments: The first author would like to give his sincere gratitude to the Japanese Government (Monbukagakusho) Scholarship for the grant for his Ph.D studies at the Tokyo Metropolitan University.

Conflicts of Interest: The authors declare no conflict of interest.

Appendix A. Fantom-Refined Model Description

The Fantom-Refined model applies 3-D Navier–Stokes equations with incompressible and Boussinesq approximation

$$\frac{\partial u}{\partial t}+u\frac{\partial u}{\partial x}+v\frac{\partial u}{\partial y}+w\frac{\partial u}{\partial z}-fu=-\frac{1}{\rho_0}\frac{\partial p}{\partial x}+\frac{\partial}{\partial x}\left(\nu_H\frac{\partial u}{\partial x}\right)+\frac{\partial}{\partial y}\left(\nu_H\frac{\partial u}{\partial y}\right)+\frac{\partial}{\partial z}\left(\nu_V\frac{\partial u}{\partial z}\right) \tag{A1}$$

$$\frac{\partial v}{\partial t}+u\frac{\partial v}{\partial x}+v\frac{\partial v}{\partial y}+w\frac{\partial v}{\partial z}+fv=-\frac{1}{\rho_0}\frac{\partial p}{\partial y}+\frac{\partial}{\partial x}\left(\nu_H\frac{\partial v}{\partial x}\right)+\frac{\partial}{\partial y}\left(\nu_H\frac{\partial v}{\partial y}\right)+\frac{\partial}{\partial z}\left(\nu_V\frac{\partial v}{\partial z}\right) \tag{A2}$$

$$\frac{\partial w}{\partial t}+u\frac{\partial w}{\partial x}+v\frac{\partial w}{\partial y}+w\frac{\partial w}{\partial z}=-\frac{1}{\rho_0}\frac{\partial p}{\partial z}+\frac{\partial}{\partial x}\left(\nu_H\frac{\partial w}{\partial x}\right)+\frac{\partial}{\partial y}\left(\nu_H\frac{\partial w}{\partial y}\right)+\frac{\partial}{\partial z}\left(\nu_V\frac{\partial w}{\partial z}\right)-\frac{g}{\rho_0}(\rho_0+\rho) \tag{A3}$$

subject to incompressibility,

$$\frac{\partial u}{\partial x}+\frac{\partial v}{\partial y}+\frac{\partial w}{\partial z}=0 \tag{A4}$$

where u, v, and w are the velocities in the x, y, and z directions, respectively. p is the pressure, ρ_0 is the reference density, $(\rho_0+\rho)$ is the density, $f=2\Omega\sin\phi$ is the Coriolis coefficient, ϕ is the latitude, Ω is the angular velocity of the earth, ν_H and ν_V are the horizontal and vertical eddy viscosity coefficients. The depth-averaged continuity equation, which is given hereafter, governs the evolution of the free surface.

$$\frac{\partial h}{\partial t}+\frac{\partial}{\partial x}\left(\int_{-d}^{h}u dz\right)+\frac{\partial}{\partial y}\left(\int_{-d}^{h}v dz\right)=0 \tag{A5}$$

where h is the free-surface height and d is the bathymetric depth. The transport equations for temperature and salinity are given by

$$\frac{\partial T}{\partial t}+u\frac{\partial T}{\partial x}+v\frac{\partial T}{\partial y}+w\frac{\partial T}{\partial z}=\frac{\partial}{\partial x}\left(K_H\frac{\partial T}{\partial x}\right)+\frac{\partial}{\partial y}\left(K_H\frac{\partial T}{\partial y}\right)+\frac{\partial}{\partial z}\left(K_V\frac{\partial T}{\partial z}\right) \tag{A6}$$

$$\frac{\partial S}{\partial t}+u\frac{\partial S}{\partial x}+v\frac{\partial S}{\partial y}+w\frac{\partial S}{\partial z}=\frac{\partial}{\partial x}\left(K_H\frac{\partial S}{\partial x}\right)+\frac{\partial}{\partial y}\left(K_H\frac{\partial S}{\partial y}\right)+\frac{\partial}{\partial z}\left(K_V\frac{\partial S}{\partial z}\right) \tag{A7}$$

where T is the temperature, S is the salinity, K_H and K_V are the horizontal and vertical eddy diffusion coefficients.

A collocated finite-volume approach was used to discretize the governing equations [51,52]. For explicit terms such as advection of momentum, horizontal diffusion, and Coriolis force, the second-order Adams–Bashforth technique was used to discretize temporal derivatives [53]. Surface elevation and vertical momentum and scalar diffusion were estimated semi-implicitly using the TRIM technique, a second-order theta approach [54]. The advection terms were spatially discretized for the momentum and scalars using the third-order ULTIMATE-QUICKEST scheme [55], while the horizontal diffusion terms were discretized using the second-order central differencing scheme.

Appendix B. Spatial Variations in Wind Quantification

To observe the spatial consistency of the two-dimension dataset, the discrepancies in subsequent increments in size selection for both wind speed and direction were calculated as:

$$\Delta f_n=f(x,y)_{n+1}-f(x,y)_n \tag{A8}$$

where $f(x,y)_{n+1}$ and $f(x,y)_n$ are the filtered images with size n+1 and n, respectively. This filtering technique will reduce the horizontal resolution of the array size. However, no padding is applied to fill the surrounding array to prevent the model from being biased. Next, the spatial variation in wind speed is described as follows:

$$C_{v_{spd},n} = \frac{1}{\overline{\Delta f}_{n,spd}} \sqrt{\frac{1}{p}\sum_{i=1}^{p}\left(\Delta f_{n,spdi} - \overline{\Delta f}_{n,spd}\right)^2} \quad \text{(A9)}$$

where $C_{v_{spd},n}$ is the coefficient of variance of spatial wind speed, p is the population size, $\Delta f_{n,spdi}$ is the value of the difference between each filter of the wind speed in each grid, $\overline{\Delta f}_{n,spd}$ is the mean wind speed. For the complexity in the wind direction, $C_{v_{dir},n}$ is estimated based on Yamartino [56].

$$C_{v_{dir},n} = \frac{\sin^{-1}(\varepsilon)\left(1.0 + 0.1547\varepsilon^3\right)}{\theta_a + 1} \quad \text{(A10)}$$

where $\varepsilon = \sqrt{1 - (S_a^2 + C_a^2)}$, $\theta_a = \tan^{-1}\left(\frac{S_a}{C_a}\right)$, $S_a = j^{-1}\sum_{i=1}^{j}\sin(\Delta f_{n,dir})_i$, and $C_a = j^{-1}\sum_{i=1}^{j}\cos(\Delta f_{n,dir})_i$, $\Delta f_{n,diri}$ is the value of the difference between each filter of the wind direction in each grid. Values of $C_{v_{spd},n}$ and $C_{v_{dir},n}$ are then averaged based on the number of layers (n) being filtered, based on the proposed method of [40].

$$\overline{C}_{v\_spd} = \frac{1}{m}\sum_{n=1}^{m} C_{v_{spd},n} \quad \text{(A11)}$$

$$\overline{C}_{v\_dir} = \begin{cases} 0.01\ if\ \overline{C}_{v\_dir} > 0.01 \\ \frac{1}{m}\sum_{n=1}^{m} C_{v_{dir},n}\ if\ \overline{C}_{v\_dir} < 0.01 \end{cases} \quad \text{(A12)}$$

References

1. Ueda, S.; Kondo, K.; Chikuchi, Y. Effects of the halocline on water quality and phytoplankton composition in a shallow brackish lake (Lake Obuchi, Japan). *Limnology* **2005**, *6*, 149–160. [CrossRef]
2. Lougee, L.A.; Bollens, S.M.; Avent, S.R. The effects of haloclines on the vertical distribution and migration of zooplankton. *J. Exp. Mach. Biol. Ecol.* **2002**, *278*, 111–134. [CrossRef]
3. Gantzer, P.A.; Bryant, L.D.; Little, J.C. Controlling soluble iron and manganese in water-supply reservoir using hypolimnetic oxygenation. *Water Res.* **2009**, *43*, 1285–1294. [CrossRef] [PubMed]
4. Yuk, J.; Aoki, S. Effect of Wind and Rainfall on Water Exchange in a Stratified Estuary. *Estuary Coasts* **2009**, *32*, 88–99. [CrossRef]
5. Yao, J.; Li, Y.; Zhang, D.; Zhang, Q.; Tao, J. Wind effects on hydrodynamics and implications for ecology in a hydraulically dominated river-lake floodplain system: Poyang Lake. *J. Hydrol.* **2019**, *571*, 103–113. [CrossRef]
6. Rafael, C.B. Wind-Induced Internal Waves in Closed Basins: A Laboratory Experiment and Field Study. Master's Thesis, Universidade Federal Do Parana, Curitiba, Brazil, 2018.
7. Maruya, Y.; Nakayama, K.; Shintani, T.; Yonemoto, M. Evaluation of entrainment velocity induced by wind stress in a two-layer system. *Hydrol. Res. Lett.* **2010**, *4*, 70–74. [CrossRef]
8. Ishikawa, T.; Tanaka, M. Diurnal stratification and its effects on wind-induced currents and water qualities in Lake Kasumigaura, Japan. *J. Hydraul. Res.* **1993**, *31*, 307–322. [CrossRef]
9. Paerl, H.W.; Hall, N.S.; Calandrino, E.S. Controlling harmful cyanobacterial blooms in a world experiencing anthropogenic and climatic-induced change. *Sci. Total Environ.* **2011**, *409*, 1739–1745. [CrossRef]
10. Kopasakis, K.; Laspidou, C.; Spiliotopoulos, M.; Kofinas, D.; Mellios, N. Numerical Investigation of Wind Driven Circulation and Horizontal Dispersion in the Surface Layer of a Re-flooded Shallow Lake. *Environ. Process.* **2016**, *3*, 693–710. [CrossRef]
11. Sprules, W.G.; Cyr, H.; Menza, C.W. Multiscale effects of wind-induced hydrodynamics on lake plankton distribution. *Limnol. Oceanogr.* **2022**, *67*, 1631–1646. [CrossRef]
12. Amadori, M.; Piccolroaz, S.; Giovannini, L.; Zardi, D.; Toffolon, M. Wind variability and Earth's rotation as drivers of transport in a deep, elongated subalpine lake: The case of Lake Garda. *J. Limnol.* **2018**, *77*, 505–521. [CrossRef]
13. Schwab, D.J.; Beletsky, D. Relative effects of wind stress curl, topography, and stratification on large-scale circulation in Laka Michigan. *J. Geophys. Res. Oceans* **2003**, *108*, 3044. [CrossRef]
14. Santo, M.A.; Toffolon, M.; Zanier, G.; Giovannini, L.; Armenio, V. Large eddy simulation (LES) of wind-driven circulation in a peri-alpine lake: Detection of turbulent structures and implications of a complex surrounding orography. *J. Geophys. Res. Oceans* **2017**, *122*, 4704–4722. [CrossRef]
15. Valerio, G.; Pilotti, M.; Marti, C.L.; Imberger, J.R. The structure of basin-scale internal waves in a stratified lake in response to lake bathymetry and wind spatial and temporal distribution: Lake Iseo, Italy. *Limnol. Oceanogr.* **2012**, *57*, 772–786. [CrossRef]

16. Amadori, M.; Giovannini, L.; Toffolon, M.; Piccolroaz, S.; Zardi, D.; Bresciani, M.; Giardino, C.; Luciani, G.; Kliphuis, M.; van Haren, H.; et al. Multi-scale evaluation of a 3D lake model forced by an atmospheric model against standard monitoring data. *Environ. Model. Softw.* **2021**, *139*, 105017. [CrossRef]
17. Li, C.; Odermatt, D.; Bouffard, D.; Wüest, A.; Kohn, T. Coupling remote sensing and particle tracking to estimate trajectories in large water bodies. *Int. J. Appl. Earth Obs. Geoinf.* **2022**, *110*, 102809. [CrossRef]
18. Okely, P.; Imberger, J.R.; Antenucci, J.P. Processes affecting horizontal mixing and dispersion in Winam Gulf, Lake Victoria. *Limnol. Oceanogr.* **2010**, *55*, 1865–1880. [CrossRef]
19. Mao, M.; Xia, M. Particle Dynamics in the Nearshore of Lake Michigan Revealed by an Observation-Modeling System. *J. Geophys. Res. Oceans* **2020**, *125*, e2019JC015765. [CrossRef]
20. Liu, S. Wind Driven Circulation in Large Shallow Lakes. Doctoral Thesis, Delft University of Technology, Delft, The Netherlands, 2020.
21. Cimatoribus, A.A.; Lemmin, U.; Barry, D.A. Tracking Lagrangian transport in Lake Geneva: A 3D numerical modeling investigation. *Limnol. Oceanogr.* **2019**, *64*, 1252–1269. [CrossRef]
22. Laval, B.; Imberger, J.; Hodges, B.R.; Stocker, R. Modelling circulation in lakes: Spatial and temporal variations. *Limnol. Oceanogr.* **2003**, *48*, 983–994. [CrossRef]
23. Podsetchine, V.; Schernewski, G. The Influence of spatial wind inhomogeneity on flow patterns in a small lake. *Water Res.* **1999**, *33*, 3348–3356. [CrossRef]
24. Rubbert, S.; Köngeter, J. Measurements and three-dimensional simulations of flow in a shallow reservoir subject to small-scale wind field inhomogeneities induced by sheltering. *Aquat. Sci.* **2005**, *67*, 104–121. [CrossRef]
25. Nakayama, K.; Shintani, T.; Shimizu, K.; Okada, T.; Hinata, H.; Komai, K. Horizontal and residual circulations driven by wind stress curl in Tokyo Bay. *J. Geophys. Res. Oceans* **2014**, *119*, 1977–1992. [CrossRef]
26. Chikita, K. Dynamic behaviours of anoxic saline water in Lake Abashiri, Hokkaido, Japan. *Hydrol. Process.* **2000**, *14*, 557–574. [CrossRef]
27. Maruya, Y.; Nakayama, K.; Sasaki, M.; Komai, K. Effect of dissolved oxygen on methane production from bottom sediment in a eutrophic stratified lake. *J. Environ. Sci.* **2023**, *125*, 61–72. [CrossRef] [PubMed]
28. Japan Meteorological Agency. Japan Meteorological Agency | AMeDAS. Available online: https://www.jma.go.jp (accessed on 10 October 2022).
29. User's Guides for the Advanced Research WRF (ARW) Modeling System, Version 3. Available online: https:/www2.mmm.ucar.edu/wrf/users/docs/user_guide_V3/contents.html (accessed on 31 May 2022).
30. Copernicus Climate Data Store. Available online: https://cds.climate.copernicus.eu/cdsapp (accessed on 10 October 2022).
31. Hersbach, H.; Bell, B.; Berrisford, P.; Hirahara, S.; Horányi, A.; Muñoz-Sabater, J.; Nicolas, J.; Peubey, C.; Radu, R.; Schepers, D.; et al. The ERA5 global reanalysis. *Q. J. R. Meteorol. Soc.* **2020**, *146*, 1999–2049. [CrossRef]
32. Solbakken, K.; Birkelund, Y.; Samuelsen, E.M. Evaluation of surface wind using WRF in complex terrain: Atmospheric input data and grid spacing. *Environ. Model. Softw.* **2021**, *145*, 105182. [CrossRef]
33. Olauson, J. ERA5: The new champion of wind power modelling? *Renew. Energy* **2018**, *126*, 322–331. [CrossRef]
34. Ramon, J.; Lledó, L.; Torralba, V.; Soret, A.; Doblas-Reyes, F.J. What global reanalysis best represents near-surface winds? *Q. J. R. Meteorol. Soc.* **2019**, *145*, 3236–3251. [CrossRef]
35. Monin, A.S.; Obukhov, A.M. Basic Laws of Turbulent Mixing in the Surface Layer of the Atmosphere. *Contrib. Geophys. Inst. Acad. Sci. USSR* **1954**, *24*, 163–187.
36. Global Map Japan. Available online: https://www.gsi.go.jp/kankyochiri/gm_japan_e.html (accessed on 10 October 2022).
37. Shintani, T. An unstructured-cartesian hydrodynamic simulator with local mesh refinement technique. *J. JSCE (Hydraul. Eng.)* **2017**, *73*, 967–972. [CrossRef]
38. Veerapaga, N.; Azhikodan, G.; Shintani, T.; Iwamoto, N.; Yokoyama, K. A three-dimensional environmental hydrodynamic model, Fantom-Refined: Validation and application for saltwater intrusion in a meso-macrotidal estuary. *Ocean Model.* **2019**, *141*, 101425. [CrossRef]
39. Kondo, J. Air-sea bulk transfer coefficients in diabatic conditions. *Bound.-Layer Meteorol.* **1975**, *9*, 91–112. [CrossRef]
40. Hieu, N.L.; Shintani, T. Numerical Investigation on Inhomogeneous Wind and Its Effects to Mass Transport. *J. JSCE* **2022**, *in press*.
41. Shintani, T.; de la Fuente, A.; de la Fuente, A.; Niño, Y.; Imberger, J. Generalizations of the Wedderburn number: Parameterizing upwelling in stratified lakes. *Limnol. Oceanogr.* **2010**, *55*, 1377–1389. [CrossRef]
42. Mortimer, C.H. Water movements in lakes during summer stratification; evidence from the distribution of temperature in Windermere. *Philos. Trans. R. Soc. Lond. Ser. B Biol. Sci.* **1952**, *236*, 355–398.
43. Spigel, R.H.; Imberger, J. The classification of Mixed-Layer Dynamics of Lakes of Small to Medium Size. *J. Phys. Oceanogr.* **1980**, *10*, 1104–1121. [CrossRef]
44. Simmons, H.; Chang, M.H.; Chang, Y.T.; Chao, S.Y.; Fringer, O.; Jackson, C.; Ko, D.S. Modeling and Prediction of Internal Waves in the South China Sea. *Oceanography* **2011**, *24*, 88–99. [CrossRef]
45. Lin, H. Hydropedology: Synergistic integration of soil science and hydrology. *Choice Rev. Online* **2013**, *50*, 4450–4451.
46. Pannard, A.; Beisner, B.E.; Bird, D.F.; Braun, J.; Planas, D.; Bormans, M. Recurrent internal waves in a small lake: Potential ecological consequences for metalimnetic phytoplankton populations. *Limnol. Ocean. Fluids Environ.* **2011**, *1*, 91–109. [CrossRef]

47. Rao, Y.R.; Schwab, D.J. Transport and Mixing between the Coastal and Offshore Waters in the Great Lakes: A Review. *J. Great Lakes Res.* **2007**, *33*, 202–218. [CrossRef]
48. Jiménez, P.A.; Dudhia, J. On the Ability of the WRF Model to Reproduce the Surface Wind Direction over Complex Terrain. *J. Appl. Meteorol. Climatol.* **2013**, *52*, 1610–1617. [CrossRef]
49. Preston, J.M.; Collins, W.T. Bottom classification in very shallow water by high-speed data acquisition. In Proceedings of the Oceans 2000 MTS/IEEE Conference and Exhibition, Providence, RI, USA, 11–14 September 2000; Volume 2, pp. 1277–1282.
50. Nakayama, K.; Okada, T.; Nomura, M. Mechanism responsible for fortnightly modulations in estuary circulation in Tokyo Bay. *Estuar. Coast. Shelf Sci.* **2005**, *64*, 459–466. [CrossRef]
51. Ferziger, J.H.; Peric, M. *Computational Methods for Fluid Dynamics*, 3rd ed.; Springer: Berlin/Heidelberg, Germany, 2002.
52. Perot, B. Conservation Properties of Unstructured Staggered Mesh Schemes. *J. Comput. Phys.* **2000**, *159*, 58–59. [CrossRef]
53. Zang, Y.; Street, R.L.; Koseff, J.R. A Non-staggered Grid, Fractional Step Method for Time-Dependent Incompressible Navier-Stokes Equations in Curvilinear Coordinates. *J. Comput. Phys.* **1994**, *114*, 18–33. [CrossRef]
54. Casulli, V. A semi-implicit finite difference method for non-hydrostatic, free-surface flows. *Int. J. Numer. Methods Fluids* **1999**, *30*, 425–440. [CrossRef]
55. Leonard, B. The ULTIMATE conservative difference scheme applied to unsteady one-dimensional advection. *Comput. Methods Appl. Mech. Eng.* **1991**, *88*, 17–74. [CrossRef]
56. Yamartino, R.J. A Comparison of Several "Single-Pass" Estimators of the Standard Deviation of Wind Direction. *J. Clim. Appl. Climatol.* **1984**, *23*, 1362–1366. [CrossRef]

water

MDPI

Article

Spatial Distribution and Source Identification of Water Quality Parameters of an Industrial Seaport Riverbank Area in Bangladesh

M. Shahidul Islam [1,2], Kei Nakagawa [3,*], M. Abdullah-Al-Mamun [2], Abu Shamim Khan [4], Md. Abdul Goni [5] and Ronny Berndtsson [6]

1 Graduate School of Fisheries and Environmental Sciences, Nagasaki University, 1-14 Bunkyo-machi, Nagasaki 852-8521, Japan; bb53421002@ms.nagasaki-u.ac.jp or msislam_chem@just.edu.bd
2 Department of Chemistry, Jashore University of Science and Technology, Jashore 7408, Bangladesh; mamunjust.03@gmail.com
3 Institute of Integrated Science and Technology, Nagasaki University, 1-14 Bunkyo-machi, Nagasaki 852-8521, Japan
4 Environmental Laboratory, Asia Arsenic Network, Arsenic Center, Benapole Road, Krishnobati, Pulerhat, Jashore 7400, Bangladesh; abuskhan2000@yahoo.com
5 Department of Chemistry, Bangladesh University of Engineering and Technology, Dhaka 1000, Bangladesh; abdulgoni@chem.buet.ac.bd
6 Division of Water Resources Engineering & Centre for Advanced Middle Eastern Studies, Lund University, P.O. Box 118, SE-221 00 Lund, Sweden; ronny.berndtsson@tvrl.lth.se
* Correspondence: kei-naka@nagasaki-u.ac.jp

Abstract: The Pasur River is a vital reservoir of surface water in the Sundarbon area in Bangladesh. Mongla seaport is located on the bank of this river. Many industries and other commercial sectors situated in this port area are discharging waste into the river without proper treatment. For this reason, geospatial analysis and mapping of water pollutant distribution were performed to assess the physicochemical and toxicological situation in the study area. We used different water quality indices such as Metal Index (*MI*), Comprehensive Pollution Index (*CPI*), and Weighted Arithmetic Water Quality Index Method (*WQI*) to improve the understanding of pollution distribution and processes determining the quality of river water. Multivariate statistical methods were used to evaluate loads and sources of pollutants in the Pasur River system. The results indicate that the sources of contaminants are both geogenic and anthropogenic, including untreated or poorly treated wastewater from industries and urban domestic waste discharge. The concentration range of total suspended solid (TSS), chloride, iron (Fe), and manganese (Mn) were from 363.2 to 1482.7, 108.2 to 708.93, 1.13 to 2.75, and 0.19 to 1.41 mg/L, respectively, significantly exceeding the health-based guideline of WHO and Bangladeshi standards. The high Fe and Mn contents are contributions from geogenic and anthropogenic sources such as industrial waste and construction activities. The average pH value was 8.73, higher than the WHO and Bangladeshi standard limit. *WQI* (ranging from 391 to 1336), *CPI* (6.71 to 23.1), and *MI* (7.23 to 23.3) were very high and greatly exceeded standard limits indicating that the Pasur River water is highly polluted. The results of this study can be used as a first reference work for developing a surface water quality monitoring system and guide decisionmakers for priorities regarding wastewater treatment.

Keywords: industrial riverbank; surface water; pollution status; Pasur River; Bangladesh

Citation: Islam, M.S.; Nakagawa, K.; Abdullah-Al-Mamun, M.; Khan, A.S.; Goni, M.A.; Berndtsson, R. Spatial Distribution and Source Identification of Water Quality Parameters of an Industrial Seaport Riverbank Area in Bangladesh. *Water* **2022**, *14*, 1356. https://doi.org/10.3390/w14091356

Academic Editor: Dimitrios E. Alexakis

Received: 2 April 2022
Accepted: 19 April 2022
Published: 21 April 2022

1. Introduction

According to the United Nations, about one-third of the world's population drinks contaminated water [1]. Clean water is essential for human health, aquatic and terrestrial ecosystems, and life-supporting activities. However, industrial, urban, and agricultural activities release untreated effluents into surface water, creating an alarming water pollution situation in Bangladesh [2]. Pathogenic bacteria (total coliform and fecal coliform) gradually degrade water quality. In the oil refinery industry, conventional oil, gas, and coal bed methane are often accompanied by large volumes of contaminated water [3]. These industries give a high load of organic pollutants such as phenols, which are potentially dangerous for the environment and human health. Industrial wastewater also contains nitrogen, phosphorus, and heavy metals such as Fe, Cr, Ni, Cd, Zn, and Mn [4]. Geogenic sources may also contribute to the pollution load in river water systems. These sources include rock–water weathering, biological activity, sediment erosion, benthic distress, and riverine system flow regime changes [5,6]. Iron and manganese exist naturally in rivers; they may also be released to water from natural geologic deposits. The Earth's crust is a major source of Mn to the atmosphere, soil, and water.

Exposure to heavy metal pollution is a significant threat to the environment and public health worldwide [7] and especially in the developing world. Heavy metals enter the food chain through biomagnification and eventually affect human health [8]. The discharge of heavy metal waste into receiving waters may result in many physical, chemical, and biological disorders such as damaged DNA and gene expression changes [9]. Heavy metals in effluents are moderately soluble in water depending on pH and may affect the total and effective exposure to humans and accumulation in soils and plants [10]. Fly ash from thermal power plants and the cement industry is either discarded of as dry disposal in landfills or discharged into natural drainage systems such as rivers. These disposal methods result in metal contamination of surface and groundwater resources that eventually will turn up in the food chain [11]. Polluted water is the main reason that several diseases such as cancer; congenital disabilities; and skin, lung, brain, kidney, and liver conditions are several times more prevalent in the investigated area than elsewhere in the country. Waterborne diseases, such as hepatitis (A, B, and C), cholera, typhoid, dysentery, and diarrhea, are also caused by polluted water [12]. To control geogenic and anthropogenic sources of pollution and prevent them from growing to levels detrimental to human beings, programs to monitor pollutants in the river water and sediments are necessary [13].

The Pasur River is one of the most important waterways for economic growth in Bangladesh. It passes through the Mongla Seaport and the Mongla shantytown of the Bagerhat district in Bangladesh. Mongla is the second largest seaport in Bangladesh. It consists of the Pasur channel beside the Sundarban mangrove forest [14]. The Sundarban mangrove forest is the largest remaining tract of mangrove in the world. Beside the river lies the coal-based Rampal power plant station.

This study was carried out on the Pasur River in the Mongla port area. A significant site of oil refinery (petroleum and vegetable oil), cement, dye and paint, leather, and shipbreaking industries is established near the Pasur River. The industry directly discharges poorly or untreated toxic effluent into the Pasur River that becomes increasingly polluted [15]. The river water is used for different functions such as industrial purposes, household activities, bathing, irrigation of fields, and cooking foods by the adjacent rural populations. In some cases, in the dry season, when the drinking water is at crisis level, local rural people use the river water for drinking purposes after boiling. The river was once an important freshwater source for drinking and domestic uses, fisheries, and agricultural irrigation. The river is still used for fishing, and fishermen use smaller or larger boats for fishing. Polluted fish is another crucial reason to assess and monitor the surface water of this river.

Given the above, this study focused on assessing the Pasur River water quality by using water quality evaluation indices such as Water Quality Index (*WQI*) [16,17], Comprehensive Pollution Index (*CPI*) [18,19], and the Metal Index (*MI*) [20,21]. Though some research has been conducted for this river [22], there are still no systematic studies focusing on both physicochemical parameters and toxic metal pollution and pollution source determination in the concerned area. Assessment of pollution status is, thus, insufficient in this area. It is also essential to identify the pollution sources. For this purpose, we used multivariate analysis to indicate sources of water pollutants, including metals [23,24]. Therefore, the objectives of this study were (i) to determine the water quality parameters of the Pasur River by using physicochemical and toxicological parameters, (ii) to use water quality indices such as *WQI*, *CPI*, and *MI* for assessing the physicochemical and toxicological properties with spatial distribution for the Pasur River, and (iii) to identify the source of pollutants including metal contamination in water by using multivariate analysis methods.

2. Materials and Methods

2.1. Study Area

The Pasur is the largest river in the Sundarban (Mangrove Forest) region, located at 89°30′0″ E and 21°45′0″ N. This river is known as Rupsha in Khulna. To the north of Khulna in the Jashore region, the river is known as Bhairab River, and to the south of Khulna, from the Chalna area, it is named Pasur River, and this location is known as Mongla Upazila. Mongla Port Municipality is the second largest seaport of the country, export processing zone, and fishing industry area on the shore of the Pasur River. The study was conducted in the Pasur River at Mongla Port Municipality area in Mongla Upazila of Bagherhat District in Khulna, Bangladesh. It is located about 55 km south of Khulna City and 131 km north of the Bay of Bengal. Land use is concentrated to manufacturing industries, forest, upland fields, and urban areas. The geology is mainly constituted by tidal deltaic deposits but the northern and western area by marsh clay and peat deposits. The southern region also contains mangrove swamp deposits and north-eastern area deltaic silt deposits. The western and central parts of the eastern region of the study area are constituted by agricultural areas and the eastern and northern area is constituted by industry mixed with urban areas (Figure 1).

Mongla Port was established in 1954 on the Pasur and Mongla River junction bank, where 1280 cargo ships were handled in the fiscal year 2019–2020. According to Bangladesh 2011 census, the municipality area is 19.4 km^2 with a population of about 40,000 [25].

The location of the present study was 22°47′ N and 22°60′ N and 89°60′ E and 89°52′ E. The river passes through the right side of the Trikona and Dubla islands and discharges into the Bay of Bengal. The river is deep, so big ships can enter the Mongla Port year-round. The river width varies. It is about 460 m wide at Rupsha, 790 m at Bajuyan, and 2.44 km at Pasur-Shibsha. The river is about 142 km long. The Pasur River and all its tributaries are tidal channels. The river is now also an effluent channel as it receives more than 80% of the wastewater generated from urban areas and industrial sites.

More than 49 small and large manufacturing and processing industries are located in the study area that discharge waste and wastewater into the river without proper treatment. Jute processing, cement manufacturing, fertilizer manufacturing, oil refinery industry, construction materials, and automobile oil storage activities are included in this area [22]. The sampling points were selected based on assumed representative connections between the river and pollution sources. The study area location and samplings points are shown in Figure 1.

2.2. Methods

The surface water quality varies with rainfall. During the rainy season, due to increased rain, the concentration of pollutants in surface water can either be decreased by dilution or increase due to surface runoff due to first flush [27]. The present research collected water samples during the rainy season at 20 sampling points. Sampling sites were selected based on functional area, including industry, urban, and agricultural areas, as shown in Figure 1.

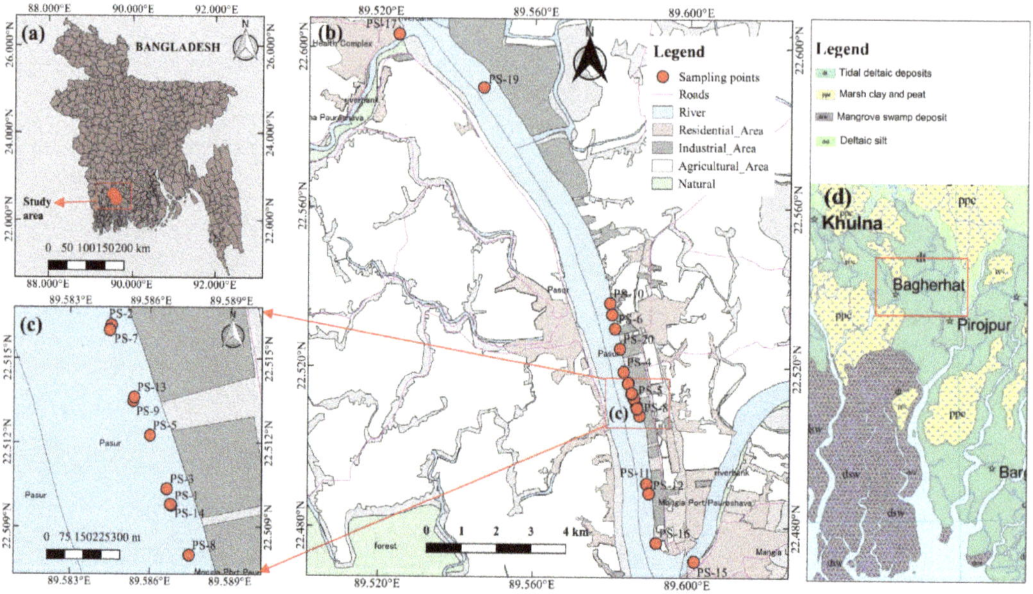

Figure 1. Location of (**a**) study area; (**b**,**c**) sampling points; (**d**) geological map of the study area [26].

Collection, storage, and transportation of water samples were performed according to standard guidelines [28]. Three water samples were randomly collected from each sampling site and then thoroughly mixed into a 1.0–1.5 L sample and transferred to clean 500 and 100 mL polyethylene bottles, respectively. This was undertaken to guarantee representative samples from each sampling point. At the same time, survey work was conducted at every sampling station to collect background information of the sampling area. The 500 mL samples were preserved with 5 mL of 55% HNO_3 per liter of water and maintained at 4 °C in the refrigerator for physical and chemical parameter analysis. The 100 mL samples were preserved with 2 mL of 6 N nitric acid for metal analysis. Before sample collection, all bottles were washed with 10% nitric acid and de-ionized water. The 100 mL samples were put in a beaker, and 5 mL concentrated HNO_3 was added and boiled at 130 °C on a hot plate until the volume was reduced to about 25 to 30 mL. HNO_3 addition was continued and repeated until the solution became clear. Then, the solution was cooled and filtered with deionized water passing through a Whatman no. 41 filter [29,30]. Ultra-pure HNO_3 was used for sample digestion. Temperature, pH, and total dissolved solids (TDS) were measured on site by a calibrated apparatus. The pH was determined by a portable calibrated pH Meter (HI 2211, HANNA Romania, Romania) and TDS was determined by calibrated multimeter (CT-676, BOECO Germany, Hamburg, Germany). Chloride, total alkalinity, total suspended solids (TSS), total hardness, iron, and manganese were measured by Mohr's titration, acid-base titration, filtration, EDTA complexometric titration, and flame-AAS method, respectively. Before analysis, all laboratory equipment was immersed in 10% nitric acid for 48 h, rinsed with distilled water, and dried in an oven. All chemicals and reagents used were of analytical grade. Deionized water with electrical conductivity less than 0.5 $\mu S\ cm^{-1}$ and resistivity ~18 MΩ cm at 25 °C was used for the preparation of all solutions. A blank sample was prepared and analyzed for water samples to ensure that the chemicals used in the preparation did not contaminate the samples. By employing approved standard solutions (HACK, Love Land, Colorado, USA), calibration curves (linearity ≥ 0.995) were created to check the quality of measurements. To prepare spiked calibration standards, 2, 5, 10, and 15 mg/L of mixed standard and known concentration solution was added. The spiked calibration standard was added after every ten samples. Recovery rates of metals spiked in water fluctuated from 93 to 100%. The detection limits were 0.006 and 0.004 mg/L for Fe and Mn, respectively. Replicate analysis (RSD less than 5%) of the traceable Certified Reference Materials (CRM) and randomly selected samples were measured to check the analyses' precision and accuracy. All samples were measured in triplicate (RSD less than 5%), and the mean was used. Maps of the study area, spatial distribution of water quality indices, and cluster groups were produced using GIS software (QGIS, version 2.18.2).

2.3. Water Quality Index (WQI)

The Water Quality Index (*WQI*) is used for the assessment of both groundwater and surface water pollution levels [31,32]. The *WQI* is calculated using the weighted arithmetic method, initially proposed by Horton [33] and developed by Brown et al. [34,35]:

$$WQI = \frac{\sum Q_n W_n}{\sum W_n}, \tag{1}$$

where Q_n = quality rating of n-th water quality parameter; W = unit weight of n-th water quality parameter. Q_n is calculated by:

$$Q_n = \left[\frac{V_n}{V_s}\right] \times 100, \tag{2}$$

For the pH, this becomes:

$$Q_{pH} = \left[\frac{V_n - V_i}{V_s - V_i}\right] \times 100, \tag{3}$$

where V_n = actual amount of n-th parameter present; V_i = ideal value of the parameter ($V_i = 0$, except for pH ($V_i = 7$)); V_s = standard permissible value for the n-th water quality parameter.

The unit weight (W_n) of the various water quality parameters is inversely proportional to the recommended standards for the corresponding parameters.

$$W_n = \frac{k}{V_s}, \tag{4}$$

where k = constant of proportionality calculated by:

$$k = \frac{1}{\sum V_s}, \tag{5}$$

The *WQI* value falls into five categories such as *WQI*: < 50 → Excellent, grade 1; 51–100 → Good, grade 2; 101–200 → Poor, grade 3; 201–300 → Very poor, grade 4; and > 300 → Likely not suitable for drinking, grade 5 [36].

2.4. Comprehensive Pollution Index (CPI)

The Comprehensive Pollution Index (*CPI*) assesses the overall water quality status [37,38]:

$$PI = \frac{C_i}{S_i}, \tag{6}$$

$$CPI = \frac{1}{n}\sum_{i=1}^{n} PI, \tag{7}$$

where *PI* is the pollution index of the i-th parameter; C_i is the measured concentration of the i-th parameter; S_i is the standard permissible concentration of the i-th parameter in the water, and n is the total number of parameters. We used health-based guidelines from the WHO and the Department of Environment (DoE), Bangladesh, standard for maximum permissible concentrations. According to Mekuria et al. [39], the *PI* is a Single Factor Evaluation Index (*SFEI*) for each water quality parameter. When the value of $PI < 1$, the water quality meets surface water quality standards. $PI > 1$, indicates that the water quality exceeds the standards, hence, the water is polluted.

The water quality can be classified into five categories based on the calculated value of *CPI* such as *CPI*: 0–0.20, Category 1, Clean; 0.21–0.40, Category 2, Sub clean; 0.41–1.00, Category 3, Slightly polluted; 1.01–2.00, Category 4, Medium polluted, and ≥2.01, Category 5, Heavily polluted [40].

2.5. Metal Index (MI)

The *MI* assesses the overall quality of surface and drinking water by [41,42]:

$$MI = \sum \frac{C_i}{(MAC)_i}, \quad (8)$$

where C_i is the mean concentration of *i*-th metal and *MAC* is the maximum allowable concentration of each metal. The maximum limit of *MI* is 1, and $MI > 1$ is a threshold of warning. The classification of water quality based on *MI* is < 0.3, class-I very pure; 0.3–1.0, class-II pure; 1.0–2.0, class-III, slightly affected; 2.0–4.0, class-IV, moderately affected; 4.0–6.0, class-V, strongly affected; and >6.0, class-VI, seriously affected [43].

2.6. Statistical Analysis

Pearson correlation, principal component analysis (PCA), and hierarchical cluster analysis (HCA) were performed to identify relationships among the examined water quality parameters in the studied area to infer sources (geogenic or anthropogenic). Pearson correlation analysis represents the strength of the relationship between different parameters. PCA and HCA are the most common multivariate statistical methods for classifying and interpreting large datasets from environmental monitoring programs that reduce the dimensionality of the data. Data were processed using SPSS 17.0 for Windows; IBM, USA, and JMP Pro 15.

3. Results and Discussion

3.1. Water Quality Guidelines

A summary of the analyses results is shown in Table 1. Concentrations were compared to standard health-based guidelines by the WHO [44,45] and the Department of Environment (DoE), Bangladesh [46]. Temperature is an essential physical water quality parameter. The photosynthesis activity of green plants, physicochemical processes, and microbial biodegradation rate are greatly influenced by temperature. According to DoE guidelines for water in Bangladesh, the temperature should be maintained between 20–30 °C. The average and median temperature in the river water of the study area were 28.7 and 29.2 °C, respectively. The maximum and minimum temperature were 31.2 and 18.3 °C at the sampling location PS-16 and PS-9, respectively. The minimum temperature is caused by tidal flows of the river, excessive rainfall (128.7–321.9 mm), or sometimes discharging cooling water from industrial sites. Samples at PS-12, PS-13, PS-14, PS-15, PS-16, PS-17, PS-18, PS-19, and PS-20 occasionally exceeded the standard limit (Table 1). In previous studies, the temperature range of the Buriganga River in Bangladesh varied from 22.8 to 31.4 °C [47].

pH is a physical water quality parameter that is crucial for aquatic life. The toxicity of metals to aquatic life, different chemical and biochemical reactions, and the suitability of water for different uses are associated with water pH [39]. The recommended pH by WHO and DoE health-based guidelines varies from 6.50 to 8.50. In the study area, the median pH was 8.72 (Figure 2a), slightly higher than the recommendations. Maximum and minimum pH was 8.97 and 8.43 at PS-11, PS-13, and PS-18, respectively (Table 1). All sampling sites showed pH higher than the WHO and Bangladesh standards except PS-18 and PS-19. The low pH value at sampling locations PS-18 and PS-19 is attributed to relatively low anthropogenic influence. The higher pH may have been influenced by dissolving alkaline waste materials from Mongla municipal domestic area and poor or untreated industrial effluents from industry near the study area.

Table 1. Comparative study of physicochemical parameters of Pasur River water with WHO and DoE Standards.

Sample ID	Temperature (°C)	pH	TH (mg/L)	TDS (mg/L)	TSS (mg/L)	Chloride (mg/L)	Alkalinity (mg/L)	Fe (mg/L)	Mn (mg/L)
PS-1	27.00	8.57	75.23	144.44	643.96	215.07	100.61	1.48	0.40
PS-2	27.33	8.57	153.80	278.12	791.58	213.13	92.85	2.25	0.68
PS-3	27.33	8.60	71.91	136.68	1482.71	250.94	88.33	1.84	0.51
PS-4	27.67	8.67	64.72	127.74	426.64	224.59	91.67	1.13	0.46
PS-5	28.00	8.80	72.97	147.77	728.27	206.48	94.67	2.10	0.70
PS-6	27.00	8.53	85.12	142.28	567.09	221.13	92.67	1.41	0.38
PS-7	27.57	8.53	78.97	146.52	988.60	241.09	94.00	2.06	0.79
PS-8	28.67	8.73	163.67	354.58	852.60	108.15	106.00	1.97	0.71
PS-9	18.33	8.93	34.80	135.18	582.34	148.61	91.00	1.15	0.34
PS-10	28.33	8.70	56.70	134.78	1004.90	170.47	91.33	2.23	0.19
PS-11	29.67	8.97	173.47	326.18	926.84	708.93	67.67	2.75	1.41
PS-12	30.57	8.87	67.28	163.81	657.54	271.51	92.00	2.32	0.86
PS-13	30.33	8.97	207.51	455.24	763.97	212.30	93.33	2.49	1.01
PS-14	31.00	8.87	79.74	153.15	363.21	233.06	87.33	1.14	0.44
PS-15	30.50	8.93	276.55	524.60	666.32	214.06	98.67	1.86	0.68
PS-16	31.17	8.93	57.69	151.67	807.87	261.32	88.67	2.32	0.97
PS-17	30.33	8.70	77.52	142.53	582.32	205.03	98.00	2.19	0.79
PS-18	31.00	8.43	76.89	144.64	674.90	191.70	93.00	2.22	0.78
PS-19	30.33	8.47	65.16	133.58	572.16	211.11	97.67	1.62	0.55
PS-20	31.00	8.80	472.64	893.27	652.45	368.87	165.33	1.72	0.61
Maximum	31.17	8.97	472.64	893.27	1482.71	708.93	165.33	2.75	1.41
Minimum	18.33	8.43	34.80	127.74	363.21	108.15	67.67	1.13	0.19
Average	28.66	8.73	120.62	241.84	736.81	243.88	96.24	1.91	0.66
STD. Dev.	±2.88	±0.18	±102.89	±193.09	±242.82	±120.8	±17.85	±0.47	±0.28
WHO STD. [a,b]		6.5–8.5	500	1000	-	250	-	0.3	0.1
DoE STD.[c]	20–30	6.5–8.5	200–500	1000	10	150–600	-	0.3–1.0	0.1

[a, b] World Health Organization (WHO) [44,45]. [c] Department of Environment (DoE), Bangladesh [46].

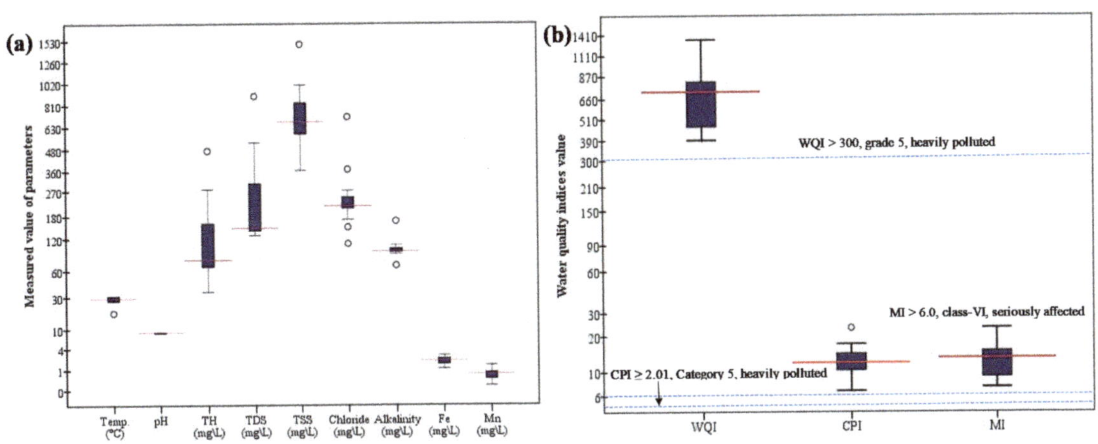

Figure 2. Boxplots for (**a**) descriptive statistics and (**b**) water quality indices.

Total hardness (TH) is influenced by contents of carbonate, bicarbonate, sulfate, and chloride salt of calcium and magnesium that decrease the water softness for cleaning, heating, and boiler systems. The WHO and national guideline for the total hardness of water is 500 mg/L, and the average observed concentration of 120.6 mg/L is lower (Table 1). The maximum and minimum concentration of total hardness was 472.6 and 34.8 mg/L at sampling station PS-20 and PS-9, respectively (Table 1). BNS Mongla (PS-20) was affected by the discharge of calcium and magnesium from household and industrial waste.

Total dissolved solids (TDS) consist of dissolved inorganic ions and suspended matter, and Bangladesh has recommended a maximum concentration of 1000 mg/L. The average concentration of TDS in the study area was 241.8 mg/L (Table 1) and the median was 147.2 mg/L (Figure 2a). The maximum and minimum concentration of TDS were 893.3 and 127.7 mg/L at sampling sites PS-20 and PS-4, respectively. TDS in the study area comes from domestic and industrial waste such as detergents, chloride, bicarbonate, fluorides, sulfate, and other ions. The chloride concentration range in samples was 108.2 to 708.9 mg/L with an average of 243.9 mg/L. The average is within the WHO and national standards (Table 1). The median value of chloride (214.6 mg/L) was also within the WHO standard limit (Figure 2a). The concentration of chloride in sampling locations PS-3, PS-12, PS-16, and PS-20 exceeded the WHO standard limit, whereas the chloride concentration of PS-11 exceeded the WHO and national guidelines of Bangladesh. Thus, the concentration of TDS and chloride at some locations did not exceed the permissible limits. A possible reason for this is the significant runoff of stormwater during the rainy season in the study area (129–322 mm) that can dilute the polluted water.

Total suspended solids (TSS) were calculated from dissolved suspended and colloidal materials that increase surface water's turbidity. TSS can affect surface water quality. Excess concentration of TSS affects light transmission and aquatic life. The standard limit of TSS should be maintained below 10.0 mg/L, which is recommended by the Bangladeshi standard (DoE). However, the average TSS concentration was 737 mg/L (Table 1). The median concentration of TSS was 671 mg/L and much higher than national standard value (Figure 2a). Maximum and minimum concentration of TSS in the river water was 1482 and 363 mg/L at the Omera Petroleum Industrial area (PS-3) and Laudobe ghat area (PS-14), respectively. All sampling sites displayed a higher TSS than permissible levels (Table 1). This is probably due to the discharge of large volumes of industrial waste and wastewater from nearby industry and local urban bazaars to the river without any treatment.

Total alkalinity is related to the contents of carbonate, bicarbonate, and hydroxyl ions in the water. The average concentration of total alkalinity in the samples was 96.2 mg/L. Maximum and minimum concentrations were 165.3 and 67.7 mg/L at sampling points PS-20 and PS-11 (Table 1), respectively. At PS-20, the total alkalinity concentration was higher than other sampling points due to alkaline household and industrial waste discharged into the river without treatment.

Iron is a common metal in surface water that may dissolve from surface sediments and suspended matter. Dissolved iron in the surface water is usually not a health hazard, but it may create a bitter taste, stain, and discolor laundry. In addition, high iron contents are not suitable for heating systems and boilers. According to the WHO standard, iron concentrations should be below 0.30 mg/L. However, the average concentration of iron in the samples was 1.91 mg/L (Table 1). The median value of Fe was 2.0 mg/L, and much higher than the standard limit (Figure 2a). The maximum and minimum concentration of iron in river water was 2.75 and 1.13 mg/L at sampling sites PS-11 and PS-4, respectively. High iron concentrations probably stem from suspended matter and industrial waste (Table 1). Thus, it can be concluded that the river water is not fit for use in household activities and industrial purposes without treatment. Iron forms several sulfides in nature, such as pyrite (FeS_2), marcasite (FeS_2), pyrrhotite ($Fe_{11}S_{12}$), troilite (FeS), and numerous other more complex compounds. Iron exists naturally in rivers; it may also be released to water from natural geologic deposits. Iron at average temperature is usually deposited from solutions such as hydrous sesquioxide, carbonates, sulfides or hydrous silicates of iron and potash known as glauconite. Anthropogenic sources such as untreated industrial effluents, improper disposal of domestic waste, and agricultural runoff are the main contributors to iron pollution in the Pasur River.

Manganese is an essential natural element in surface water but may adversely affect aquatic ecosystems [48]. The manganese concentration should be below about 0.1 mg/L according to the WHO and Bangladeshi standards for water. The average manganese concentration in the study area was 0.66 mg/L. The median value was 0.68 mg/L, and consequently much higher than recommended standards (Figure 2a). The maximum and minimum manganese concentration in the study area was 1.41 and 0.19 mg/L at sampling points PS-11 and PS-10, respectively (Table 1). The concentration of metals in the water of the Pasur River is compatible with that of surface water of other aquatic systems in Bangladesh and worldwide [29,49–51]. The manganese concentration in the Pasur River is alarming for aquatic environments, most likely caused by dissolved, suspended solids and industrial waste. Manganese sources can be geogenic or anthropogenic. Manganese forms two sulfides, alabandite (MnS) and hauerite (MnS_2). Both minerals are scarce and so unstable that they rapidly oxidize on exposure. Alabandite is the less rare form and usually occurs as a subordinate constituent of metalliferous veins or allied deposits. The anthropogenic sources of Mn are industrial effluents, runoff from agricultural activities, and uncontrolled release or leakage from landfill sites. The Sela River at Sundarbans area, an ecologically important river like the Pasur River, has drawn global attention after an oil spill accident in December 2014, when about 94,000 gallons of heavy fuel were released into the river, causing instant damage to the mangrove habitat and wildlife. Thus, chemical accidents or disasters in adjacent rivers are also sources of physicochemical and toxicological pollution.

3.2. Water Quality Indices

The water quality indices in the analyzed water were expressed by box plots. The three water quality indices are shown in Figure 2b. The *WQI* was within the range of 391–1336. The median was 725 (Figure 2b) and greatly exceeded the maximum limit of 300, which indicates the pollution status of grade 5. Thus, water in all locations falls under the "very poor water" and "likely not suitable for drinking purposes" categories. The highest *WQI* was found at PS-11 (Table 2). The metal contamination in this location is very high, which is a major concern. The high *WQI* in the study area is due to both geogenic sources of pollutants and discharge of municipal wastewater and industrial effluents, fishing boats, agricultural runoff, loading–unloading, and construction activities in the Mongla seaport area.

The *CPI* was applied to understand the overall status of Pasur River water pollution. This varied between 6.7–23.1, with an average of 12.8 (Table 2), and a median of 12.1 (Figure 2b). The median *CPI* (12.1) indicates high pollution levels. All sampling locations exceeded the maximum limit of $\geq$2.0, indicating Category 5 and heavy pollution. The highest *CPI* was recorded at PS-3. The reason is discharge of domestic waste, sewerage, and septic waste. At the same time, metals are generated from garages, vehicle battery maintenance shops, and other adjacent industries.

To assess and evaluate the combined effects of metals, *MI* was used. The *MI* ranged from 7.2–23 with a median of 13 (Figure 2b). The median exceeded the maximum limit of 1 ($MI > 1$ is a threshold of warning). All sampling locations were clearly above the maximum limit of >6.0 (class-VI, seriously affected). Highest *MI* was found at PS-11 (Table 2).

The source of metals may be anthropogenic or geogenic. Anthropogenic sources of metal contamination are probably due to industrial wastewater from workshops and garages, vehicle batteries, paints and pigment, fishing boats, fuel stations, and agrochemicals.

3.3. Spatial Distribution of Water Quality Indices

GIS analysis was applied to improve the understanding of the spatial distribution of the different water quality indices (*WQI*, *CPI*, and *MI*) (Table 2). The QGIS (version 2.18.2) was used for this purpose [52]. The distribution maps, thus, show the concentration variation of water quality parameters within the region. Lower values of quality indices are represented by light red color while deep red indicates a polluted area (Figures 3–5).

Table 2. Water quality index (*WQI*), Comprehensive Pollution Index (*CPI*), Metal Index (*MI*), and pollution status.

Sample ID	*WQI*	*CPI*	*MI*	Pollution Level
PS-1	466.4	10.80	8.93	High
PS-2	746.6	13.72	14.30	High
PS-3	638.8	23.13	11.23	High
PS-4	466.1	7.62	8.37	High
PS-5	744.8	12.73	14.00	High
PS-6	440.2	9.65	8.50	High
PS-7	826.5	16.58	14.77	High
PS-8	750.6	14.46	13.67	High
PS-9	391.3	9.64	7.23	High
PS-10	399.9	15.99	9.33	High
PS-11	1336.1	17.24	23.27	High
PS-12	875.6	12.09	16.33	High
PS-13	1008.1	13.96	18.40	High
PS-14	447.9	6.71	8.20	High
PS-15	706.1	11.82	13.00	High
PS-16	967.9	14.39	17.43	High
PS-17	807.5	10.82	15.20	High
PS-18	808.9	12.13	15.20	High
PS-19	582.9	10.05	10.90	High
PS-20	641.9	11.65	11.83	High
Average	702.7	12.76	13.0	High

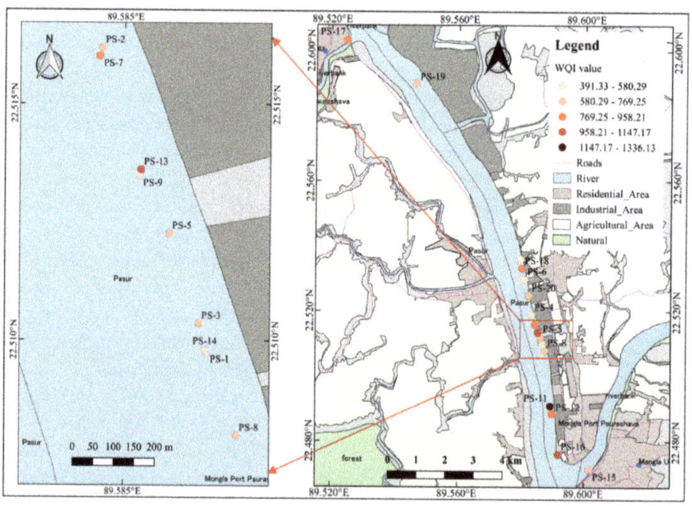

Figure 3. Spatial distribution of Water Quality Index (*WQI*).

The spatial variation of *WQI* is shown in Figure 3. The *WQI* ranged from 391 to 1336, calculated using pH, TDS, chloride, total hardness, TSS, Fe, and Mn. It can be concluded that the southeast part of the study area had the highest values. The sampling points were close to the Mongla Seaport associated with intense urban activities, including trunk roads with many landfills, garbage dumps, and municipal waste. Probably these are the primary pollution sources responsible for increasing the pollution and hence *WQI*. The lowest *WQI* was at the center of the area and decreasing when moving from south to north. High *WQI* is probably related to domestic waste, effluents from industry and nearby local markets, and agricultural runoff. Tidal processes and rainy period stormwater runoff are likely important for the transport characteristics of *WQI* pollutants. Runoff can also wash away topsoil particles and contribute to riverbank erosion. As the flow rate increases, resuspension can mobilize bottom sediments, further raising TSS concentrations.

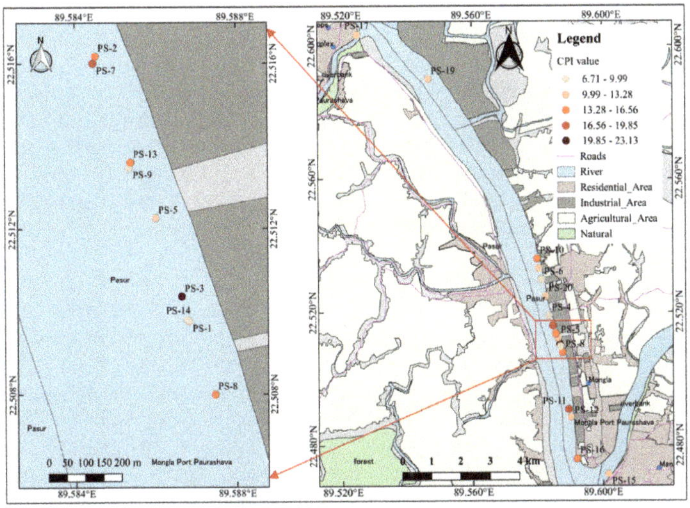

Figure 4. Spatial distribution of Comprehensive Pollution Index (*CPI*).

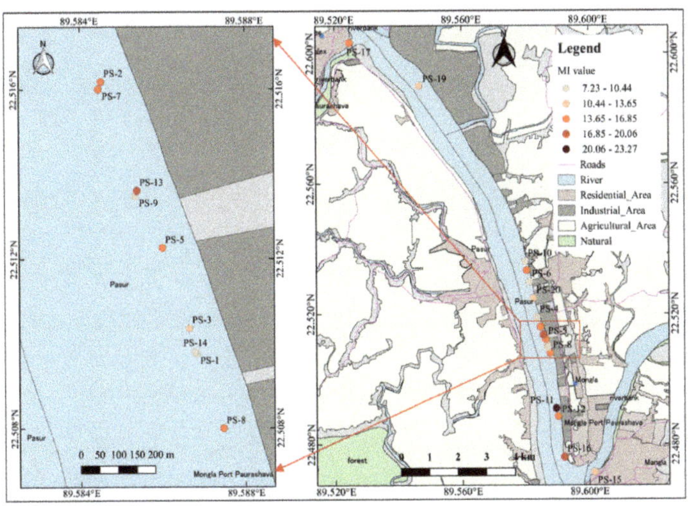

Figure 5. Spatial distribution of Metal Index (*MI*).

The spatial distribution of *CPI* is shown in Figure 4. The trend of *CPI* distribution is somewhat similar to the distribution for *WQI*. The *CPI* gradually decreases from south to north in the study area. Highest *CPI* was found in the south and south-eastern parts, whereas the north region is associated with low *CPI* (Figure 4). The sites with high pollution load are densely populated urban areas including a trunk road, bazaars, shops, car washing garages, and many small and heavy industries situated beside the bank of Pasur River. The pollution from industrial effluents, urban and agricultural waste, municipal and household waste in some rivers in Bangladesh has reached alarming levels. High water temperature and hardness increase the heavy metal toxicity.

The spatial variation of *MI* is shown in Figure 5. The spatial distribution indicates a trend with increasing concentrations from north to south. The southern study area thus indicates higher pollution pressure (Figure 5). The southern region is close to Mongla Seaport and urban areas with heavy industries. The highest value of *MI* was found in sampling location PS-11, which is closely associated with the Mongla seaport area. Many ships gather here for extensive loading and unloading activities. This may be the probable reason for higher metal pollution at this site. The *MI* index exceeded the standard limit ($MI > 1$) for all sampling points. Thus, it can be concluded that the entire study area is seriously threatened by metal pollution. Rivers transport large amounts of domestic sewage, industrial wastewater, and seasonal runoff from agricultural fields. The high *MI* is probably due to industrial, seaport, and construction activities, fishing boats, and domestic waste.

3.4. Multivariate Analysis

3.4.1. Pearson's Correlation Matrix

Pearson correlation was used to study relationships between surface water contaminants with significance levels at $p < 0.01$ and $p < 0.05$ [53–55]. Table 3 presents the Pearson correlation for physicochemical parameters. Parameters that correlated positively with one another included Mn with Fe ($r = 0.757$), $p < 0.01$); total hardness (TH) with alkalinity ($r = 0.739$, $p < 0.01$) and TDS ($r = 0.992$, $p < 0.01$), and TDS with alkalinity ($r = 0.735$, $p < 0.01$). TSS was moderately correlated with Fe ($r = 0.483$) and chloride was moderately related to Fe ($r = 0.641$, $p < 0.05$). Higher correlation between variables may indicate common sources, mutual dependence, and similar or nearly identical metal accumulation properties in surface water.

Table 3. Pearson's correlation matrix.

Parameter	Temp.	pH	TH	TDS	TSS	Chloride	Alkalinity	Fe	Mn
Temp.	1								
pH	0.344	1							
TH	0.104	0.292	1						
TDS	0.137	0.365	0.992 **	1					
TSS	−0.259	−0.097	−0.022	−0.021	1				
Chloride	0.068	0.328	0.32	0.287	0.165	1			
Alkalinity	0.015	−0.062	0.739 **	0.735 **	−0.159	−0.123	1		
Fe	0.127	0.2	0.133	0.146	0.483 *	0.391	−0.205	1	
Mn	0.239	0.391	0.226	0.236	0.17	0.641 **	−0.227	0.757 **	1

** Correlation is significant at the 0.01 level (2-tailed). * Correlation is significant at the 0.05 level (2-tailed).

The strong correlation between total hardness with alkalinity indicates a common source of contamination. Hardness is mainly caused by calcium and magnesium salts. These salts are dissolved from geologic deposits through which water travels. Most alkalinity in surface water comes from calcium carbonate, $CaCO_3$, being leached from rocks and soil. The anthropogenic sources for both are industrial effluents, municipal wastewater discharge, or excessive application of lime to the soil in agricultural areas. The strong correlation between TDS and alkalinity indicates a similar source. TDS in surface water may come from agricultural and residential (urban) runoff, leaching of soil contamination, and point source water pollution discharge from industrial or sewage treatment plants.

Fe and Mn were strongly correlated. Both have a similar geogenic source. Carbonates of Fe and Mn are isomorphous with each other, hence a possible cause of their association, such as is seen in almost all manganiferous spathic iron ores, whether these ores are formed by direct precipitation or by replacement of carbonate of lime. The oxidation of such a mixture would give a combined iron and manganese ore of the common form. Common anthropogenic sources of Fe and Mn are industrial effluents and local urban wastewater discharge. TSS did not show a strong correlation with any variable. Thus, the source is probably different as compared to other pollutants.

3.4.2. Principal Component Analyses

The sources of pollutants were further investigated using PCA and HCA. The results of the PCA are shown in Table 4. The total number of components (common factors) in the PCA was determined based on the Kaiser criterion [56]. Under this criterion, the only component with eigenvalues ≥ 1 should be accepted as a possible source of variance in the data, with the highest priority ascribed to components with the highest eigenvector sum. Scree plots were used to identify the number of PCs to be retained to comprehend the underlying data structure [57]. This indicated that the first three components capture the most significant variation in the data. Thus, three PCs were extracted. The eigenvalues for these PCs ranged from 1.16 to 3.29, explaining 76.1% of the total variance. PC 1, PC 2, and PC 3 explained 36.6, 26.6, and 12.8% of the total variance, respectively (Table 4).

Table 4. Results of principal component analysis.

	PC 1	PC 2	PC 3
Temperature	0.57	0.10	0.24
pH	0.48	0.10	−0.70
TH	0.81	−0.54	0.06
TDS	0.81	−0.54	0.02
TSS	0.14	0.43	0.63
Chloride	0.63	0.38	−0.22
Alkalinity	0.38	−0.83	0.25
Fe	0.59	0.63	0.30
Mn	0.70	0.59	−0.12
Eigen values	3.29	2.40	1.16
% of variance	36.59	26.65	12.84
Cumulative %	36.59	63.23	76.07

PC 1, accounting for 36.6% of the total variance, had positive loadings for all parameters but especially high loading for total hardness, TDS, Mn, and chloride (r = 0.63–0.81), and moderately associated with pH, alkalinity, temperature, and Fe. The Pearson correlation matrix showed that total hardness was strongly correlated with TDS. These pollutants have similar geogenic and anthropogenic sources such as untreated industrial effluents, agricultural runoff, municipal waste, and landfills from nearby urban areas. PC 2 was strongly associated with Fe, Mn, and TSS also confirmed from the correlation matrix. The strong association between Fe and Mn indicates that similar sources are at hand. PC 3 had high association with TSS and pH (Table 4).

3.4.3. Hierarchical Cluster Analysis (HCA)

The HCA was based on Ward's method with squared Euclidean distance [58,59] and performed on standardized data based on the three PC scores outlined above. The 20 samples were classified into four distinct groups, clusters A, B, C, and D. The average concentration of each cluster group is shown in Table 5. The table shows that cluster A was not strongly related with any parameters. Cluster B was related to high TSS, Fe, and Mn. Cluster B contained higher indices values (*WQI*, *CPI*, and *MI*) than cluster A and D. Cluster C and D, in which only one sample was classified of each. High TSS, pH, chloride, Fe, and Mn were classified in cluster C. Cluster C was ranked as the most polluted area among the four clusters with respect to *WQI*, *CPI*, and *MI*. Cluster D was related to pH, TDS, and chloride. This cluster's *WQI*, *CPI*, and *MI* values were lower than cluster B and C but higher than cluster A. The decreasing order of indices was cluster C > cluster B > cluster D > cluster A (Table 5). Thus, cluster C was most polluted, and cluster A the least.

Table 5. Average concentration of water quality parameters in each cluster group.

	Cluster A	Cluster B	Cluster C	Cluster D
No. of samples	5	13	1	1
Temperature (°C)	26.2	29.34	29.67	31.0
pH	8.71	8.71	8.97	8.8
TH (mg/L)	67.92	109.74	173.47	472.64
TDS (mg/L)	140.56	224.19	326.18	892.27
TSS (mg/L)	516.65	813.36	926.84	652.45
Chloride (mg/L)	208.49	212.10	708.93	368.87
Alkalinity (mg/L)	92.66	94.5	67.67	165.33
Fe (mg/L)	1.26	2.11	2.75	1.72
Mn (mg/L)	0.40	0.71	1.41	0.61
WQI	442.4	758.8	1336.1	641.9
CPI	8.88	14.0	17.2	11.7
MI	8.25	14.14	23.3	11.8

The scatter plot of the 20 samples described by principal components (A: PCs 1 and 2; B: PCs 1 and 3; C: PCs 2 and 3) and classified into four clusters is shown in Figure 6. If a PC score is greater than 0, the water quality parameter characteristics influence the component at the site. Conversely, if a PC score is less than 0, it means that the component was not significantly affected by the water chemistry at the site [60,61]. In PC 1, cluster C and D are separated from cluster A, and B. PC 1 has smaller scores for cluster A indicating that it had less concentration of the quality parameters than cluster B, C, and D. Thus, it can be confirmed that samples of cluster A were less polluted than others. Cluster B contained higher PC 1 scores than cluster A. In addition, due to positive PC 2 scores of cluster B compared to cluster A, samples of this cluster had a higher concentration of water quality parameters than cluster A. The higher concentrations are due to geogenic and anthropogenic sources such as untreated industrial wastewater, domestic and municipal wastewater discharge from the urban areas, and agricultural runoff. Both cluster C and the majority sample of cluster B showed a positive score for PC 1 and 2, indicating that both clusters contained high concentration pollutants from industrial, agricultural, and urban areas. This is also confirmed from the average *WQI*, *CPI*, and *MI* value of Table 5. Both cluster C and cluster B showed positive scores for PCs 1 and 2, but cluster C was more affected than cluster B (Figure 5). PC 3 revealed that cluster B is significantly more influenced than cluster D. Thus, the sample of cluster B was more affected and polluted than cluster D.

The spatial distribution of each cluster can be observed in Figure 7. Cluster C is generally located nearer the Sundarban mangrove forest in the west part of the study area and the southeast region in the urban area. Cluster C is the study area's most contaminated sampling point containing the highest metal contamination and water quality indices (*WQI*, *CPI*, *MI*). The location of cluster C is associated with Mongla Seaport. Many ships and vessels are taking part in loading and unloading activities for different types of goods. Sometimes they discharge waste, including used oil and oily substances, coal, asbestos, and chemicals directly into the river. The high TSS may come from local bazaars in urban areas and municipal wastewater and landfills.

Cluster B is located south to north at different study area sites (Figure 7). The samples of this cluster contained high metal concentrations and high water quality indices (*WQI*, *CPI*, *MI*), indicating a metal contamination area. The contamination of this area is higher than for cluster D and cluster A. Sample of cluster D is in the southeast region and near the residential area. The sample of these points contained high calcium and magnesium carbonates, bicarbonates and sulfates, and combined content of all inorganic and organic substances present in a liquid in molecular, ionized, or micro-granular (colloidal sol) suspended form. For this reason, it contains the highest value of total hardness and TDS.

The source of these pollutants is agricultural runoff and residential (urban) runoff. This area had high metal contents and TSS.

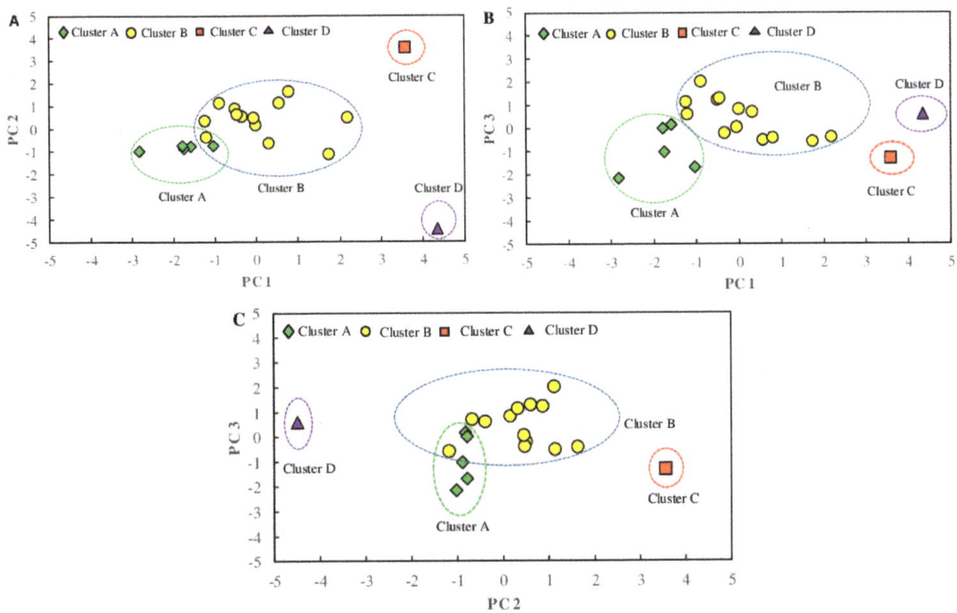

Figure 6. Scatter plots for two principal components with respect to clusters; (**A**) PCs 1 and 2, (**B**) PCs 1 and 3, and (**C**) PCs 2 and 3.

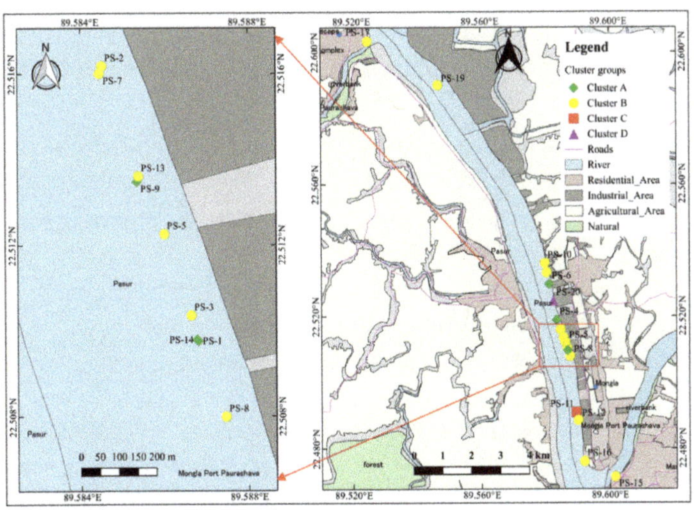

Figure 7. Distribution of each cluster in the study area.

All cluster A samples are in the southeast area but far from cluster C. Samples of cluster A contain the lowest value of metals and other physicochemical parameters. Thus, these samples were less polluted than other clusters. To conclude, this study shows that the pollution level of the study area is divided into four main areas.

4. Conclusions

Rivers are essential sources of water supply for humans and environment with critical conditions for water quality. This study concluded that iron, manganese, total suspended solids, chloride at some locations, and pH had much higher concentrations in the river water than recommended by WHO and Bangladesh Environmental standards (DoE) and thus are not safe for household use or aquatic ecosystems. We used water quality indices, spatial distribution, and multivariate statistical analyses to evaluate the pollution level and source determination. Multivariate analysis was used to improve the knowledge regarding the source of pollution. Based on the PCA results, four distinct groups were obtained by HCA. The concentration of different water quality parameters was different regarding land use and the importance of the three PC scores. The cluster groups revealed that the sample of cluster C was most polluted, and the samples of cluster A were the least polluted in the study area. The severe enrichment of pollutants in this area is primarily due to anthropogenic sources related to industrial, agricultural, urbanization, and fishing activities.

Mongla Seaport Authority, Mongla Export Processing Authority, Mongla Municipal Corporation, District Administration, and the Department of Environment can take the initiative to protect the river water from pollution and untreated industrial and municipal waste. Monitoring of coastal activities is essential to save the coastal ecosystems. This monitoring system can give policymakers and stakeholders an interest in the coastal environment and resources. A systematic and periodic inspection of each industry located beside the river should be performed before certificates of compliances are issued by the Department of Environment (DoE), Bangladesh. Short- and long-term scientific studies should be immediately started to assess the impacts of industrial activities on coastal water, soil, and fishery resources, as well as human health. Thus, this study recommends that continuous monitoring of the pollution level of the Pasur River as well as adjoining agricultural areas should be assessed regarding the risk for human health and ecosystems in the vicinity of the river. To avoid and alleviate environmental contamination, effective approaches for extracting harmful heavy metals from sewage and industrial effluents are urgently needed before the effluent is released into the environment. As well, traditional treatment techniques are necessary for water used for domestic purposes. Finally, public awareness of the impacts and remedies of pollution should be raised so that they can play a key role in pollution reduction.

Author Contributions: Conceptualization: M.S.I. and M.A.G.; Methodology: M.S.I. and M.A.G.; Formal analysis and investigation: M.S.I., M.A.-A.-M. and A.S.K.; Visualization; M.S.I. and K.N.; Writing—original draft preparation: M.S.I. and K.N.; Writing—review and editing: K.N., R.B., and M.S.I.; Resources: M.S.I.; Supervision: K.N. All authors have read and agreed to the published version of the manuscript.

Funding: This research received no external funding.

Institutional Review Board Statement: Not applicable.

Informed Consent Statement: Not applicable.

Data Availability Statement: The data presented in this study are available on request from the corresponding author.

Acknowledgments: The authors wish to express their sincere thanks to Department of Chemistry at "Jashore University of Science and Technology" for the laboratory support to complete this research and the "Asian Arsenic Network" laboratories for their generous cooperation in analyzing samples for elemental analysis by AAS.

Conflicts of Interest: The authors declare that they have no competing financial interest or personal relationships that could have appeared to influence the work reported in this paper.

References

1. Yeh, G.; Lin, C.; Nguyen, D.H.; Hoang, H.G.; Shern, J.C.; Hsiao, P.J. A five-year investigation of water quality and heavy metal mass flux of an industrially affected river. *Environ. Sci. Pollut. Res.* **2021**, *29*, 12465–12472. [CrossRef]
2. Proshad, R.; Zhang, D.; Idris, A.M.; Islam, M.S.; Kormoker, T.; Sarker, M.N.I.; Khadka, S.; Sayeed, A.; Islam, M. Comprehensive evaluation of chemical properties and toxic metals in the surface water of Louhajang River, Bangladesh. *Environ. Sci. Pollut. Res.* **2021**, *28*, 49191–49205. [CrossRef] [PubMed]
3. Pichtel, J. Oil and Gas Production Wastewater: Soil Contamination and Pollution Prevention. *Appl. Environ. Soil Sci.* **2016**, *2016*, 2707989. [CrossRef]
4. Pavlidou, A.; Anastasopoulou, E.; Dassenakis, M.; Hatzianestis, I.; Paraskevopoulou, V.; Simboura, N.; Rousselak, E.; Drakopoulou, P. Effects of olive oil wastes on river basins and an oligotrophic coastal marine ecosystem: A case study in Greece. *Sci. Total Environ.* **2014**, *497–498*, 38–49. [CrossRef] [PubMed]
5. Smedley, P.L.; Kinniburgh, D.G. A review of the source, behaviour and distribution of arsenic in natural waters. *Appl. Geochem.* **2002**, *17*, 517–568. [CrossRef]
6. Apollaro, C.; Di Curzio, D.; Fuoco, I.; Buccianti, A.; Dinelli, E.; Vespasiano, G.; Castrignano, A.; Rusi, S.; Barca, D.; Figoli, A.; et al. A multivariate non-parametric approach for estimating probability of exceeding the local natural background level of arsenic in the aquifers of Calabria region (Southern Italy). *Sci. Total Environ.* **2022**, *806*, 150345. [CrossRef]
7. Lin, Y.; Gritsenko, D.; Feng, S.; Teh, Y.C.; Lu, X.; Xu, J. Detection of heavy metal by paper-based microfluidics. *Biosens. Bioelectron.* **2016**, *83*, 256–266. [CrossRef]
8. Aradpour, S.; Noori, R.; Naseh, M.R.V.; Hosseinzadeh, M.; Safavi, S.; Rozegar, F.G.; Maghrebi, M. Alarming carcinogenic and non-carcinogenic risk of heavy metals in Sabalan dam reservoir, Northwest of Iran. *Environ. Pollut. Bioavailab.* **2021**, *33*, 278–291. [CrossRef]
9. Sorlini, S.; Palazzini, D.; Sieliechi, J.M.; Ngassoum, M.B. Assessment of Physical-Chemical Drinking Water Quality in the Logone Valley (Chad-Cameroon). *Sustainability* **2013**, *5*, 3060–3076. [CrossRef]
10. Mahmoud, E.K.; Ghoneim, A.M. Effect of polluted water on soil and plant contamination by heavy metals in El-Mahla El-Kobra, Egypt. *Solid Earth* **2016**, *7*, 703–711. [CrossRef]
11. Ramachandra, T.V.; Bhat, S.P.; Mahapatra, D.M.; Krishnadas, G. Impact of indiscriminate disposal of untreated effluents from thermal power plant on water resources. *Indian J. Environ. Prot.* **2012**, *32*, 705–718. Available online: http://wgbis.ces.iisc.ernet.in/energy/water/paper/ijep_water_resources/ijep_water_resources.pdf (accessed on 7 October 2021).
12. Khan, M.H.; Nafees, M.; Muhammad, N.; Ullah, U.; Hussain, R.; Bilal, M. Assessment of Drinking Water Sources for Water Quality, Human Health Risks, and Pollution Sources: A Case Study of the District Bajaur, Pakistan. *Arch. Environ. Contam. Toxicol.* **2021**, *80*, 41–54. [CrossRef] [PubMed]
13. Zereen, F.; Islam, F.; Habib, M.A.; Begum, D.A.; Zaman, M.S. Inorganic pollutants in the Padma River, Bangladesh. *Environ. Geol.* **2011**, *39*, 1059–1062. [CrossRef]
14. Hossain, F.; Islam, M.A.; Al-Mamun, A.; Naher, K.; Khan, R.; Das, S.; Tamim, U.; Hossain, S.M.; Nahid, F.; Islam, M.A. Assessment of Trace Contaminants in Sediments of the Poshur River Nearby Mongla Port of Bangladesh. *Nucl. Sci. Appl.* **2016**, *25*, 7–11. Available online: http://baec.portal.gov.bd/sites/default/files/files/baec.portal.gov.bd/page/1f00cd0e_737d_4e2e_ab9f_08183800b7a2/2=2504-F.pdf (accessed on 15 October 2021).
15. *Environment Review Report*; Mongla Economic Zone, Bangladesh Export Zone Authority: Mongla, Bagerhat, 2014; pp. 1–33. Available online: https://www.beza.gov.bd/wp-content/uploads/2015/10/ERR-of-Mongla-EZ.pdf (accessed on 13 June 2021).
16. Chowdhury, R.M.; Muntasir, S.Y.; Hossain, M.M. Water quality index of water bodies along Faridpur-Barisal Road in Bangladesh. *Glob. Eng. Technol. Rev.* **2012**, *2*, 1–8. Available online: https://citeseerx.ist.psu.edu/viewdoc/download?doi=10.1.1.681.1779&rep=rep1&type=pdf (accessed on 13 June 2021).
17. Tabrez, S.; Zughaibi, T.A.; Javed, M. Water quality index, Labeo rohita, and Eichhornia crassipes: Suitable bio-indicators of river water pollution. *Saudi. J. Biol. Sci.* **2022**, *29*, 75–82. [CrossRef]
18. Yan, C.A.; Zhang, W.; Zhang, Z.; Liu, Y.; Deng, C.; Nie, N. Assessment of Water Quality and Identification of Polluted Risky Regions Based on Field Observations & GIS in the Honghe River Watershed, China. *PLoS ONE* **2015**, *10*, e0119130. [CrossRef]
19. Xiong, B.; Li, R.; Johnson, D.; Luo, Y.; Xi, Y.; Ren, D.; Huang, Y. Spatial distribution, risk assessment, and source identification of heavy metals in water from the Xiangxi River, Three Gorges Reservoir Region, China. *Environ. Geochem. Health* **2021**, *43*, 915–930. [CrossRef]
20. Bakan, G.; Ozkoc, H.B.; Tulek, S.; Cuce, H. Integrated environmental quality assessment of Kızılırmak River and its coastal environment. *Turk. J. Fish. Aquat. Sci.* **2010**, *10*, 453–462. [CrossRef]
21. Goher, M.E.; Hassan, A.M.; Abdel-Moniem, I.A.; Fahmy, A.H.; El-sayed, S.M. Evaluation of surface water quality and heavy metal indices of Ismailia Canal, Nile River, Egypt. *Egypt. J. Aquat. Res.* **2014**, *40*, 225–233. [CrossRef]
22. Ali, M.M.; Ali, M.L.; Islam, M.S.; Rahman, M.Z. Assessment of toxic metals in water and sediment of Pasur River in Bangladesh. *Water Sci. Technol.* **2018**, *77*, 1418–1430. [CrossRef] [PubMed]
23. Algul, F.; Beyhan, M. Concentrations and sources of heavy metals in shallow sediments in Lake Bafa, Turkey. *Sci. Rep.* **2020**, *10*, 11782. [CrossRef] [PubMed]

24. Krishna, A.K.; Satyanarayanan, M.; Govil, P.K. Assessment of heavy metal pollution in water using multivariate statistical techniques in an industrial area: A case study from Patancheru, Medak District, Andhra Pradesh, India. *J. Hazard. Mater.* **2009**, *167*, 366–373. [CrossRef] [PubMed]
25. Mongla Municipality in Khulna Division. 2011. Available online: https://www.citypopulation.de/en/bangladesh/khulna/admin/bagerhat/015899 (accessed on 20 June 2021).
26. Persits, F.M.; Wandrey, C.J.; Milici, R.C.; Manwar, A. Digital geologic and geophysical data of Bangladesh. *U.S. Geol. Surv. Open-File Rep.* **2001**. [CrossRef]
27. Ling, T.Y.; Gerunsin, N.; Soo, C.L.; Nyanti, L.; Sim, S.F.; Grinang, J. Seasonal Changes and Spatial Variation in Water Quality of a Large Young Tropical Reservoir and Its Downstream River. *J. Chem.* **2017**, *2017*, 8153246. [CrossRef]
28. APHA. *Standard Methods for the Examination of Water and Wastewater*, 23rd ed.; American Public Health Association: Washington, DC, USA, 2017; pp. 1–1546. Available online: http://dl.mozh.org/upload/StandardMetods23RD.pdf (accessed on 1 April 2022).
29. Bhuyan, M.S.; Bakar, M.A.; Nabi, M.R.U.; Senapathi, V.; Chung, S.Y.; Islam, M.S. Monitoring and assessment of heavy metal contamination in surface water and sediment of the Old Brahmaputra River, Bangladesh. *Appl. Water Sci.* **2019**, *9*, 125. [CrossRef]
30. Ali, M.M.; Ali, M.L.; Islam, M.S.; Rahman, M.Z. Preliminary assessment of heavy metals in water and sediment of Karnaphuli River, Bangladesh. *Environ. Nanotechnol. Monit. Manag.* **2016**, *5*, 27–35. [CrossRef]
31. Molekoa, M.D.; Avtar, R.; Kumar, P.; Thu Minh, H.V.; Dasgupta, R.; Johnson, B.A.; Sahu, N.; Verma, R.L.; Yunus, A.P. Spatio-Temporal Analysis Spatio-Temporal Analysis of Surface Water Quality in Mokopane Area, Limpopo, South Africa. *Water* **2021**, *13*, 220. [CrossRef]
32. Noori, R.; Berndtsson, R.; Hosseinzadeh, M.; Adamowski, J.F.; Abyaneh, M.R. A critical review on the application of the National Sanitation Foundation Water Quality Index. *Environ. Pollut.* **2019**, *244*, 575–587. [CrossRef]
33. Horton, R.K. An index number system for rating water quality. *J. Water Pollut. Control Fed.* **1965**, *373*, 303–306.
34. Brown, R.M.; McClelland, N.I.; Deininger, R.A.; O'Connor, M.F. A Water Quality Index—Crashing the Psychological Barrier. In *Indicators of Environmental Quality. Environmental Science Research*; Thomas, W.A., Ed.; Springer: Boston, MA, USA, 1972; Volume 1. [CrossRef]
35. Howladar, M.F.; Hossain, M.N.; Anju, K.A.; Das, D. Ecological and health risk assessment of trace metals in water collected from Haripur gas blowout area of Bangladesh. *Sci. Rep.* **2021**, *11*, 15573. [CrossRef] [PubMed]
36. Wu, H.; Yang, W.; Yao, R.; Zhao, Y.; Zhao, Y.; Zhao, Y.; Zhang, Y.; Yuan, Q.; Lin, A. Evaluating surface water quality using water quality index in Beiyun River, China. *Environ. Sci. Pollut. Res.* **2021**, *27*, 35449–35458. [CrossRef] [PubMed]
37. Bi, S.; Wang, L.; Li, Y.; Zhang, Z.; Wang, Z.; Ding, X.; Zhou, J. A Comprehensive Method for Water Environment Assessment considering Trends of Water Quality. *Adv. Civ. Eng.* **2021**, *2021*, 5548113. [CrossRef]
38. Kumar, A.; Mishra, S.; Taxak, A.K.; Pandey, R.; Yu, Z.G. Nature rejuvenation: Long term (1989–2016) vs short-term memory approach-based appraisal of water quality of the upper part of Ganga River, India. *Environ. Technol. Innov.* **2020**, *20*, 101164. [CrossRef]
39. Mekuria, D.M.; Kassegne, A.B.; Asfaw, S.L. Assessing pollution profiles along Little Akaki River receiving municipal and industrial wastewaters, Central Ethiopia: Implications for environmental and public health safety. *Heliyon* **2021**, *7*, e07526. [CrossRef]
40. Son, C.T.; Giang, N.T.; Thao, T.P.; Nui, N.H.; Lam, N.T.; Cong, V.H. Assessment of Cau river water quality assessment using a combination of water quality and pollution indices. *J. Water Supply Res. Technol.* **2020**, *69*, 160–172. [CrossRef]
41. Ojekunle, O.Z.; Ojekunle, O.V.; Adeyemi, A.A.; Taiwo, A.G.; Sangowusi, O.R.; Taiwo, A.M.; Adekitan, A.A. Evaluation of surface water quality indices and ecological risk assessment for heavy metals in scrap yard neighborhood. *Springerplus* **2016**, *5*, 560. [CrossRef]
42. Piroozfar, P.; Alipour, S.; Modabberi, S.; Cohen, D. Using multivariate statistical analysis in assessment of surface water quality and identification of heavy metal pollution sources in Sarough watershed, NW of Iran. *Environ. Monit. Assess.* **2021**, *193*, 564. [CrossRef]
43. Withanachchi, S.S.; Ghambashidze, G.; Kunchulia, I.; Urushadze, T.; Ploeger, A. Water Quality in Surface Water: A Preliminary Assessment of Heavy Metal Contamination of the Mashavera River, Georgia. *Int. J. Environ. Res. Public Health* **2018**, *15*, 621. [CrossRef]
44. WHO. Guidelines for Drinking-Water Quality. 1984. Available online: https://apps.who.int/iris/bitstream/handle/10665/252072/9241541687eng.pdf?sequence=1&isAllowed=y (accessed on 30 December 2021).
45. WHO. Guidelines for Drinking-Water Quality, 4th ed. 2011. Available online: http://apps.who.int/iris/bitstream/handle/10665/44584/9789241548151_eng.pdf (accessed on 2 January 2022).
46. Department of Environment, Peoples' Republic of Bangladesh. Environmental Conservation Rules. 1997. Available online: https://www.elaw.org/system/files/Bangladesh+--+Environmental+Conservation+Rules,+1997.pdf (accessed on 30 December 2021).
47. Fatema, K.; Begum, M.; Zahid, M.A.; Hossain, M.E. Water quality assessment of the river Buriganga, Bangladesh. *J. Biodivers. Conserv. Bioresour. Manag.* **2018**, *4*, 47–53. [CrossRef]
48. U.S. Department of Health and Human Services, Public Health Service, Agency for Toxic Substances and Disease Registry. Toxicological Profile for Manganese. 2012. Available online: https://www.atsdr.cdc.gov/toxprofiles/tp151.pdf (accessed on 18 July 2021).

49. Islam, M.S.; Nakagawa, K.; Mamun, M.A.A.; Siddique, M.A.B.; Berndtsson, R. Is road-side fishpond water in Bangladesh safe for human use? An assessment using water quality indices. *Environ. Chall.* **2022**, *6*, 100434. [CrossRef]
50. Aradpour, S.; Noori, R.; Tang, Q.; Bhattarai, R.; Hooshyaripor, F.; Hosseinzadeh, M.; Haghighi, A.T.; Klove, B. Metal contamination assessment in water column and surface sediments of a warm monomictic man-made lake: Sabalan Dam Reservoir, Iran. *Hydrol. Res.* **2020**, *51*, 799–814. [CrossRef]
51. Peluso, J.; Coll, C.S.P.; Cristos, D.; Rojas, D.E.; Aronzon, C.M. Comprehensive assessment of water quality through different approaches: Physicochemical and ecotoxicological parameters. *Sci. Total. Environ.* **2021**, *800*, 149510. [CrossRef] [PubMed]
52. Nakagawa, K.; Amano, H.; Berndtsson, R.; Takao, Y.; Hosono, T. Use of sterols to monitor surface water quality change and nitrate pollution source. *Ecol. Indic.* **2019**, *107*, 105534. [CrossRef]
53. El Sorogy, A.S.; Youssef, M. Pollution assessment of the Red Sea Gulf of Aqaba seawater, northwest Saudi Arabia. *Environ. Monit. Assess.* **2021**, *193*, 141. [CrossRef]
54. Jahan, S.; Strezov, V. Water quality assessment of Australian ports using water quality evaluation indices. *PLoS ONE* **2017**, *12*, e0189284. [CrossRef]
55. Huang, Z.; Zheng, S.; Liu, Y.; Zhao, X.; Qiao, X.; Liu, C.; Zheng, B.; Yin, D. Distribution, toxicity load, and risk assessment of dissolved metal in surface and overlying water at the Xiangjiang River in southern China. *Sci. Rep.* **2021**, *11*, 109. [CrossRef]
56. Nakagawa, K.; Amano, H.; Asakura, H.; Berndtsson, R. Spatial trends of nitrate pollution and groundwater chemistry in Shimabara, Nagasaki, Japan. *Environ. Earth Sci.* **2016**, *75*, 234. [CrossRef]
57. Hossain, M.B.; Shanta, T.B.; Ahmed, A.S.S.; Hossain, M.K.; Semme, S.A. Baseline study of heavy metal contamination in the Sangu River estuary, Chattogram, Bangladesh. *Mar. Pollut. Bull.* **2019**, *140*, 255–261. [CrossRef]
58. Khan, M.H.R.; Liu, J.; Liu, S.; Li, J.; Cao, L.; Rahman, A. Anthropogenic effect on heavy metal contents in surface sediments of the Bengal Basin river system, Bangladesh. *Environ. Sci. Pollut. Res.* **2020**, *27*, 19688–19702. [CrossRef]
59. Sharma, G.; Lata, R.; Thakur, N.; Bajala, V.; Kuniyal, J.C.; Kumar, K. Application of multivariate statistical analysis and water quality index for quality characterization of Parbati River, Northwestern Himalaya, India. *Discov. Water* **2021**, *1*, 5. [CrossRef]
60. Amano, H.; Nakagawa, K.; Berndtsson, R. Surface water chemistry and nitrate pollution in Shimabara, Nagasaki, Japan. *Environ. Earth Sci.* **2018**, *77*, 354. [CrossRef]
61. Yu, Z.Q.; Amano, H.; Nakagawa, K.; Berndtsson, R. Hydrogeochemical evolution of groundwater in a Quaternary sediment and Cretaceous sandstone unconfined aquifer in Northwestern China. *Environ. Earth Sci.* **2018**, *77*, 629. [CrossRef]

water

MDPI

Article

Topographical Characteristics of Frequent Urban Pluvial Flooding Areas in Osaka and Nagoya Cities, Japan

Daisuke Komori [1,*], Kota Nakaguchi [2], Ryosuke Inomata [1], Yuika Oyatsu [1], Ryohei Tachikawa [1] and So Kazama [1]

1 Department of Civil and Environmental Engineering, Tohoku University, Sendai 980-8579, Japan
2 Public-Use Systems Department 1, Kobe Works, Mitsubishi Electric Corporation, Kobe 652-8555, Japan
* Correspondence: daisuke.komori.e8@tohoku.ac.jp; Tel.: +81-22-795-5007

Abstract: Flooding area records have been available since 1993 in Japan; however, there have been no studies that have utilised these records to elucidate urban pluvial flooding formation mechanisms. Therefore, frequent urban pluvial flooding areas using 20 years of urban pluvial flooding area records during 1993–2012 were identified and analysed using the principal component analysis of their topographical characteristics in Osaka and Nagoya Cities, Japan. The results showed that the topographical characteristics of the frequent urban pluvial flooding areas in both cities were different, with particularly conflicting trends in principal component 1. Furthermore, the urban pluvial flooding in Osaka City could not be described solely by topographical characteristics, and the influence of anthropogenic factors such as dominant structures that may influence inundated water flows in and around frequent urban pluvial flooding areas and stormwater drainage improvements on the occurrence of urban pluvial flooding were shown to be influential. In addition, most of the frequent urban pluvial flooding areas in Nagoya City were located on almost no gradient with a slope of less than 1 degree, and thus, the mere presence of dominant structures around it would dam up the inundated water and cause urban pluvial flooding. The results of this study quantitatively showed the paradigm shift of urban pluvial flooding factors from topographical characteristics to anthropogenic characteristics by the statistical analysis of newly defined urban pluvial flooding frequency areas.

Keywords: frequent urban pluvial flooding; topographical characteristics; anthropogenic characteristics; urbanisation; GIS; principal component analysis

Citation: Komori, D.; Nakaguchi, K.; Inomata, R.; Oyatsu, Y.; Tachikawa, R.; Kazama, S. Topographical Characteristics of Frequent Urban Pluvial Flooding Areas in Osaka and Nagoya Cities, Japan. *Water* **2022**, *14*, 2795. https://doi.org/10.3390/w14182795

Academic Editor: Akira Kawamura

Received: 8 August 2022
Accepted: 6 September 2022
Published: 8 September 2022

1. Introduction

Urban flooding has had a strong negative impact on many cities around the world for most of human history and certainly in recent decades [1–3]. More than half of the global population lives in urbanised areas, and the frequency as well as the intensity of hydro-meteorological extremes are on the rise [4,5]. Urban flooding is therefore likely to cause greater losses in the coming decades. For example, urban flooding and associated property damage accounted for 73% of the total damage caused by flooding in the USA from 1960 to 2016, amounting to USD 107.8 billion [6]. Of the total damage caused by flooding in Japan's three largest cities, namely Tokyo's 23 wards, Osaka City, and Nagoya City, from 2006 to 2013, 82% of the total damage was caused by urban pluvial flooding [7]. Urban pluvial flooding is one of the typical urban floodings and flood damage that occurs when rainwater is not discharged to mainstream or tributary rivers because rainfall exceeds the design capacity of a drainage facility. In recent years, severe urban pluvial flooding has been caused by the reduction of rainwater soil penetration amounts because of changes in land use and increased strong downpour occurrence frequency. Furthermore, according to the IPCC assessment report [8], the risk of urban pluvial flooding in urban areas will increase in the future as a result of increased strong downpours due to climate change. Therefore, it is important to clarify the characteristics of past urban pluvial flooding areas in large cities and urbanising cities, where damage from urban pluvial flooding will become more apparent in the future, and take efficient countermeasures against urban pluvial flooding.

Managing urban flood risk is a high priority worldwide, as suggested by a large number of cities, from all continents, taken as case studies in recent research papers dedicated to urban flood modelling [9]. In Japan, it has become mandatory for local authorities to publish urban pluvial flooding hazard maps as a measure against urban pluvial flooding. On the other hand, the spatial and temporal characteristics of flooding in urban areas are complex due to extensive changes to land use [10] that introduce micro-urban features such as buildings, roads, and drainage networks [11,12], and most inundation simulations cannot accurately simulate flooding in urban areas [13]. Thus, flooding simulations using physical models represent the flooding mechanism through known physical analyses. Therefore, if the unknown factors are mainly important for the inundation mechanism, the accuracy will be low. On the other hand, the statistical analytical approach is considered to be able to elucidate the distribution of areas where urban pluvial flooding frequently occurs (hereafter referred to as frequent urban pluvial areas) based on past flooding area records and to quantitatively evaluate their characteristics, including unknown factors that cannot be considered in flooding simulations [14].

A statistical analytical approach has been greatly developed by utilising several appropriate factors, such as bedrock geology, soil properties, land use, drainage networks, road networks, building, and precipitation. Some previous studies have identified flood mechanisms. For example, Fariza et al. (2019) used fuzzy multi-criteria decision making (FMCDM) to assess urban flood risk levels in Sidoarjo, Indonesia [15]. Sato and Hayashi (2014) used principal component analysis (PCA) to analyse the main topographic characteristics of inundation in the Musashino Plateau of Tokyo and Saitama [16]. However, no report of an earlier study describes the quantitative assessment of the universal characteristics of flood-prone areas because they analysed areas where inundation has occurred at least once [14], and most of the earlier studies were limited to case studies of one city or one case study.

Djamres et al. (2021) identified the frequent urban pluvial flooding areas using seven years of urban pluvial flooding area records during 2008–2015 and analysed, using the PCA, their topographical characteristics in Tangerang, Indonesia [14]. Results showed that frequent urban pluvial flooding areas in Tangerang emerged because of a slope in the upstream condition, the correlation between concave and flow length conditions, the correlation of the slope condition and distance to a river, and relationships among flow length in upstream characteristics and distance to a pond. Furthermore, 29% of frequent urban pluvial flooding areas had low topographical similarity because of anthropogenic factors such as changes in overland flow directions due to the change of a slope in the upstream condition by land-use change and trapping of flood water by "dominant structures" that may influence inundated water flows in and around frequent urban pluvial flooding areas. Mignot and Dewals (2022) argued that a particular impediment to progress in urban flood modelling science is that the conclusions of most studies remain genuinely site-specific formulations and that significant progress could be made by attempting to extract general knowledge from the collection of existing case studies [9]. They also suggested that this may be achieved by designing appropriate metrics for classifying and standardizing the definition of flooding scenarios, investigated processes, and effects of analysed factors [9]. As the newly defined urban pluvial flooding frequency areas contain primary factors related to flooding mechanisms, statistical analysis of frequent urban pluvial flooding areas may reveal previously unknown factors such as anthropogenic factors and their relative influence on the flooding mechanism.

In Japan, flooding area records have been available since 1993, and Kakehashi et al. (2014) [17] used these records to identify frequent river flooding areas throughout Japan and to elucidate their formation mechanisms. However, there have been no studies that have utilised these records for urban pluvial flooding. Therefore, the objectives of this study were to identify the frequent urban pluvial flooding area using these records for 20 years during 1993–2012 in Osaka and Nagoya Cities, the two cities with the largest proportion of urban pluvial flooding damage from 2006 to 2013 [7], and to analyse the topographical characteristics of frequent urban pluvial flooding areas and their distribution at both cities by applying methods reported by Djamres et al. (2021) [14]. This study also aimed to elucidate anthropogenic characteristics of urban pluvial flooding from the views of the location of "dominant structures" and the impact of drainage system improvement.

2. Materials and Methods

2.1. Study Area

In this study, two cities, namely Osaka and Nagoya Cities in Japan, were selected as the study area. The location and detailed information of the target cities in Japan are shown in Figure 1 and Table 1. These cities were selected because their urbanised areas cover more than 90% of the city area, while they have differences in urban population, urban area population, and other factors related to the urban scale of the cities, and therefore, they are suitable for this study to compare characteristics of urban pluvial flooding areas due to differences in an urban scale. Furthermore, there were relatively more urban pluvial flooding area records than in other cities, and it was thought that more reliable analysis results could be obtained (in detail, see Section 2.2). In addition, the amount of capital expenditure on sewerage projects in each city was large, and it was considered that cities with larger investments in sewerage projects were more likely to have experienced more urban pluvial flooding in the past (Table 1).

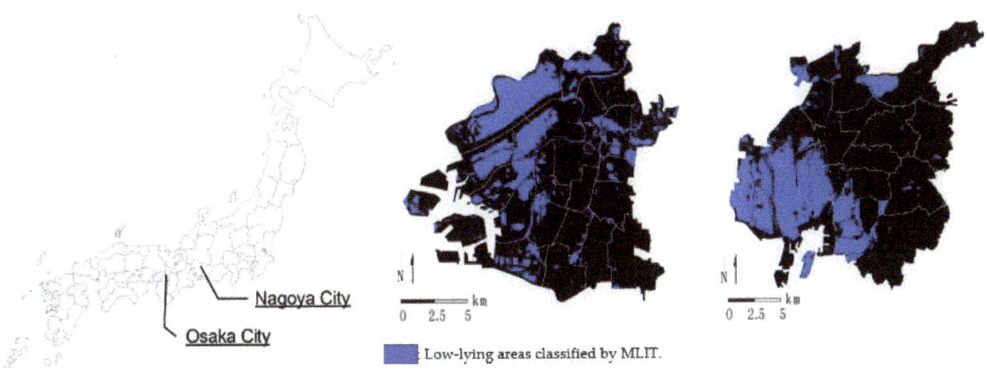

Figure 1. Location of the target cities in Japan and low-lying areas in the target cities. Low-lying areas are classified by the Ministry of Land, Infrastructure, Transport, and Tourism (MLIT) as areas where the elevation is lower than the surrounding [18].

Table 1. Detailed information on Osaka and Nagoya Cities.

	Osaka City	Nagoya City
Population (person)	2,717,000	2,311,000
Population density (person/km^2)	12,000	7000
Urbanised area ratio (%)	93.9	92.6
Gross domestic product (trillion yen)	20.0	14.4
Urban pluvial flooding area records	155 [1]	302 [1]

Note: [1] In detail, see Section 2.2.

The annual maximum one-hour rainfall data as short-time extreme rainfall causing urban pluvial flooding in the study area for 20 years during 1993–2012 in Osaka and Nagoya Cities are shown in Table 2. One-hour precipitation data from JMA AMeDAS stations (Osaka: 34°40.9′ N, 135°31.1′ E; Nagoya: 35°10.0′ N, 136°57.9′ E) were used [19]. The average annual maximum one-hour rainfall was 33.2 mm in Osaka City and 39.4 mm in Nagoya City, respectively. Above-average annual maximum one-hour rainfalls were recorded seven times in Osaka and nine times in Nagoya, with no bias towards increased annual maximum one-hour rainfall through the 20-year period 1993–2012.

Table 2. Annual maximum one-hour rainfall during 1993–2012 in Osaka and Nagoya Cities. The shaded values are greater than the average for each city.

Rank	Osaka City		Nagoya City	
	Date	Annual Maximum One-Hour Rainfall (mm)	Date	Annual Maximum One-Hour Rainfall (mm)
1	16:00 27 Aug. 2011	73	19:00 18 Sep. 2000	93
2	15:00 18 Aug. 2012	52	6:00 08 Oct. 2009	67
3	1:00 29 Aug. 1996	49.5	0:00 29 Aug. 2008	55
4	11:00 03 July 1995	45.5	0:00 28 July 1998	49.5
5	15:00 05 Aug. 1977	45.5	15:00 05 Sep. 2004	49.5
6	18:00 20 Oct. 2004	38	23:00 21 Aug. 1999	49
7	2:00 11 Aug. 1999	36	21:00 17 Sep. 1994	46.5
8	15:00 24 Sep. 2003	31	16:00 20 Sep. 2011	43.5
9	15:00 28 July 2008	30.5	18:00 25 July 1993	40.5
10	2:00 16 June 2010	29.5	14:00 20 July 2012	37
11	3:00 05 July 1993	28.5	5:00 23 Aug. 2007	33.5
12	15:00 30 Aug. 2005	27	5:00 11 July 1997	33
13	8:00 08 Sep. 1994	26.5	9:00 21 July 1995	28.5
14	5:00 13 July 2002	26.5	18:00 26 Aug. 2003	27
15	0:00 17 July 2007	26	9:00 06 Sep. 2002	26
16	22:00 16 May 1998	23	9:00 20 Aug. 2005	24.5
17	18:00 11 Sep. 2000	22	20:00 30 Aug. 2001	23.5
18	8:00 01 Oct. 2001	21	1:00 16 June 2006	23.5
19	6:00 19 July 2006	19	11:00 08 Sep. 2010	21.5
20	1:00 30 July 2009	13.5	6:00 09 May 1996	17

2.2. Identification of Frequent Urban Pluvial Flooding Areas

An example of a flooding area record in the case of urban pluvial flooding (urban pluvial flooding area records) is shown in Figure 2. Flooding area records are produced by each municipality as part of flood damage statistical surveys. Flooding area records since 1993 are stored as image data (pdf format) by MLIT. In this study, a total of 457 urban pluvial flooding area records for 20 years from 1993 to 2012 in Osaka and Nagoya Cities were obtained from MLIT, and frequent urban pluvial flooding areas were identified.

To confirm the reliability of urban pluvial flooding area records in the study area, these records were compared with the presence or absence of flood damage caused by urban pluvial flooding in the flood damage statistics. It was confirmed that 89.5% of urban pluvial flooding in Osaka City and 97.4% in Nagoya City were correctly recorded as the urban pluvial flooding area records. Here, these ratios were calculated by dividing the number of urban pluvial flooding area records by the number of urban pluvial flooding events recorded in the flood statistics. The reason why the flood pluvial flooding area record did not reach 100% might be that the flooding area record has been allowed not to record cases where the flooded area was less than 1000 m^2, and the number of flooded houses was less than 10 [17]. Namely, it was possible that although urban pluvial flooding damage was recorded in the flood damage statistics, flooding area was not recorded in the urban pluvial flooding area record.

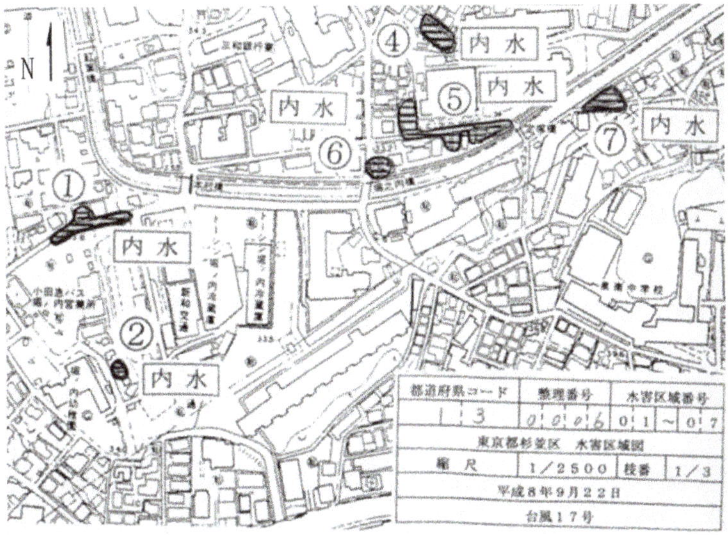

Figure 2. An example urban pluvial flooding area record (in Japanese). The seven shaded areas indicate the area where urban pluvial flooding occurred by Typhoon No. 17 on 22 September 1996. The scale of the map is 1/2500.

These urban pluvial flooding area records for the 20 years from 1993 to 2012 in Osaka and Nagoya Cities were input into GIS, and vector data were created for each year for the identification of frequent urban pluvial flooding areas. The areas where urban pluvial flooding occurred in Osaka and Nagoya Cities each year were converted into raster data of 10 m, 30 m, 50 m, 100 m, 200 m, and 400 m meshes. If even a small amount of urban pluvial flooding occurred within a mesh, that mesh was regarded as having been subject to urban pluvial flooding. The total years of urban pluvial flooding that occurred in each mesh over the 20 years from 1993 to 2012 were calculated by adding up the raster data for each year.

2.3. *Topographical Characteristics of Frequent Urban Pluvial Flooding Area*

2.3.1. Topographical Factors

Referring to Djamres et al. (2021) [14], the topographical factors of each mesh in the target city were quantified. The topographical factors used were "Elevation", "Slope", "Depth of concave", "Capacity of concave", "Catchment area", "Slope of upstream", "Slope of downstream", "Difference of slope", "Flow length of upstream", "Flow length of downstream", and "Difference of flow length". The method used to create each topographical factor is shown in Table 3. In this study, the 5 m mesh digital elevation model (DEM) provided by the Geospatial Information Authority of Japan (2013) [20] was used as the elevation data. This digital elevation model is produced based on Lidar data.

2.3.2. Visualisation of the Topographical Characteristics of Urban Pluvial Flooding

Based on Djamres et al. (2021) [14], it was attempted to identify areas with similar topographical characteristics to those of frequent urban pluvial flooding areas as areas that are strongly influenced by topographical factors of urban pluvial flooding, based on the results of the PCA. First, from the principal component loadings evaluated in the PCA, each principal component score for the whole area mesh of each city was calculated using the following formula:

$$P_{n(i,j)} = \sum_{m=1}^{12} \left(a_{m(i,j)} \cdot l_m \right) \quad (1)$$

where $P_{n(i,j)}$ is the principal component score of number n at $mesh\ (i,j)$; $a_{m(i,j)}$ is the standardised value of factor m at $mesh\ (i,j)$; and l_m is the factor loading of factor m. To assess similarity rates of topographical characteristics in frequent urban pluvial flooding areas and other areas in the city, we computed the deviation value of the averaged principal component score of urban pluvial flooding areas and scores of every mesh in the targeted areas using the following formula:

$$H_{(i,j)} = \sum_{n=1}^{4}\left(\left|P_{n(i,j)} - \overline{P'_n}\right| \cdot w_n\right) \quad (2)$$

where $H_{(i,j)}$ is the indicator of how similar topographical characteristics with frequent urban pluvial flooding areas at $mesh\ (x,y)$; $P_{n(i,j)}$ is the principal component score of number n at $mesh\ (i,j)$; $\overline{P'_n}$ is the mean principal component score of number n in the frequent urban pluvial flooding area; w_n is the contribution rate of the principal component of number n. In other words, the smaller $H_{(i,j)}$ is, the more the location has similar topographical characteristics to the frequent urban pluvial flooding area, and the greater the influence of topographical factors on urban pluvial flooding.

Table 3. The method used to create each topographical factor.

Topographical Factor	Unit	Methodology for Creating Dataset
Elevation	m	DEM is converted to raster data.
Slope	%	The slope at each mesh is calculated from "Elevation".
Depth of concave	m	The area where the elevation is lower than the surrounding area and the contour lines are closed is extracted as a concave area, and the depth and volume of the concave area are calculated.
Capacity of concave	m^3	
Catchment area	m^2	Flow direction data are created from "Elevation", and the area is calculated from the number of meshes upstream of each mesh.
Slope of upstream	%	The average of the slope in the catchment area of each mesh is calculated.
Slope of downstream	%	The average of the slopes in the downstream channel of each mesh is calculated.
Difference of slope	%	The difference between "Slope of upstream" and "Slope of downstream" at each mesh is calculated.
Flow length of upstream	m	The length of longest upstream channel in each mesh is extracted from the flow direction data.
Flow length of downstream	m	The length of the downstream channel to a sink or outlet of the catchment area in each mesh is extracted from the flow direction data.
Difference of flow length	m	The difference between "Flow length of upstream" and "Flow length of downstream" at each mesh is calculated.

3. Results

3.1. Identification of Frequent Urban Pluvial Flooding Areas

There are two points to consider in identifying frequent urban pluvial flooding areas: the frequency and extent of urban pluvial flooding. First, the frequency of urban pluvial flooding was examined. The relationship between the total years of urban pluvial flooding and the number of urban pluvial flooding areas in the raster data for each mesh size is shown in Table 4. The maximum total years of urban pluvial flooding in Osaka City and Nagoya City was 5–7 years. It is natural that when the criterion for the total years of urban pluvial flooding is small, areas that are not suitable for frequent urban pluvial flooding areas are selected. However, when the criterion is too large, the number of identified urban pluvial flooding areas becomes small, and it is not possible to obtain sufficient areas for analysis. In this study, urban pluvial flooding areas were identified as areas where urban pluvial flooding occurred more than four years (more than once every five years) in the 20 years from 1993 to 2012, considering more than half of the maximum total number of years in Osaka and Nagoya Cities.

Table 4. (**a**) The relationship between the total years of urban pluvial flooding and the number of urban pluvial flooding areas in the raster data for each mesh size in Osaka City. (**b**) The relationship between the total years of urban pluvial flooding and the number of urban pluvial flooding areas in the raster data for each mesh size in Nagoya City.

(a)

Total Years of Urban Pluvial Flooding (Year)	Osaka City					
	Mesh Size					
	10 m	30 m	50 m	100 m	200 m	400 m
1	38,687	7648	4089	1879	898	337
2	3820	1082	692	462	323	210
3	645	261	189	138	116	99
4	70	36	45	49	42	45
5	4	8	10	20	23	18
6		1		1	4	5
7						2
More than 4	74	45	55	70	69	70

(b)

Total Years of Urban Pluvial Flooding (Year)	Nagoya City					
	Mesh Size					
	10 m	30 m	50 m	100 m	200 m	400 m
1	632,915	73,366	27,436	7406	2093	542
2	102,193	13,330	5540	1920	743	303
3	11,737	1931	929	435	261	161
4	1243	256	153	83	54	51
5	172	35	22	16	17	19
6	39	16	12	9	5	6
7					1	2
More than 4	1454	307	187	108	77	78

Next, the mesh size of frequent urban pluvial flooding areas was examined. In Table 4b, the number of frequent urban pluvial flooding areas in Nagoya City decreased as the mesh size increased in terms of the reference total years (more than four years) of frequent urban pluvial flooding. This could be since large urban pluvial flooding areas that were duplicated and over-accounted for in smaller meshes were accounted for as a single area when the mesh size increased. Furthermore, this was also considered to be since some urban pluvial flooding areas that were closely located were accounted as one area with increasing the mesh size. On the other hand, in Osaka City, the number of identified frequent urban pluvial flooding areas increased when the mesh size was larger than 30 m and was almost the same when the mesh size was larger than 100 m (Table 4a). This suggested that the urban pluvial flooding areas in Osaka City were smaller and less closely located than in Nagoya City and that a mesh size of 100 m or larger would enable the frequent urban pluvial flooding areas to be identified.

Since the objective of this study is to elucidate the specific distribution of frequent urban pluvial flooding areas and their topographical characteristics, the smaller the mesh size to be identified, the better. It is considered that the applicable mesh size to be identified was about 30 m mesh because the flooding area record has been allowed not to record cases where the flooded area was less than 1000 m^2, and the number of flooded houses was less than 10 [17], as mentioned above. However, the urban pluvial flooding area record has been produced manually, and some errors may be included between the actual urban pluvial flooding area and its record. For example, a discrepancy of about 1 m to 50 m between them was recognised in the field survey, and if the mesh size was too small, the frequent urban pluvial flooding area could not be identified properly because flooded areas that occurred at the same location might be accounted for separately. Therefore, the mesh size was set to 100 m × 100 m in this study.

The frequent urban pluvial flooding areas in Osaka and Nagoya Cities identified according to the above-mentioned identifications are shown in Figure 3. In total, 70 frequent urban pluvial flooding areas were identified in Osaka City and 108 in Nagoya City. The proportion of frequent urban pluvial flooding areas in the area of each city was 0.34% in Osaka City and 0.33% in Nagoya City.

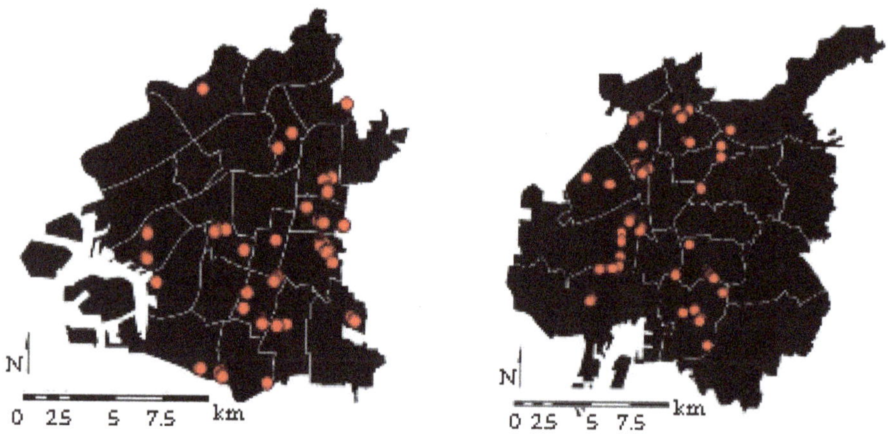

Figure 3. The distribution of frequent urban pluvial flooding areas (red circle) in (**a**) Osaka City and (**b**) Nagoya City.

3.2. *Topographical Characteristics of Frequent Urban Pluvial Flooding Areas*

Principle Component Analysis (PCA) of Topographical Characteristics

The PCA of frequent urban pluvial flooding areas on topographical characteristics was performed using the topographical factors quantified in the previous chapter. Detailed results are shown in Table 5. The principal components with eigenvalues exceeding 1 were up to the fourth principal component (PCs) in both cities, with cumulative contribution rates of 85.7% for Osaka City and 88.6% for Nagoya City. This indicated that the topographical factors used in this study were appropriate for describing frequent urban pluvial flooding areas.

In Osaka City, the first principal component (PC1) accounted for 34.3% of the total variance. The factors that correlated the most with the PC1 were "Slope upstream" (0.404) and the difference of slope (0.390) in positive values. One can infer that frequent urban pluvial flooding areas had an upstream slope that was higher than the downstream slope. Principal component 2 (PC2) was negatively correlated with "Slope of downstream" and "Difference of flow length" and positively correlated with "Flow length of downstream". Therefore, PC2 was considered to be the component that aggregates the downstream situation, and one can conclude that frequent urban pluvial flooding areas had a downstream slope gentler than upstream, downstream flow length longer than upstream, and the location at a low elevation. In contrast, the third component (PC3) showed strong positive correlations with "Flow length of upstream" and "Catchment area". Therefore, PC3 was considered to be the component that aggregates the upstream situation. Principal component 4 (PC4) had negative correlations with "Depth of concave" and "Capacity of concave" and a positive correlation with "Slope of downstream". Therefore, PC4 was considered to be the component that aggregates the concave situation.

Table 5. (**a**) The PCA result for Osaka City. The shaded values are the factors that correlated the most in each PC. "-" indicates that no topographical factors in the principal components satisfied the 95% confidence interval. (**b**) The PCA result for Nagoya City. The shaded values are the factors that correlated the most in each PC.

(a)

Factor Loading	PC1	PC2	PC3	PC4
Standard deviation (eigenvalue)	1.94	1.63	1.39	1.04
Contribution ratio	0.343	0.241	0.176	0.098
Cumulative contribution ratio	0.343	0.584	0.759	0.857
Elevation	0.373	−0.017	−0.016	0.162
Slope	0.354	−0.387	-	0.008
Depth of concave	−0.278	−0.311	−0.051	−0.539
Capacity of concave	−0.308	−0.331	−0.049	−0.401
Catchment area	−0.027	0.060	0.683	0.018
Slope of upstream	0.404	−0.295	0.102	−0.212
Slope of downstream	0.213	−0.435	−0.043	0.391
Difference of slope	0.390	−0.210	0.125	−0.337
Flow length of upstream	−0.070	0.014	0.687	−0.018
Flow length of downstream	0.310	0.409	0.045	−0.333
Difference of flow length	−0.319	−0.389	0.163	0.315

(b)

Factor Loading	PC1	PC2	PC3	PC4
Standard deviation (eigenvalue)	1.93	1.77	1.34	1.05
Contribution ratio	0.338	0.285	0.164	0.100
Cumulative contribution ratio	0.338	0.623	0.786	0.886
Elevation	−0.387	0.193	0.078	−0.249
Slope	−0.366	0.195	−0.297	−0.322
Depth of concave	0.352	−0.172	−0.126	−0.581
Capacity of concave	0.383	−0.120	−0.131	−0.547
Catchment area	−0.147	−0.463	0.201	−0.073
Slope of upstream	−0.434	−0.217	−0.025	−0.172
Slope of downstream	−0.317	0.200	−0.416	−0.180
Difference of slope	−0.349	−0.309	0.128	−0.118
Flow length of upstream	−0.104	−0.495	0.212	−0.040
Flow length of downstream	−0.054	0.206	0.647	−0.271
Difference of flow length	−0.012	−0.443	−0.422	0.204

In Nagoya City, PC1 had negative correlations for all factors except "Depth of concave" and "Capacity of concave", showing the opposite trend to that of Osaka City. PC2 was negatively correlated with three factors: "Flow length of upstream", "Catchment area", and "Difference of flow length". Therefore, PC2 was considered to be the component that aggregates the upstream situation and the factors related to the flow channel. PC3 had a strong positive correlation with "Flow length of downstream" and negative correlations with "Slope of downstream" and "Difference of flow length". Therefore, PC3 was considered to be the component that aggregates the downstream situation. PC4 had negative correlations with "Depth of concave" and "Capacity of concave", as did Osaka City. The results of the PCA for Osaka and Nagoya Cities showed that the topographical characteristics of the frequent urban pluvial flooding areas in both cities were different, with particularly conflicting trends in PC1.

The distribution of $\boldsymbol{H}_{(i,j)}$ and the urban pluvial flooding areas of each city is shown in Figure 4. Many urban pluvial flooding areas were located in areas with small $\boldsymbol{H}_{(i,j)}$, which has similar topographical characteristics to the frequent urban pluvial flooding areas. On the other hand, especially in Osaka City, the urban pluvial flooding areas were also distributed in areas with relatively large $\boldsymbol{H}_{(i,j)}$, which do not have similar topographical characteristics to the frequent urban pluvial flooding areas. This suggested that factors other than topographical characteristics that caused urban pluvial flooding were largely responsible for such areas in many parts of Osaka City and some parts of Nagoya City.

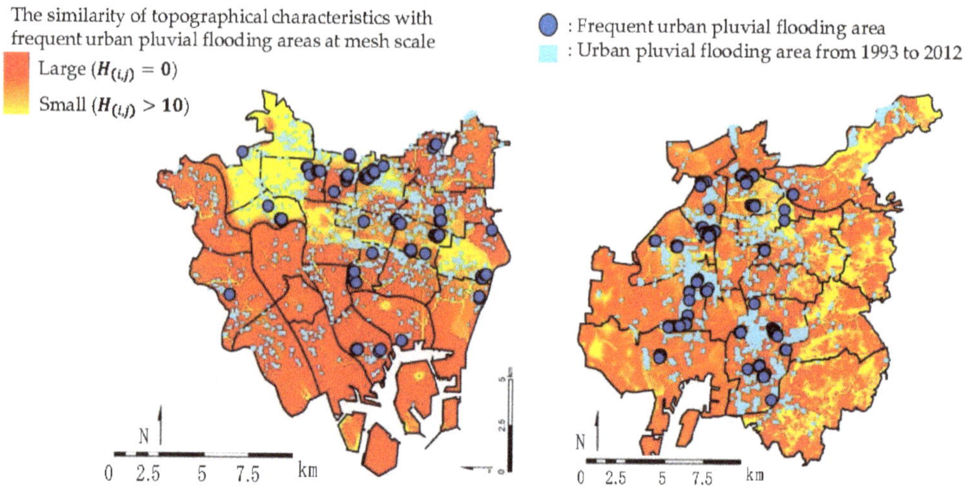

Figure 4. The distribution of $H_{(i,j)}$ and the urban pluvial flooding areas (light blue dot) in (**a**) Osaka City and (**b**) Nagoya City.

In addition, the average $H_{(i,j)}$ in the inundated areas by urban pluvial flooding from 1993 to 2012 (areas shown in the light blue dot in Figure 4) was calculated to be 3.26 in Osaka City and 2.81 in Nagoya City. This suggested that the distribution of urban pluvial flooding areas in Nagoya was better described by topographical characteristics than in Osaka City. Furthermore, the average $H_{(i,j)}$ for the whole area of each city, which was 2.64 in Osaka City and 3.18 in Nagoya City, was smaller in Osaka City than the average $H_{(i,j)}$ for the inundated area. This indicated that the urban pluvial flooding in Osaka City could not be described solely by topographical characteristics.

3.3. *Other Characteristics of Frequent Urban Pluvial Flooding Areas*

3.3.1. Impact of Structures in Frequent Urban Pluvial Flooding Areas

Referring to Djamres et al. (2021) [14], the location of "dominant structures" was examined as another characteristic of these areas. Here, "dominant structures" are roads that divide sewers, railway lines (excluding subways and elevated lines), embankments, and structures with a site of more than 100 m per side. In addition, a road dividing a sewer is not a road with a sewer pipe buried directly under the centre of the road but a road with a sewer pipe buried under each side of the road. These are roads with a median strip or large road widths, such as dual carriageways in one direction, which were included in the analysis in this section because they influenced the flow of water.

Frequent urban pluvial flooding areas were classified according to the relationship between the location of "dominant structures" and the direction of inclination to "dominant structures". A conceptual diagram of the classification conditions and the proportions of the classified frequent urban pluvial flooding areas are shown in Figure 5 and Table 6. Figure 5a,d,g,h (the red line enclosure in Figure 5) show areas where "dominant structures" were located in the direction of inclination (direction of inundated water flow). In Osaka City, there were 90% of the total frequent urban pluvial flooding areas in which one or more "dominant structures" existed within 8 mesh around the area. Of the frequent urban pluvial flooding areas with "dominant structures" within 8 mesh of the perimeter, 74% of the areas had "dominant structures" in the direction of inundated water flow (Figure 5a,d,g,h).

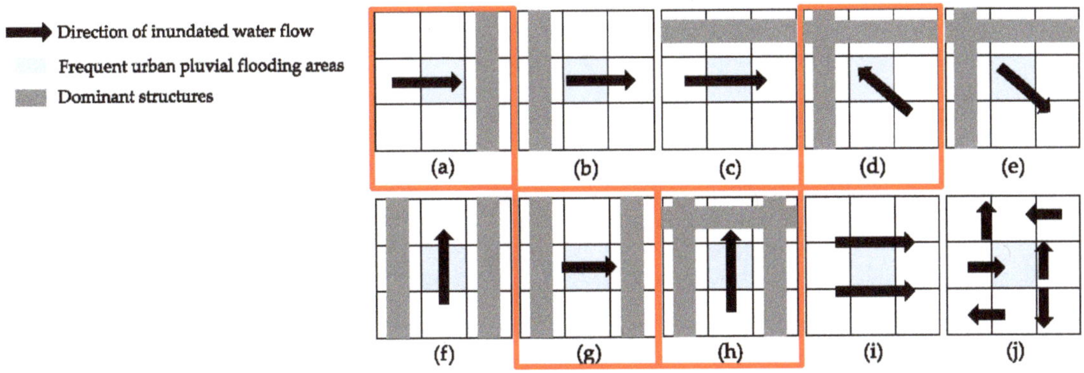

Figure 5. A conceptual diagram of the classification conditions of the classified frequent urban pluvial flooding areas. (**a**–**h**) are areas where "dominant structures" were located within 8 mesh around the urban pluvial flooding area, while (**i**,**j**) are areas where "dominant structures" were not located within 8 mesh around the urban pluvial flooding area. (**h**) is the area where "dominant structures" were located in three directions around the frequent urban pluvial flooding area, and "dominant structures" were located in the front of the slope direction. The red line enclosure are areas where "dominant structures" was located in the direction of inclination (direction of inundated water flow).

Table 6. The proportions of the classified frequent urban pluvial flooding areas in Figure 5. The grey values are areas where "dominant structures" were located in the direction of inundated water flow.

Classified Frequent Urban Pluvial Flooding Areas in Figure 5	Osaka City	Nagoya City
(a)	29%	14%
(b)	7%	21%
(c)	3%	19%
(d)	20%	4%
(e)	9%	6%
(f)	4%	2%
(g)	1%	1%
(h)	17%	0%
(i)	9%	22%
(j)	1%	11%
Total	100%	100%

In Nagoya City, there were 67% of the all-frequent urban pluvial flooding areas in which one or more "dominant structures" existed within 8 mesh around the area. Of the frequent urban pluvial flooding areas where "dominant structures" existed within 8 mesh of the perimeter, 28% of the areas had "dominant structures" in the direction of inundated water flow (Figure 5a,d,g,h).

A comparison of the results for Osaka and Nagoya Cities showed that the proportion of frequent urban pluvial flooding areas with "dominant structures" in Nagoya City was smaller than that in Osaka City, and in particular, the proportion of frequent urban pluvial flooding areas with "dominant structures" in the direction of the inundated water was much smaller. Therefore, the distribution of elevation and slope in frequent urban pluvial flooding areas in Osaka and Nagoya Cities was calculated based on the assumption that the topographical differences in frequent urban pluvial flooding areas might influence the characteristics of frequent urban pluvial flooding areas in both cities. The average elevation of the frequent urban pluvial flooding areas was 3.30 m in Osaka City and 3.61 m in Nagoya City, respectively. Although the elevation was higher in Nagoya City, there was no significant difference. On the other hand, the average slopes of all frequent urban pluvial flooding areas in Osaka City and Nagoya City were 2.20% and 0.33%, respectively, indicating that the slopes of the frequent urban pluvial flooding areas in Nagoya City were much smaller than those in Osaka City. The frequency distribution of slopes at the frequent urban pluvial flooding areas in Osaka and Nagoya Cities is shown in Figure 6. Most of the frequent urban pluvial flooding areas in Nagoya City were located on almost no gradient with a slope of less than 1 degree. It was inferred that in such areas, regardless of the location of "dominant structures", the mere presence of "dominant structures" around it would dam up the inundated water and cause urban pluvial flooding. In addition, this was considered to be one of the reasons why "Slope" and "Slope of upstream" were not correlated as PCs in the PCA as the topographical characteristics of the frequent urban pluvial flooding area in Nagoya City described in the previous section.

3.3.2. Impact of Drainage System Improvements in Frequent Urban Pluvial Flooding Areas

To examine the impact of the improvement of drainage systems as another characteristic of frequent urban pluvial flooding areas, the occurrence trends of urban pluvial flooding in the urban pluvial flooding areas of Osaka and Nagoya Cities were investigated. The difference in urban pluvial floodings between the first 10 years (1993–2002) and the second 10 years (2003–2012) of the 20 years (1993–2012) is shown in Figure 7. Of the areas classified as green (1993–2002), 64 (91%) were in Osaka City compared with 8 (7%) in Nagoya City. On the other hand, red areas (2003–2012) were only 1 (1%) in Osaka City compared with 56 (52%) in Nagoya City. In addition, blue areas (1993–2012) were also only 5 (7%) in Osaka City compared with 44 (41%) in Nagoya City.

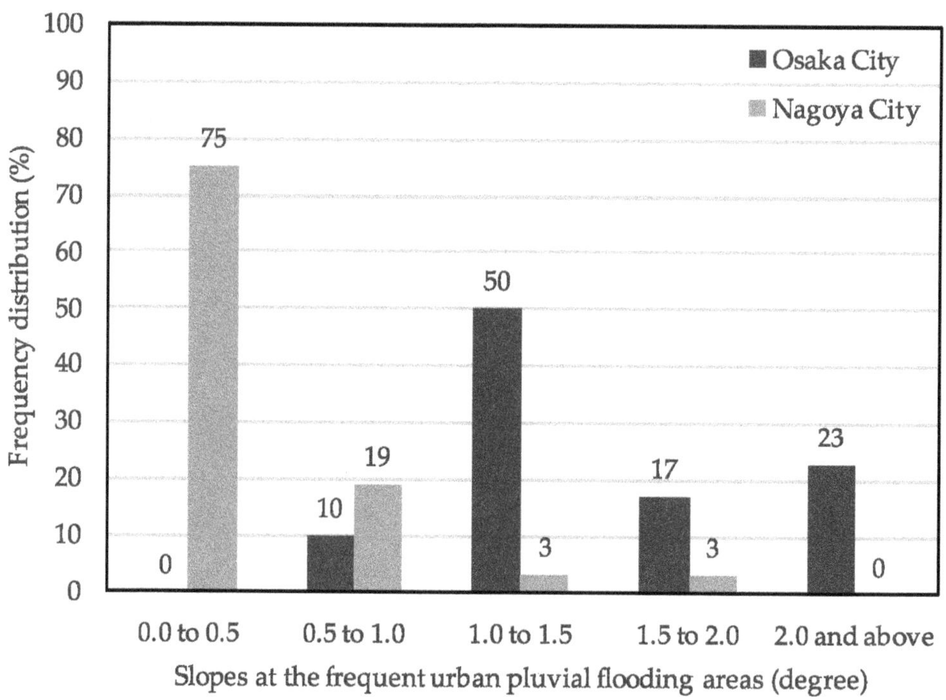

Figure 6. The frequency distribution of slopes at the frequent urban pluvial flooding areas in Osaka and Nagoya Cities.

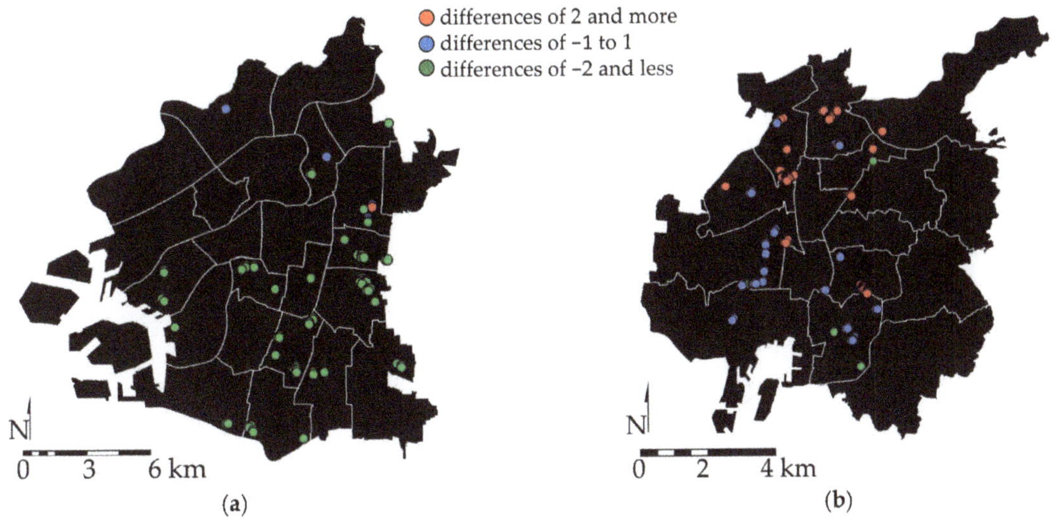

Figure 7. The difference in urban pluvial floodings between the first 10 years (1993–2002) and the second 10 years (2003–2012) of the 20 years (1993–2012) in (**a**) Osaka City and (**b**) Nagoya City. The differences of 2 or more are shown in red, −1 to 1 in blue, and −2 or less in green. Red indicates areas where urban pluvial flooding occurred more frequently in the second 10 years, blue indicates areas where urban pluvial flooding occurred continuously throughout the 20 years, and green indicates areas where urban pluvial flooding occurred more frequently in the first 10 years.

Regarding the improvement of drainage systems, in Osaka City, the Naniwa Underground Discharge Channel (total length of 12.2 km), a large-scale drainage system to drain rainwater from the south-eastern area of Osaka City into the Sumiyoshi River, began to be constructed in 1984 and was completed in 2000, following the large-scale urban pluvial flooding caused by Typhoon No. 19 in September 1979 and Typhoon No. 10 in August 1982. As a result of such progress made in countermeasures against urban flooding, it can be inferred that urban pluvial flooding ceased to occur in frequent urban pluvial flooding areas during 2003–2012. On the other hand, Nagoya City, which is on a smaller urban scale than Osaka City, has been implementing emergency drainage system improvements in response to the large-scale urban pluvial flooding caused by the torrential rains of September 2002 and September 2006. However, it was found that most of the frequent urban pluvial flooding areas during 2003–2012 were located in areas where the improvement had not yet been completed (Figure 8). These results indicated that anthropogenic factors such as "dominant structures" and drainage system improvements influence the occurrence of urban pluvial flooding as characteristics other than topographical characteristics.

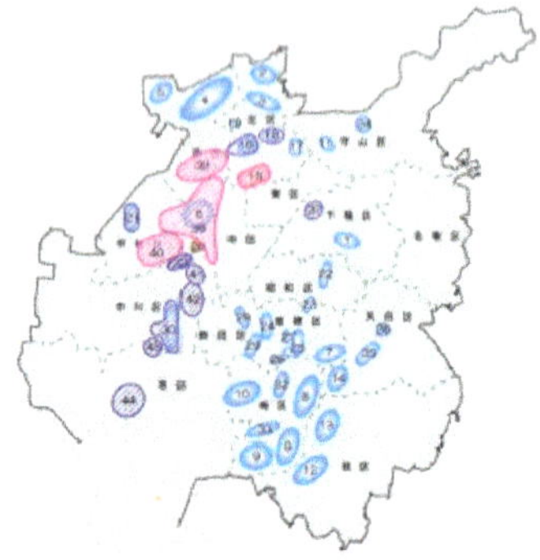

Figure 8. Urban flood management plan in Nagoya City (Progress as at the end of 2015) [21].

4. Discussion and Summary

This study clarified the distribution of frequent urban pluvial flooding areas in Osaka and Nagoya Cities by using urban pluvial flooding area records. The identified frequent urban pluvial flooding areas were 70 in Osaka City and 108 in Nagoya City, and their proportion to the area of each city was 0.34% in Osaka City and 0.33% in Nagoya City. Analyses of their characteristics revealed the following:

- The PCA of frequent urban pluvial flooding areas using the eleven topographical factors showed high cumulative contribution rates for Osaka and Nagoya Cities, which indicated that the topographical factors used in this study were appropriate for describing frequent urban pluvial flooding areas. The results of the PCA quantitively showed that the topographical characteristics of the frequent urban pluvial flooding areas in both cities were different.
- Using the results of PCA of topographical characteristics in the frequent urban pluvial flooding area, the similarity with these topographical characteristics at a 100 m mesh scale in both cities was quantified. Although many urban pluvial flooding areas were located in areas with similar topographical characteristics, especially in Osaka City, the urban pluvial flooding areas were also distributed in areas without similar topographical characteristics. This suggested that factors other than topographical characteristics that caused urban pluvial flooding were largely responsible for such areas in many parts of Osaka City and some parts of Nagoya City.
- Anthropogenic factors such as "dominant structures" and drainage system improvements as characteristics other than topographical characteristics on the occurrence of urban pluvial flooding were shown to be influential. The results also showed that the impact of anthropogenic factors was greater in Osaka City, which is on a larger urban scale than Nagoya City.

In general, urbanisation increases the accumulation of assets in topographically flood-prone areas, and the risk of urban pluvial flooding increases. On the other hand, urban pluvial flood risk has been reduced through the improvement of drainage systems in such areas; however, urban pluvial flooding has also been observed in topographically less-flood-prone areas due to changes in land use and land cover. In other words, as urbanisation progresses, the main cause of urban pluvial flooding is likely to shift from topographical factors to anthropogenic factors. The results of this study quantitatively showed this paradigm shift of urban pluvial flooding factors by the statistical analysis of newly defined urban pluvial flooding frequency areas. This is demonstrated as one of the methods for a major advance in urban flood modelling science proposed by Mignot and Dewals [9]. Furthermore, this study showed that it is difficult to describe past urban pluvial flooding areas in Osaka and Nagoya Cities solely based on topographical characteristics. This was consistent with the findings of the numerical experiment [22] and statistical analyses [14] and showed that in some cases that artificial structures formed areas vulnerable to urban pluvial flooding even outside topographically flood-prone areas.

In addition, it is particularly difficult to collect flow velocity and depth observations during urban floods, as they are usually of short duration. The recent proliferation of mobile phones and online video-sharing platforms gives access to countless amateur videos [23–25], which are being used in most geophysical sciences, but difficulties with retrieving the location and time of the scenes impair the use of these data for detailed model validation [9]. The urban pluvial flooding area record used in this study and the newly defined frequent urban pluvial flooding area could be important sources for this validation.

On the other hand, although this study succeeded in quantitatively assessing the differences in the impact of topographical characteristics on the formation of frequent urban pluvial flooding areas due to urban scale by comparative analyses in Osaka and Nagoya Cities, this study only compared two cities, and the relationship between urban scale and frequent urban pluvial flooding areas has not been quantitively clarified. Further quantitative comparison with the characteristics of frequent urban pluvial flooding areas in other cities of different urban scales would help to understand the characteristics of urban pluvial flooding and their transition from topographical factors to anthropogenic factors, which is associated with urbanisation. Furthermore, socio-hydrology [26,27], which deals with the interaction between water systems and human activities, has been recently applied worldwide for flooding and water resources management [28]. Further quantitative comparative analysis of the characteristics of frequent urban pluvial flooding areas among different urban scale cities would provide a quantitative understanding of the system dynamics of urban pluvial flooding interacting with urbanisation, namely human activities.

Author Contributions: Conceptualization, D.K.; methodology, D.K. and K.N.; formal analysis and investigation, D.K., K.N., R.I., Y.O. and R.T.; writing—original draft preparation, D.K. and K.N.; writing—review and editing, D.K. and S.K.; visualization, D.K. and K.N.; supervision, D.K. and S.K.; funding acquisition, D.K. and S.K. All authors have read and agreed to the published version of the manuscript.

Funding: This research was partly funded by the Nomura Foundation for Membrane Structure Technology and the Obayashi Foundation, and the APC was funded by the Masaki Sawamoto Research Publication Grant.

Institutional Review Board Statement: No applicable.

Informed Consent Statement: No applicable.

Data Availability Statement: No applicable.

Acknowledgments: This research was supported by CSIS Joint Research No. 1123, University of Tokyo. The urban pluvial flooding area record was provided by the River Planning Division, Water Management and Land Conservation Bureau, Ministry of Land, Infrastructure and Transport. The authors would like to express their gratitude.

Conflicts of Interest: The authors declare no conflict of interest.

References

1. Hammond, M.J.; Chen, A.S.; Djordjević, S.; Butler, D.; Mark, O. Urban flood impact assessment: A state-of-the-art review. *Urban Water J.* **2015**, *12*, 14–29. [CrossRef]
2. Lowe, R.; Urich, C.; Domingo, N.S.; Mark, O.; Deletic, A.; Arnbjerg-Nielsen, K. Assessment of urban pluvial flood risk and efficiency of adaptation options through simulationse a new generation of urban planning tools. *J. Hydrol.* **2017**, *550*, 355–367. [CrossRef]
3. Wang, Y.; Chen, A.S.; Fu, G.; Djordjević, S.; Zhang, C.; Savić, D.A. An integrated framework for high-resolution urban flood modelling considering multiple information sources and urban features. *Environ. Model. Softw.* **2018**, *107*, 85–95. [CrossRef]
4. Economic Lossess from Climate-Related Extremes in Europe. Available online: https://www.eea.europa.eu/data-and-maps/indicators/river-floods-3/assessment (accessed on 5 August 2022).

5. Hirabayashi, Y.; Mahendran, R.; Koirala, S.; Konoshima, L.; Yamazaki, D.; Watanabe, S.; Kanae, S. Global flood risk under climate change. *Nat. Clim. Chang.* **2013**, *3*, 816–821. [CrossRef]
6. National Academies of Sciences, Engineering, and Medicine. *Framing the Challenge of Urban Flooding in the United States*; National Academies Press: Washington, DC, USA, 2019.
7. Statistics of Flood Damage. Available online: https://www.mlit.go.jp/river/toukei_chousa/kasen/suigaitoukei/index.html (accessed on 5 August 2022).
8. The Fifth Assessment Report (AR5) WG2-Climate Change 2014: Impacts, Adaptation, and Vulnerability. Available online: https://www.ipcc.ch/report/ar5/wg2/ (accessed on 5 August 2022).
9. Mignot, E.; Dewals, B. Hydraulic modelling of inland urban flooding: Recent advances. *J. Hydrol.* **2022**, *609*, 127763. [CrossRef]
10. Jacobson, C.R. Identification and quantification of the hydrological impacts of imperviousness in urban catchments: A review. *J. Environ. Manag.* **2011**, *92*, 1438–1448. [CrossRef] [PubMed]
11. Zhang, W.; Villarini, G.; Vecchi, G.A.; Smith, J.A. Urbanization exacerbated the rainfall and flooding caused by hurricane Harvey in Houston. *Nature* **2018**, *563*, 384–388. [CrossRef] [PubMed]
12. Shuster, W.D.; Bonta, J.; Thurston, H.; Warnemuende, E.; Smith, D. Impacts of impervious surface on watershed hydrology: A review. *Urban Water J.* **2005**, *2*, 263–275. [CrossRef]
13. Wang, X.; Kinsland, G.; Poudel, D.; Fenech, A. Urban flood prediction under heavy precipitation. *J. Hydrol.* **2019**, *577*, 123984. [CrossRef]
14. Djamres, E.K.D.; Komori, D.; Kazama, S. Topographical Characteristics of Frequent Inland Water Flooding Areas in Tangerang City, Indonesia. *Front. Water* **2021**, *3*, 661299. [CrossRef]
15. Fariza, A.; Basofi, A.; Prasetyaningrum, I.; Pratiwi, V.I. Urban flood risk assessment in sidoarjo, indonesia, using fuzzy multi-criteria decision making. *J. Phys. Conf. Ser.* **2019**, *1444*, 012027. [CrossRef]
16. Sato, R.; Hayashi, T. Topographical and topographical characteristics of flooded areas and visualization of potential for pluvial flooding. *J. Jpn. Soc. Hydrol. Water Resour.* **2014**, *27*, 158–169. [CrossRef]
17. Guidelines for Statistics of Flood Damage. Available online: http://www.mlit.go.jp/river/toukei_chousa/kasen/suigaitoukei/shiryou.html (accessed on 5 August 2022).
18. National Land Information Division. Available online: https://nlftp.mlit.go.jp/ksj/index.html (accessed on 25 August 2022).
19. Japan Meteorological Agency AMeDAS. Available online: https://www.jma.go.jp/jp/amedas/ (accessed on 25 August 2022).
20. Geospatial Information Authority of Japan. Available online: https://fgd.gsi.go.jp/download/menu.php (accessed on 5 August 2022).
21. Nagoya City Water and Sewerage Bureau: Countermeasures in Response to the Torrential Rains in Tokai, Japan. Available online: https://www.water.city.nagoya.jp/category/saigaisonae/2174.html (accessed on 5 August 2022).
22. de Almeida, G.A.; Bates, P.; Ozdemir, H. Modelling urban floods at submetre resolution: Challenges or opportunities for flood risk management? *J. Flood Risk Manag.* **2018**, *11*, S855–S865. [CrossRef]
23. Le Coz, J.; Patalano, A.; Collins, D.; Guillén, N.F.; García, C.M.; Smart, G.M.; Braud, I. Crowdsourced data for flood hydrology: Feedback from recent citizen science projects in Argentina, France and New Zealand. *J. Hydrol.* **2016**, *541*, 766–777. [CrossRef]
24. Kankanamge, N.; Yigitcanlar, T.; Goonetilleke, A.; Kamruzzaman, M. Determining disaster severity through social media analysis: Testing the methodology with South East Queensland Flood tweets. *Int. J. Disaster Risk Reduct.* **2020**, *42*, 101360. [CrossRef]
25. Zhang, Y.; Chen, Z.; Zheng, X.; Chen, N.; Wang, Y. Extracting the location of flooding events in urban systems and analyzing the semantic risk using social sensing data. *J. Hydrol.* **2021**, *603*, 127053. [CrossRef]
26. Sivapalan, M.; Savenjie, H.H.G.; Blöschl, G. Sociohydrology: A new science of people and water. *Hydrol. Process.* **2012**, *26*, 1270–1276. [CrossRef]
27. Di Baldassarre, G.; Sivapalan, M.; Rusca, M.; Cudennec, C.; Garcia, M.; Kreibich, H.; Konar, M.; Mondino, E.; Mård, J.; Pande, S.; et al. Sociohydrology: Scientific challenges in addressing the sustainable development goals. *Water Resour. Res.* **2019**, *55*, 6327–6355. [CrossRef] [PubMed]
28. Kreibich, H.; Schröter, K.; Di Baldassarre, G.; Van Loon, A.; Mazzoleni, M.; Abeshu, G.W.; Agafonova, S.; AghaKouchak, A.; Aksoy, H.; Alvarez-Garreton, C.; et al. Panta Rhei benchmark dataset: Socio-hydrological data of paired events of floods and droughts. *GFZ Data Serv.* **2022**. [CrossRef]

water

MDPI

Article

Willingness to Pay for Improved Urban Domestic Water Supply System: The Case of Hanoi, Vietnam

Nuong Thi Bui [1,2], Stephen Darby [2], Trang Quynh Vu [1], Jean Margaret R. Mercado [3], Thao Thi Phuong Bui [4], Komali Kantamaneni [5], Thuong Thi Hoai Nguyen [1], Tu Ngoc Truong [6], Hue Thi Hoang [1] and Duong Du Bui [7,*]

1 Faculty of Environment, Hanoi University of Natural Resources and Environment (HUNRE), No. 41A Phudien Street, North Tuliem District, Hanoi City 100000, Vietnam; btnuong@hunre.edu.vn (N.T.B.); trangvqt101999@gmail.com (T.Q.V.); nththuong.mt@hunre.edu.vn (T.T.H.N.); hthue@hunre.edu.vn (H.T.H.)

2 Department of Geography and Environmental Science, University of Southampton, Southampton SO17 1BJ, UK; s.e.darby@soton.ac.uk

3 Department of Tourism Management, University of Santo Tomas, España Blvd, Sampaloc, Manila 1008, Philippines; jrmercado@ust.edu.ph

4 Disaster Prevention Research Institute, Kyoto University, Kyoto 606-8501, Japan; bui.thaothiphuong.3e@kyoto-u.ac.jp

5 Faculty of Science and Technology, University of Central Lancashire, Preston PR1 2HE, UK; kkantamaneni@uclan.ac.uk

6 TNTU Sciences & Technology Cooperation, Hanoi City 100000, Vietnam; truongtu13@hotmail.com

7 National Center for Water Resources Planning and Investigation, Ministry of Natural Resources and Environment, Hanoi City 100000, Vietnam

* Correspondence: duongdubui@gmail.com; Tel.: +84-966-214-316

Citation: Bui, N.T.; Darby, S.; Vu, T.Q.; Mercado, J.M.R.; Bui, T.T.P.; Kantamaneni, K.; Nguyen, T.T.H.; Truong, T.N.; Hoang, H.T.; Bui, D.D. Willingness to Pay for Improved Urban Domestic Water Supply System: The Case of Hanoi, Vietnam. *Water* **2022**, *14*, 2162. https://doi.org/10.3390/w14142161

Academic Editors: Guy Howard and Marco Franchini

Received: 1 May 2022
Accepted: 14 June 2022
Published: 8 July 2022

Abstract: Access to a reliable and safe domestic water supply is a serious challenge for many developing countries worldwide. In the capital of Vietnam, Hanoi, the municipal government is facing a number of difficulties in providing sufficient water in a sustainable manner due to the increasing urban population and the serious pollution of both surface and groundwater resources, but this is also due to a lack of resources to invest in the supply system. This study aimed to investigate water users' willingness to pay for the improvement of Hanoi's domestic water supply system. A contingent valuation process based on a survey of 402 respondents was used to explore citizens' willingness to pay (WTP) for the improvement of their urban water supply. The results show that Hanoi's urban communities (more than 90%) were generally satisfied with the quantity of their water supply, but tended to be dissatisfied with its quality, with 80% of the respondents using advanced water purifiers before drinking and cooking. Respondents were also concerned about the overall reliability of the service, with 40% of respondents indicating that they received no check and maintenance service. A WTP regression model was developed based on the survey findings. The average WTP is 281,000 dong/household/month (approximately 12.2 USD at the exchange rate of 1 USD to about 23,000 VND), equivalent to 1.4% of the average household income at the end of 2019, indicating the level of affordability of monthly water payments among Hanoi citizens.

Keywords: urban domestic water supply; willingness to pay; CVM-based process; theoretical regression framework; Hanoi; Vietnam

1. Introduction

A shortage of clean water is one of the world's most pressing concerns. According to United Nations Water, in 2014 water scarcity affected approximately 700 million people worldwide, with this figure likely to increase to approximately 1.8 billion people by 2025 [1]. Furthermore, two-thirds of the world's population live in areas with severe water scarcity [1], particularly those living in urban areas [2,3]. Aside from physical water scarcity, there is also economic water scarcity, which means that water supply is inadequate due to the lack of, or poor water infrastructure, management, and policy [4]. There is, therefore, a strong imperative to improve domestic water supplies in many urban areas around the world [5]. Several studies have demonstrated that investment in the domestic water supply can deliver a range of direct and indirect economic co-benefits, including lower healthcare expenditure, time-saving non-health-related benefits (e.g., no queuing at shared water facilities or walking a distance to a water collection site), and water resource protection, among others [6]. According to the definition of the Joint Monitoring Programme, improved water sources include piped water into dwellings, plots or yards, public taps or standpipes, boreholed, a protected dug well, a protected spring, or rainwater collection, all of which are likely to provide safe drinking-water for communities. Thus, improved domestic water supplies in urban areas should seek to deliver three basic goals, namely: sufficiency in quantity, safe quality of water, and reliable management and service.

Although many households in developing countries lack access to improved domestic water supplies, governments in these countries often cannot afford to offer substantially subsidized improved domestic water to all, or even to the majority of their populations [7]. As a result, improvements in the domestic water supply in these countries frequently relies mainly on financial contributions (payment of water bills) from households. However, such contributions depend not only on each household's willingness to pay (WTP), i.e., the maximum amount that households are able to pay for their water supply [8], but also on each household's capacity to pay, i.e., the total household income minus the amount to cover basic needs [9,10]. WTP information may be used by planners at all levels (national, provincial, city, and rural) to evaluate a project's economic feasibility, set affordable tariffs, evaluate policy alternatives, assess financial sustainability, and design socially equitable subsidies. Moreover, a cost–benefit analysis would be inadequate without such WTP data; the net economic benefits of an improved domestic water supply are calculated as the difference between the consumers' maximum WTP for better services and the actual cost of the services [11]. To estimate the WTP, the Contingent Valuation Method (CVM) is often used. CVM is an economic, non-market valuation method which is particularly useful for determining human preferences for public goods that have no monetary value in the market. CVM is an established method and has found many applications in water-related fields, such as assessing the social value of increasing water quality, reducing risks from drinking water and groundwater contamination, and the provision of drinking water services in developing countries [12]. In CVM applications to water supply services [13–19] the main objectives are to estimate the WTP to improve current water supply services and to explore the factors controlling WTP values via empirical statistical modelling. Mostly, the WTP regression models in these CVM applications are similar in terms of their selection of the WTP as a dependent variable and typically employ demographic factors (usually age, gender, education, income, family size, etc.) as independent variables.

Similar to other developing countries, Vietnam has limited improved domestic water supply, especially in urban areas [20]. According to the National Environment Report [21], it is estimated that only about 70% of the population has access to potable water. Hanoi, Vietnam's capital city, is now facing several water scarcity issues connected to its urban water supply. The fast rate of urbanization and rapid increase in the city's population (~3.4%/year) are significantly inflating the demand for clean water [22]. Meanwhile, the quality of the water resources that are being used to supply Hanoi is decreasing. Domestic and industrial effluents have polluted surface water sources in river basins of many major rivers, such as the Red, Nhue, and Day Rivers located in the Red River Delta. It is estimated that between 100,000 and 150,000 m^3/day of untreated industrial wastewater flows directly into the rivers in Hanoi alone [23]. Furthermore, upstream of the administrative area of Hanoi, the quality and quantity of water resources are significantly affected, affecting abstraction possibilities. Hanoi's water issues are exacerbated further by high water loss rates, averaging 23% [24]. As a result, it is critical to improve the water supply system in Hanoi by investing in the necessary supporting infrastructure, adjusting water allocations to fulfill residents' demands reasonably, and by improving the quality of water supplied so that it meets the Vietnam Ministry of Health Quality Standard QCVN 02-BYT. Thus, the socialization of investment capital is critical to offer financial support for these activities, especially in developing countries [25].

This study explores these issues within the context of the case study of Hanoi by developing a CVM-based process that focuses on three important points: (1) a naturally exploring WTP technique aligning with the way in which humans think, (2) determining whether the results of the CVM's social investigation are consistent and reliable, and (3) finding the appropriate variables to include in the WTP regression model based on the groups of key factors that significantly affect WTP and the current circumstances of Hanoi's urban domestic water supply system's (HUDWSS) performance. With the application of the developed CVM-based process to Hanoi, it is possible to examine the factors affecting residents' WTP, thus providing an essential first step in the improvement of the domestic water supply and community expectations in Hanoi.

2. Study Area

Figure 1 displays the study location, consisting of ten urban districts, and the main rivers and lakes of Hanoi. As the economic, political, and social center of the country, Hanoi's population and its density are very high compared to other developed cities in Vietnam. In 2020, the total population of Hanoi was approximately eight million people; female and male residents accounted for a similar proportion, and almost half of the total population was comprised of urban residents, according to the General Statistics Office of Vietnam. Average monthly income per capita in Hanoi was an estimated 6.3 million dong (about 280 US dollars).

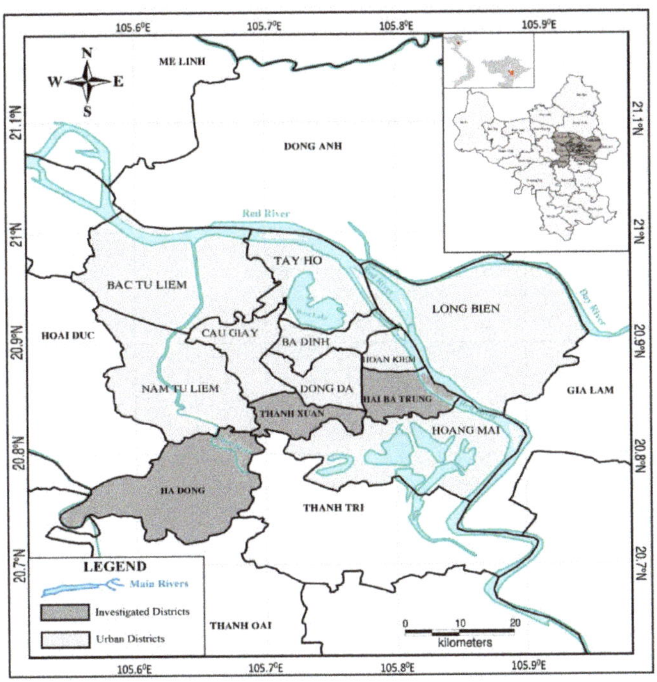

Figure 1. Study area and main rivers in Hanoi.

Hanoi's Urban Domestic Water Supply System (HUDWSS) faces a number of significant challenges in delivering its plans to supply improved water to all residents before 2030. Currently, the per capita average urban water demand is approximately 200 L/day [26] However, the municipal government is facing several challenges in providing sufficient water sustainably due to: (i) the ever-increasing urban population and density; (ii) serious pollution of both the surface and groundwater resources that serve as the main input water sources for HUDWSS's operation, and especially; (iii) the large losses of water (due to leakage) that can then not be used to generate any revenue.

Regarding the first challenge, the rapid urban growth places great pressure on natural resources and the environment, putting particular stress on HUDWSS to meet the growing water demand of customers, causing the fragmentation of Hanoi's urban water infrastructure [27]. In recent years, Hanoi's population has increased rapidly; the density of the nine urban districts has increased up to 11,759 people/km^2, while the urbanization rate reached 40.5% in 2013 [26]. In recent years, the population grew by approximately 3.4% per year. In just two years, from 2015 to 2017 the population grew from 7.2 to 7.7 million people, and the population density rose from 2000 to 2209 people/km^2 [28]. The city's water distribution system, including new and old networks, has a long history of construction and rehabilitation. Hanoi Water Limited Company (HAWACO) is the city's largest water distribution enterprise and has legal status under the Hanoi Transportation Department. The company was established under Decision No. 546/QDUB dated 4 April 1994 of the People's Committee of Hanoi, with a history that can be traced back to the nineteenth century. Currently, Hanoi has 12 main water plants and, (including the district of Ha Dong) eight water supply zones managed by HAWACO, Vietnam's Freshwater Business and Construction Investment Joint Stock Company (VIWACO), and Ha Dong Waterworks [29]. As reported by HAWACO, the average water supply capacity is 1,462,000 m^3/month, of which 35% is distributed to the old network, which mainly serves the Old Quarter's communities; and 65% is distributed to the new network, which covers the inner, the west, and the southeast regions of Hanoi, see Figure 1 [30]. HAWACO is failing to meet current water needs. With the per capita water demand in the urban districts at approximately 130 L/person/day, the public water utilities failed to supply urban districts approximately every two days per month in 2016 [30]. Consequently, as mentioned by HAWACO [31], just 55% of the city's population has access to HUDWSS even though the public distribution network fully covers all of the urban districts. However, around 30% of urban households use freely accessible well water sources [32].

Regarding the second challenge, both surface and groundwater resources for HUDWSS are seriously degraded and polluted. This critical situation challenges the water enterprises in terms of how to provide a high-quality water supply for local communities. The majority of input water sources for HUDWSS consist of groundwater harvested from Pleistocene aquifers. Groundwater resources are distributed unevenly, with the largest recharge of 700,000 m^3/day in the south and the smallest of 66,000 m^3/day in the Soc Son district (Figure 1). The disastrous situation of seriously degraded groundwater quantity and quality as a consequence of inappropriate usage and management has been comprehensively presented in a number of previous studies [33–35]. In addition, the surface water sources in rivers and lakes are also seriously contaminated due to solid waste dumping, and domestic and industrial wastewater flowing directly into the water bodies without treatment. The total domestic and industrial wastewater volume in the central area of Hanoi is approximately 600,000–700,000 m^3/day; the combined capacity of all the water treatment plants in Hanoi, however, is only around 245,000 m^3/day. Hence, two-thirds of the generated wastewater is not treated before being discharged into rivers and lakes [12]. Consequently, water quality parameters are far below the national water quality standards. For instance, as reported by HAWACO in 2016, the observed values of an important water quality parameter of chemical oxygen demand of Quynh lake, which is located in our targeted urban district Hai Ba Trung, is extremely high at 136 mg/L, compared to the recommended value of 10 mg/L, which is the national standard for good surface water quality. Moreover, major rivers such as the Red River, Da River, and Day River are inter-provincial water resources, and they have a great impact on the quality and reserves from the watersheds; thus, it is difficult to use these water resources.

Regarding the third challenge, according to HAWACO [30], the most pressing issue for the Hanoi water supply sector is the large volume of non-revenue water, which is approximately 23%. In fact, considering around 600,000 customers and the total capacity of plants of 534,500 m^3/day, it is estimated that the quantity of per day non-revenue water is approximately 2.8 times the existing supply capacity. The rate of non-revenue water loss was substantially reduced from 38% in 2007 to 23% in 2015 as a result of the efforts of HAWACO's management. The average water price as of 2020 is 7000 dong/m^3 (~USD 0.31/m^3) in urban districts; thus, this massive wasted budget could be estimated as more than 1 billion dong per day (~USD 43,700/day), while the residents still lack a water supply. The main reasons for this huge loss include the poor maintenance of water pipelines, inaccurate water meters leading to the incorrect recording of water use, or even water theft or illegal water tapping, as observed in other developing countries [36]. This critical circumstance threatens the sustainability of the financial situation of HUDWSS and thus challenges the ability of HUDWSS to meet the goal of achieving a sufficient water supply for all, as mentioned in the global sustainable development goals.

3. Contingent Valuation Method-Based Process for Exploring the Factors Affecting the WTP of Hanoi's Urban Domestic Water System

The CVM used here employed a survey of Hanoi water supply service customers (see Appendix A) that consisted of five main sections: (1) a section asking questions concerning respondents' public awareness of the water supply service situation, with the aim to understand the community's response to the service; (2) the presentation of the CVM scenario; (3) several questions for those who do not use HUDWSS; (4) a question asking about the respondent's WTP for the improvement of the water supply service, and; (5) a series of demographic questions. This survey outline follows the general process of the application of CVM to elicit WTP based on a range of previous studies [37–39]. We further developed these general sections to apply them to the specific situation of HUDWSS. The main objectives addressed in the questionnaire were (i) to understand the current situation, public perception, and satisfaction regarding the domestic water supply among local communities; (ii) to explore the WTP of local residents for water supply improvement; and (iii) to enable the subsequent analysis of the key factors affecting the WTP. Conducting a comprehensive investigation, we finally had 402 respondents, which as discussed in Section 3.3, is sufficient to give adequate statistical power.

In the application of the CVM developed in this study, we developed three methodological innovations. First, we improved the technique of determining the respondents' WTP values step by step, logically and naturally, in line with natural thinking patterns; the so-called 'naturally exploring' WTP technique. Secondly, we checked whether the obtained responses from the CVM social investigation were acceptably consistent and reliable, and this step is essential since the social investigation is usually carried out in diverse situations. Thus, its reliability and consistency are uncertain, depending on awareness, the convenience of the interview process, and even the personalities of both interviewers and respondents. We then clarified the groups of main factors affecting WTP, which have not been mentioned in the previous literature. This step is also crucial because it provides us with the background to propose a list of appropriate variables that should be included in the WTP regression model. The details of these three points are explained in the following subsections.

3.1. Questionnaire Design

In order to gain a better understanding of the current domestic water use situation in urban communities in Hanoi, this study conducted a social investigation based on a questionnaire survey and a face-to-face interview. The questionnaire was set up and completed in two phases. The first phase was a pilot survey, in which 10 samples were collected to test how the respondents understood the primary list of questions and how much information the interviewers could collect from the face-to-face interviews. After the pilot survey, a few questions were changed, making them easier to understand for the respondents and thereby increasing the effectiveness in approaching the problem.

The final questionnaire consisted of five parts, as we needed to deal with respondents with and without a water supply from HUDWSS. The first part sought to understand which water sources were used for domestic purposes; the second part aimed to provide a list of questions for those who used water from HUDWSS. Here, we attempted to measure the satisfaction of the communities in terms of the quantity, quality, and management of the current domestic water supply; the third part addressed communities that did not use water from HUDWSS; the fourth part enquired about which aspects the communities wished to improve regarding the performance of HUDWSS and the amount that the communities were willing to pay for this improvement in the future; and the final part focused on the residential demographic variables such as age, gender, education level, and monthly income. The demographic information provided the basis for understanding the factors affecting the respondents' WTP for the HUDWSS improvement.

3.2. The Naturally Exploring WTP Technique

To determine respondents' WTP, one of the most widely used WTP questioning techniques is open-ended questions [37,39–44]. In an open-ended question, the respondents are asked to state the maximum amount they could pay for water-targeted improvement. The advantage of this technique is that the question is easy to understand and gives the respondents freedom in giving their WTP values. However, as highlighted in a previous study [38], the open-ended question technique can result in several "zero bids", i.e., respondents indicating a WTP of zero. Therefore, to avoid this zero-WTP situation in our investigation, before asking about the residents' WTP, we provided them with information about the price they are currently paying for 1 m^3 of water supplied by HUDWSS. By doing this, their WTP values were at least equal to their current price for water use, the so-called WTP_0. Thus, in practice, we used three consecutive questions to determine WTP values as the maximum amount that they were willing to contribute. The first question was "To improve the current HUDWSS to the level of your expectations, are you willing to support the water price?". Upon answering YES to this question, they were asked the second question, "The current price of water supply is 7000 dong/m^3, how much do you think that this current price could be increased to have a better budget for HUDWSS improvement?". When answering the second question, respondents were offered several levels by which the current water price WTP_0 could be increased, resulting in pre-WTP values. The third question, "What is the maximum amount which you are willing to support to improve HUDWSS's performance?", was used to determine the maximum amount that the respondents were willing to pay, which is referred as their actual WTP values.

3.3. Sampling

The minimum sample size was determined according to Krejcie and Morgan [45]; for large areas, the number of participants required for the survey was calculated as follows:

$$s = \frac{X^2NP(1-P)}{d^2(N-1) + X^2P(1-P)} \tag{1}$$

in which S: minimum sample size; X^2: the table value of Chi-square for 1 degree of freedom at the desired confidence level (3.841); N: the population size; P: the population proportion (assumed to be 0.5 since this would provide the maximum sample size); d: the degree of accuracy expressed as a proportion (0.05).

The estimated population in these Hanoi urban communities was 3,962,310 in 2019. Thus, the minimum sample size was calculated as 384 samples. In this study, we first conducted a pre-test survey of ten samples of several staff working at the Hanoi University of Natural Resources and Environment and the residents living near this university in order to determine whether our questionnaire was understandable and appropriate. We then finalized the questionnaire and started conducting the comprehensive social investigation in the three most urbanized district representatives in Hanoi urban communities, including the newly urbanized Ha Dong and two old inner-city districts of Thanh Xuan and Hai

Ba Trung (Figure 1). After three months of surveying from November and December of 2019 to January of 2020, we had randomly collected 454 samples. Among these, we could not use 52 samples, mainly due to missing information for more than 30% of the questions. Hence, the final data set comprised 402 samples, of which 91 samples were from the Ha Dong district (353,200 residents in 2019), 155 samples were from the Thanh Xuan district (286,700 residents in 2019), and 156 samples were from the Hai Ba Trung district (311,800 residents in 2019).

3.4. The Theoretical WTP Regression Model

In the literature, we found that 21 main factors/variables that could be considered to affect the WTP for water supply and water-related service improvement. These factors/variables could be divided into four groups: demographic factors, water supply quantity, quality, and service, as shown in Table 1. Therefore, the theoretical WTP regression model could be formulated as in Equation (2).

$$WTP = f(De; Quan; Qual; Ser) = C_{De} \times De + C_{Quan} \times Quan + C_{Qual} \times Qual + C_{Ser} \times Ser + e \quad (2)$$

in which *WTP* is the dependent variable; *De* are demographic variables; *Quan*, *Qual*, and *Ser* are variables regarding water supply quantity, quality, and service, respectively; C_{De}, C_{Quan}, C_{Qual}, C_{Ser} are coefficients; and *e* is the random error. The IBM Statistical Package for the Social Sciences (SPSS) was used to generate the WTP regression model and perform data analysis.

Table 1. Classification of 21 factors/variables considered to affect WTP according to the literature.

Classification	Factors/Variables	Publications
Demographic Factors (8)	Gender	[13–19] *
	Age	[46–50] *
	Education level	[25,39,51,52] *
	Family size	[19,53–55] *, [56]
	Children	[57], [58–60] *
	Occupation	[48,54], [61,62] *
	Income	[36,37,56,58,63] *
	Wealth of the household	[14] *, [64]
Water Supply Quantity (3)	Water source	[51], [65–67] *
	Water reliability	[50,68–70] *
	Water use quantity	[13,47,48,55] *, [71]
Water Supply Quality (4)	Clean water awareness	[43] *, [71], [72,73] *
	Water quality care	[37], [39,44,49] *, [74]
	Water-borne diseases	[38,66,75] *
	Water treatment measures	[39,68,76–78]
Water Supply Service (6)	Bid value	[15,50] *, [65], [70] *
	Monthly water bill	[55], [61] *, [77]
	Household location	[48] *, [51], [67] *, [71]
	Distance to water source	[16] *, [25]*, [77]
	Time for water connection	[41,47,53], [55] *
	Water connection charges	[54,57,68] *, [78]

Note: The studies with "*" are the ones in which the corresponding factors/variables are significant at a 0.05 confidence level.

Selection of factors/variables in each group: Depending on the targeted water supply issues and the public preferences in a specific study area, the factors/variables of water supply quantity, quality, and service groups could be selected. In terms of demographic factors, this selection depended on the characteristics of the communities that the researchers wished to test for their correlations with the WTP.

3.5. Reliability Check

To validate the reliability of the responses in the questionnaire, we used a statistical measure of confidence consistency, namely Cronbach's alpha coefficient, for a given sample. The coefficient formula is according to Equation (3), as indicated in [79].

$$\alpha = \frac{K}{K-1}\left[1 - \frac{\sum_{i=1}^{K} \sigma_{Y_i}^2}{\sigma_X^2}\right] \quad (3)$$

in which K is the number of components (usual questions); σ_X^2 is the variance of the observed total test scores, and $\sigma_{Y_i}^2$ is the variance of component i for the sample.

Reliability scale: As mentioned in [79], the coefficient values in the range of (0.9; 1.0), (0.8; 0.9), (0.7; 0.8), (0.6; 0.7), (0.5; 0.6), and (0; 0.5) indicate respectively excellent, good, acceptable, questionable, poor, and unacceptable internal consistency.

3.6. Application for HUDWSS

Applying our CVM-based process, we obtained 402 useable samples from the three urban districts. Using Equation (3), we find a value for α of 0.879 for the responses to the group of questions regarding the type of water used daily; 0.863 for the responses to the questions regarding the residential satisfaction towards HUDWSS's water consumption; 0.943 for the responses to the questions regarding the residential satisfaction towards HUDWSS's water quality; and 0.865 for the responses to the questions regarding the residential satisfaction towards HUDWSS's service and management. These coefficients are all higher than a threshold value of 0.8, indicating that the obtained responses are internally consistent and reliable. Therefore, the data set obtained from these 402 reliable responses was used for further analysis.

Table 2 provides further details about the characteristics and status of the survey sample. From the total of 402 people interviewed, 208 (51.7%) were female, and 194 (48.3%) were male. The majority of the respondents' ages were in the range of 24 to 55 years (66.9%), and only one respondent was under 18 years old (0.2%). Of the five occupation groups mentioned in the questionnaire, business was the most common, accounting for 46.3% of all occupations. The level of education of the surveyed subjects was partly reflected through their occupation. The more educated subjects were aware of the importance of clean water and the impact of inadequate water quality on their health. It was observed during the pilot survey that interviewees did not wish to disclose their actual incomes due to various personal reasons. Consequently, dividing the total family income into ranges of values in this study made the respondents more comfortable in indicating their relative income. The number of respondents with a total family income of between 10 to 30 million VND (approximately between 434 to 1304 USD) accounted for half (51.7%) of the sample.

Table 2. Basic characteristics of the sample (N = 402).

Characteristics		In Number (Persons)	In Percentage (%)
Age	Less than 18 years old	1	0.2%
	From 18 to 24 years old	50	12.4%
	From 24 to 55 years old	269	66.9%
	Over 55 years old	82	20.4%
Gender	Female	208	51.7%
	Male	194	48.3%
Occupation	Student	29	7.2%
	Employees and officers	64	15.9%
	Business	186	46.3%
	Worker	52	12.9%
	Work at home	71	17.7%

Table 2. *Cont.*

Characteristics		In Number (Persons)	In Percentage (%)
Total income	Less than 3 million VND	1	0.2%
	3–5 million VND	9	2.2%
	5–10 million VND	77	19.2%
	10–30 million VND	208	51.7%
	Over 30 million VND	107	26.6%

4. Results and Discussion

Here we first present the current public awareness, satisfaction, and expectations of HUDWSS quantity, quality, and service that are described to allow for a better understanding of the urban community perceptions and to evaluate the use of the naturally exploring WTP technique on Hanoi's dataset. This is followed by a comparison of the usual statistic variation in both Pre-WTP and WTP values. Finally, we describe the establishment of the WTP regression model based on the theoretical regression framework presented in the methodology to assess how the community's maximum willingness to pay was affected by demographic and domestic water-related factors.

4.1. Social Satisfaction of the HUDWSS's Quantity, Quality, and Service

We divided the sample into two groups: households using water from HUDWSS, comprising 331 respondents (82.3%), and households not using water from HUDWSS, comprising 71 respondents (17.7%). Hence, water from HUDWSS, or "tap water", is most widely used in the inner city of Hanoi (Figure 2). The percentage of those using tap water for cooking and eating purposes accounted for 81.8%, while some households still used well water (12.4%) and rainwater (5.7%). For other purposes, such as bathing, sanitation, gardening, and car washing, the percentage of water usage remained at the highest levels, at 78.4% and 72.6%, respectively. The reason for this was that there were 35 households (approximately 8.7% of the total 331 households using water from HUDWSS) using two or more water sources along with water from HUDWSS. Using alternative water sources is considered to be more advantageous because of their convenience and the low cost of installation. Around 17.7% of households did not have access to tap water for various reasons. There were three subjective reasons discovered through the survey: (i) unstable, dripping, and limited water supply; (ii) no water quality guarantee; and (iii) the limited maintenance service of the supply system.

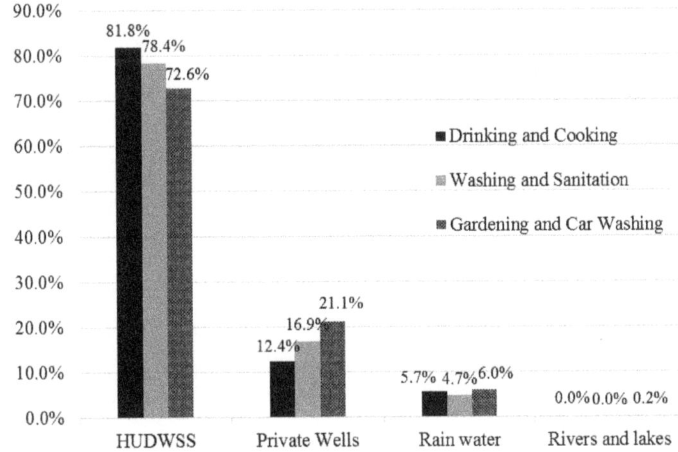

Figure 2. The main water sources used for basic purposes among the studied households.

4.2. Residential Satisfaction of Domestic Water Quantity

The ability to provide a sufficient amount of water by HUDWSS to consumers depends on many factors and was examined objectively through direct interviews with the residents. The survey results show that monthly water consumption and water cut-off status directly influenced the extent to which residents' consumption needs were met. Most (55%) of the surveyed households usually used 20 to 30 m^3 daily, and water consumption was gradually reduced with the number of members in the household and the use purpose. However, there were still a small number (7.9%) of households with a high water demand of over 30 m^3 per day, usually for business and production purposes. In general, the amount of water consumption was relatively high, and this is expected to rise in the future. Therefore, a calculation based on the number of days and monthly times of water cut-off was performed to evaluate whether the water supply was adequate. The amount of water supplied by the city water supply companies was considered to be relatively sufficient. In particular, 77.3% of the surveyed respondents said that the water was sufficiently supplied. In addition, the households with a water cut-off of 1 day, 2 days, and 3 days per month accounted for 15.4%, 6.6%, and 0.6%, respectively (Table 3).

Table 3. Household responses regarding domestic water quantity.

In Terms of	Number of Responses	Percentage (%)
	Monthly water consumption	
Less than 10 m^3	33	10.0%
From 10 to 20 m^3	90	27.2%
From 20 to 30 m^3	182	55.0%
More than 30 m^3	26	7.9%
	Domestic water cut-off frequency	
1 day/month	51	15.4%
2 day/month	22	6.6%
>3 day/month	2	0.6%
No water cut-off	256	77.3%
	Domestic water use quantity satisfaction	
<50%	0	0.0%
50–60%	6	1.8%
60–80%	18	5.4%
80–100%	245	74.0%
>100%	62	18.7%
Total	331	100%

Most of the subjects were relatively satisfied with the amount of water provided. The ability to meet from 80% to 100% of the demand for users accounted for 74% of the total population surveyed. In addition, the amount of water meeting more than 100% of the water consumption demand accounted for 18.7% of the total population surveyed. The remaining residents considered that the supplied water was insufficient for use. The survey results show that households that used less water had enough water to use. As for some cases where households required large amounts of water (>30 m^3) or had frequent water cut-offs, respondents were not satisfied and underestimated the water supply capacity of HUDWSS (Table 3).

4.3. Residential Satisfaction of Domestic Water Quality

According to the survey results, the households using water from HUDWSS responded that they used clean water. Approximately 9.4% of the households mentioned that the water sometimes had a different color or taste, and others commented that the water was cloudy (3.9%). Besides the general public response regarding domestic water supply quality, the water quality nevertheless still could not meet the residents' standards because, according to our results, 80.4% of households use advanced water purifiers before cooking and drinking. For the rest, depending on the household income and also the perceptions of the household decision-makers, households boiling their water before use accounted for 15.4%, and households that did not use any kind of treatment accounted for only 4.2%. The results reveal that there is a need to improve HUDWSS to meet the standards of the local communities. For further information on health impacts related to water supply quality, we also considered waterborne diseases such as diarrhea, skin diseases, gynecological diseases, dengue fever, Japanese encephalopathy, and helminthic infection in the three urban districts. As shown in Table 4, the proportion of residents who had not suffered from any waterborne disease accounted for 90%, but 10% of the respondents had been affected by the aforementioned water-borne diseases.

Table 4. Household responses regarding domestic water quality.

In Terms of	Number of Responses	Percentage (%)
Public awareness of domestic water quality		
Clean	279	84.3%
Color/smell/taste sometimes	31	9.4%
Cloudy	13	3.9%
Strange color/smell/taste	8	2.4%
Water treatment used		
Advanced water purifiers	266	80.4%
Just boiling	51	15.4%
None	14	4.2%
Water-borne diseases affected		
Not yet	297	89.7%
Slightly	9	2.7%
Skin disease	12	3.6%
Diarrhea/Gynecological diseases/Dengue fever/Japanese encephalopathy/Helminthic infection	13	3.9%
Total	331	100%

4.4. Residential Satisfaction of Domestic Water Service

The questionnaire survey for households using the water supply from HUDWSS showed that the service is relatively acceptable. In terms of the water bills, approximately 60% of respondents' monthly payments ranged from 200,000 to 500,000 dong/month (approximately 9 to 22 USD at the exchange rate of 1 USD to about 23,000 VND on the first day of 2020); 35% had to pay less than 200,000 dong/month (mostly households with few members and households who used water from alternative water sources, such as private wells, rivers, and lakes); and approximately 5% were mainly business-based households and paid more than 500,000 dong/month.

Convenient and multiple payment methods are crucial in increasing residents' satisfaction with HUDWSS' service quality. As investigated in our study, there were three main payment methods, with those utilizing smartphone apps and computers accounting for 38.1%, those using home-visiting staff accounting for 35%, and those using supermarkets/post offices accounting for 26.9% (Table 5). This shows that modern payment methods via apps and third-party stakeholders have taken a strong foothold over the traditional home-visiting payment method. The transition is quite important for busy urban households in Hanoi, where modern residents spend most of their time away from their houses.

Table 5. Household responses regarding domestic water management and service.

In Terms of	Number of Responses	Percentage (%)
Payment method		
Via apps on smartphone/computer	126	38.1%
Via home-visiting staff	116	35.0%
Via Supermarket/Post office	89	26.9%
Receiving water cut-off notifications		
From HAWACO website	5	1.5%
From HAWACO document	2	0.6%
From the community's radio	8	2.4%
From the community's bulletin	38	11.5%
No notification	22	6.6%
No water cut-off	256	77.3%
How long has the local water supply system been installed?		
Less than 5 years	38	11.5%
5–10 years	94	28.4%
10–15 years	154	46.5%
Over 15 years	38	11.5%
Don't know	7	2.1%
Maintenance service		
Don't know	23	6.9%
Regularly checked and maintained	176	53.2%
No maintenance service	132	39.9%
Total	331	100%

Regarding water scarcity in urban districts and the limitation of the water supply system's capacity, especially during the summer season, urban districts in Hanoi sometimes face several days without water supply. We found that the urban communities in Hanoi wish to receive water cut-off notifications from the suppliers. In this study, water cut-off notification refers to the interruption announcement of water supply. Most households (77.3%) did not experience days without a water supply. However, around 11.5% of households received water cut-off notifications from the community's bulletin and only 0.6% of them received written notices from the water companies. Among the residents who experienced days without a water supply, around 6.6% of households said that they did not receive any notification before this happened (Table 5).

Furthermore, we found a lack of regular checks and maintenance for HUDWSS even though these services are essential to ensure that the system is still working well. As observed in our study, almost two-thirds of the respondents (58%) confirmed that the local water system had existed for more than ten years, and 11.5% relied on a system that had even been installed more than fifteen years ago. New systems, which had been installed less than five years ago, accounted for only about 10% of the respondents. However, almost 40% of the respondents complained that there was no maintenance service. Only half of the local systems were regularly checked and maintained (Table 5). The lack of maintenance service explains why the non-revenue water proportion in Hanoi remains high, despite recent efforts to reduce it, as shown in Section 2.

4.5. Social Needs Priorities for HUDWSS Improvement

Given the current state of HUDWSS, with its many shortcomings, an important issue is how residents' would like the quality of the water supply services to be improved. We examined the work needed to improve the HUDWSS in the questionnaire, which the respondents evaluated based on the priority of the tasks to be carried out (Figure 3). According to the community's assessment, "improving the quality of water supply" should be prioritized first, followed by "the amount of water needs to be stabilized regularly".

Other issues related to the management and service quality were not a priority but still need to be dealt with in the future. Therefore, to realize these improvements, the question is as follows: Are residents willing to pay higher water prices to obtain more funding in order to improve the quality and quantity of the water supply?

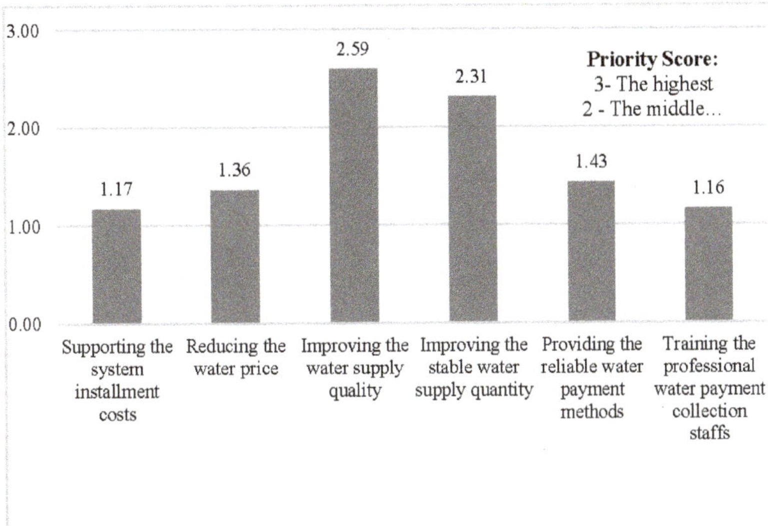

Figure 3. Prioritizing the tasks needed in order to improve HUDWSS.

4.6. Household Positive Pre-WTP and WTP for Improved HUDWSS

Among the 402 survey respondents corresponding to 402 households, 331 used water from HUDWSS and the remaining 71 did not. Therefore, resulting from the naturally exploring WTP technique, both the Pre-WTP values—resulting from the answer to the question on how many of the respondents think the current price could be increased—and WTP values—resulting from the question on the maximum willingness to pay—were evaluated based on the interview results of the 331 households that used water from HUDWSS. The changes in the respondents' willingness from Pre-WTP to WTP are shown in Table 6 and Figure 4. When answering the question about Pre-WTP, around 71% of households stated that their willingness was equal to the current water supply price of WTP_0 = 7000 dong/m^3 (approximately 0.3 USD). Around half of these households increased their willingness to pay higher values when stating their maximum levels of WTP. This change indicates that 35.6% of households were willing to pay 14,000 dong/m^3 (approximately 0.6 USD). The highest pre-WTP value of 13,000 dong/m^3 was replaced by 20,000 dong/m^3 as the highest value of WTP. By multiplying the obtained pre-WTP and WTP values per m^3 of water supply with the corresponding average monthly water use of the households, the pre-WTP and WTP values per month could be evaluated. Thus, as seen in Figure 4, the pre-WTP mean also increased from 164,000 dong/household/month (approximately 7 USD) to a mean WTP of 281,000 dong/household/month (about 12.2 USD). A substantial proportion (22.1%) of households were willing to pay the highest amount, equivalent to 350,000 dong/household/month (about 15 USD). For each targeted district, the mean WTP in the Thanh Xuan district was approximately 278,000 dong/household/month (about 12.1 USD), equivalent to 1.4% of the total average income of households; in Hai Ba Trung district, the mean WTP was the highest, approximately 286,000 dong/household/month (about 12.4 USD), equivalent to 1.5%; and the mean WTP in Ha Dong district was the lowest, at approximately 270,000 dong/household/month (about 11.7 USD), equivalent to 1.4% of the total average household income. Overall, the mean WTP of all three districts was equivalent to approximately 281,000 dong/household/month (about 12.2 USD), which is equivalent to 1.4% of the total average income of households. In this case, our proposed WTP exploring technique helped the respondents to determine their maximum willingness values, offering them the freedom to provide their figures rather than selecting from a set of proposed values. This allowed the respondents to carefully consider their needs in order to obtain water at their recommended rates. The average estimated WTP value of the three districts in general, and the value of each district, in particular, did not exceed 2.5%, as established by the United States Environmental Protection Agency (US EPA) as an indicator of the affordability of monthly water payments among citizens.

Table 6. Pre-WTP and WTP of the respondents.

Pre-WTP Value (dong/m^3/Household)	Number of Responses	Percentage (%)	WTP Value (dong/m^3/Household)	Number of Responses	Percentage (%)
7000	234	70.7	7000	118	35.6
8000	2	0.6	14,000	118	35.6
9000	5	1.5	15,000	2	0.6
10,000	19	5.7	16,000	3	0.9
12,000	16	4.8	17,000	19	5.7
13,000	55	16.6	19,000	16	4.8
			20,000	55	16.6
Total	331	100	Total	331	100

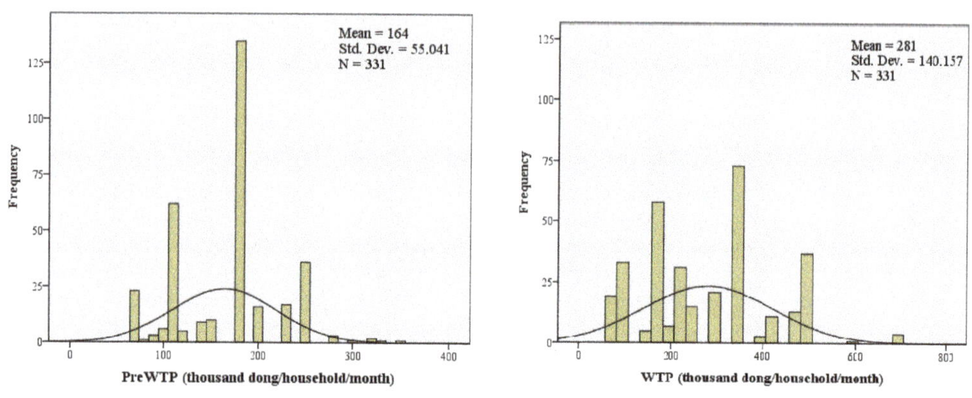

Figure 4. Pre-WTP and WTP histograms.

4.7. Determining and Analyzing the Factors Affecting the WTP Values of the Respondents

Determining the variables included in the WTP regression model is important to understand the factors that affect WTP. In this study, the variables were determined and selected based on two criteria. First, based on the synthesis of 21 common factors grouped into four categories, as shown in Table 1, this study also clarified the main factors from these four common groups. In the first group of demographics, we considered the five variables of gender, age, income, family size, and occupation, as these variables are usually measured in the literature regarding CVM applications. Second, based on the actual situation of HUDWSS' performance, the variables of water supply quantity, quality, and service were selected and proposed in this study. For the group of water supply quantity, the water use variable referred to the average monthly amount of water consumption in households; regarding the quantity variable, this measured how satisfied the respondents were with the water supply quantity. The water supply quality group included the quality variable, which measured whether the water met the quality standards of the communities, and the treatment variable, which referred to the water purification methods that the households usually used before cooking and drinking. The water supply service here consisted of three variables measuring how well and conveniently the HUDWSS' service could assist their customers in using the water supply. The payment method variable considered the way in which the households paid their monthly water bills. The maintenance variable evaluated how often the household's water supply equipment was checked and maintained to prevent non-revenue water. The notification variable measured whether the communities were informed before the water supply was cut off. As a result, 12 independent variables were selected and proposed to affect the WTP for improving the quality of the HUDWSS service. The descriptions of these variables are also given in Table 7. Moreover, the dataset obtained from any social questionnaire survey is usually complex, causing skewing problems. In this study, we thus applied the log transformation method for the values obtained for the income and WTP variables in order to reduce the data complexity. The transformation rules for all the dependent and independent variables of the WTP regression model are described in detail in Table 7.

Table 7. Description of variables used in regression analysis.

Variable		Description
Dependent variable		
Log_WTP		Continuous variable denoting the natural log value of each respondent WTP
Independent variables		
Demographic variable	Gender	Dummy variable equal to 1 for male and 0 for female
	Age	Dummy variable equal to 1 if age of respondent is in the range of 24 to 55 years old and 0 otherwise
	Log_Income	Continuous variable denoting the natural log value of each household's monthly income
	Family Size	Dummy variable equal to 1 if household has more than normal size of four members and 0 otherwise
	Occupation	Discrete variable denoting the respondent's occupation type
Water supply quantity	Water use	Dummy variable equal to 1 if the monthly household water consumption is in the range of 20 to 30 m^3 and 0 otherwise
	Quantity	Dummy variable equal to 1 if the water supply meets more than 80% of the household's water needs and 0 otherwise
Water supply quality	Quality	Dummy variable equal to 1 for clean water response and 0 otherwise
	Treatment	Dummy variable equal to 1 if household uses water treatment and 0 if household uses no water treatment
Water supply service	Payment Method	Discrete variable denoting the household's water payment method
	Maintenance	Dummy variable equal to 1 if household's water supply system is regularly maintained and 0 otherwise
	Notification	Dummy variable equal to 1 if household receives a notification before water cut-off day and 0 otherwise

Regression analysis for 331 households using water from HUDWSS was conducted. The results of the regression model are shown in Table 8. The specified regression function fit the estimated mean WTP of individual respondents, as indicated by an R^2 estimated at 0.314, and the standard error of this estimation is approximately 0.208. The results show that age significantly (at 0.1 significance level) affects the mean WTP of respondents. More specifically, the older the respondent, the higher their mean WTP. As expected, similar to the situation in many other developing countries [19,36,37,56,58,63], income affects the mean WTP of respondents significantly (at 0.05 significance level); the higher the household income, the higher the mean WTP of the respondent. The results also confirm economic theory, which states that an individual/household's demand for a particular commodity depends on his/her income [38]. The effects of gender are significant (at 0.05) and negative, implying that men are willing to pay more for improved HUDWSS than women. This is contrary to the assumption that women are more likely to pay more because they invest more time in household activities and have a greater need for water for domestic purposes, as suggested by the findings of Ayanshola et al. [80]. The effect of water use is significant (at 0.01 level) and positive, and the greater the amount of water used by a household influences the WTP for the improvement of the HUDWSS service. It is interesting to note that the payment methods and maintenance variables, which were proposed for the first time in this study for the targeted HUDWSS, appear to positively affect the respondents' WTP and are also respectively significant, at 0.1 and 0.01 levels. This implies that households that used modern payment methods were willing to pay more to improve the water supply service, and households that have regular water supply system maintenance checks by staff are also willing to pay more to improve the water supply service than other households whose water supply systems are not maintained.

Table 8. WTP regression results for HUDWSS improvement.

Independent Variable	Estimated Coefficient	*p*-Value	Standard Error
(Constant)		0.000	0.376
Gender	−0.101	0.032 **	0.023
Age	0.089	0.079 *	0.027
Log_Income	0.126	0.013 **	0.052
Family Size	−0.054	0.259	0.024
Occupation	0.009	0.855	0.011
Water_use	0.319	0.000 ***	0.027
Quantity	0.044	0.391	0.049
Quality	0.046	0.376	0.035
Treatment	−0.011	0.819	0.058
Payment Method	0.092	0.053 *	0.015
Maintainance	0.228	0.000 ***	0.028
Notification	0.003	0.953	0.034

Dependent variable: Log_WTP; R^2 = 0.314; Adjusted R^2 = 0.288; Standard error of the estimate: 0.208; *** statistically significant at 0.01, ** statistically significant at 0.05, and * statistically significant at 0.1.

The other variables seem to be insignificant in the level of $p = 0.1$ influencing respondents' WTP. Specifically, the occupation of respondents does not seem to have a significant impact. This finding is similar to other related studies conducted using CVM for the improvement of the domestic water supply system in Nigeria [38] and in Palestine [49]. The number of people in each household (family size) usually appears to positively affect WTP, as found by Byambadorj and Han [54]; Akeju et al. [38] and Fujita et al. [55]. However, in our study, this variable negatively affected the respondents' WTP and is statistically insignificant at (p-value = 0.1). The water quantity and water quality satisfaction seem to positively affect the respondents' WTP but are also not significant at (p-value = 0.1). Moreover, the effect of the treatment variable was negative and is statistically insignificant at (p-value = 0.1), as similarly found by Rodríguez-Tapia et al. [39]; Guilfoos et al. [75]; Orgill et al. [76] and Odwori [68]. The effects of the water cut-off notifications were positive and do not seem to be significant at the level of p-value = 0.1.

5. Conclusions

This study successfully proposes a CVM-based process and tests its effectiveness in an investigation of Hanoi urban communities WTP regarding the improvement of HUDWSS. Our results show that Hanoi urban communities were essentially satisfied with the water supply quantity, as our investigation found that more than 90% of surveyed respondents considered their water supply to be mostly sufficient. The water quality was still lower than the quality expectations of the urban residents since most households (80.4%) used advanced purifiers to treat the tap water before drinking and cooking. Almost half of the respondents complained about the lack of maintenance services. From the regression model results, we found that significant factors (at p-value = 0.1, 0.05, and even 0.01) affecting the WTP are gender, age, income, water use, payment method, and maintenance;

meanwhile, occupation, family size, quantity and quality satisfaction, treatment, and notification were found to be insignificant factors (at p-value = 0.1). These findings reveal the crucial role of understanding the target problems in selecting and proposing appropriate variables to increase the effectiveness of the WTP regression model. Our results also show that the naturally exploring WTP technique proposed in this study makes it easier for both respondents and interview conductors in determining the WTP values. The average WTP is approximately 1.4% of the average household income, well below the 2.5% threshold established by the US EPA as an indicator of the affordability of monthly water payments among citizens. However, the implementation of policies will take several years, especially in developing countries such as Vietnam, where the average household income is considered low compared to other developed countries, and should take into account the varying income levels among households. This shows the practicality of a future social investment fund contributed to by the communities that is used for upgrading and improving the quality of Hanoi's urban water supply services. The willingness to pay (WTP) technique offered genuine results that helped to make realistic recommendations to the policy- and decision-makers without any complications. The methodology that we developed for this study can be applied to any similar area, without any geographical limitations, in several STEM and social science subjects and multidisciplinary fields. These characteristics make this model unique and easy to use in all circumstances.

Following are several remarks about how to get highly reliable data in such CVM applications. One thing was asking the cooperative residents who were willing to spend more than 15 min to finish the questionnaire. This thing mainly depended on how the interview conductors started asking the questions. Another factor involved about the investigation approach, which significantly affected the reliability of the collected data. Using email and Google Forms i could reduce the survey time and make it more convenient to complete, but the reliability of the obtained data is usually low in comparison to the face-to-face interviewing approach. In this study, the face-to-face interviewing approach was employed to maximize the possibility of obtaining highly reliable data. The reason is that the respondents usually do not fully understand all the questions, thus the interview process should be like a friendly discussion. Particularly in the CVM investigation, the respondents should imagine the unreal market and give the most proper payment for the goods (domestic water supply, in this case). The friendly face-to-face discussion was thus crucial to help the respondents in finding their appropriate WTPs. Moreover, as experienced from our investigation, mentioning the current water price of WTP_0 = 7000 dong when asking about the WTP really affected the respondent's opinions. All of the obtained Pre-WTPs and WTPs were higher than WTP_0. That means that the respondents thought of the WTP_0 as their acceptable minimum payment for 1 m^3 of water. In addition, the WTP0 could also give us the possible variation range of respondent WTPs. In our case, no respondents gave a WTP of more than three times the WTP_0 (i.e., about 21,000 dong). Therefore, in order to obtain a highly reliable data set, it is possible to eliminate the response where the WTPs are higher than 3^*WTP_0. Regarding future research, we would like to apply the same methodology by combining physical (water quality and quantity) and social science (people income, employment, gender) components in other similar areas in Vietnam. Moreover, we would like to expand our study area to other South Asian countries, such as India, the Philippines, Sri Lanka, and China. Our collaborators have also shown interest in applying this unique method to the countries mentioned above.

Author Contributions: Conceptualization, N.T.B., D.D.B., J.M.R.M., S.D., K.K. and T.N.T.; methodology, N.T.B., T.T.H.N. and H.T.H.; software, T.Q.V.; validation, N.T.B. and T.Q.V.; formal analysis, N.T.B. and T.Q.V.; investigation, N.T.B. and T.Q.V.; resources, N.T.B. and T.Q.V.; data curation, N.T.B. and T.Q.V.; writing—original draft preparation, N.T.B., J.M.R.M., T.T.H.N., T.T.P.B. and T.Q.V.; writing—review and editing, J.M.R.M., S.D. and K.K.; visualization, N.T.B., T.N.T. and T.Q.V.; supervision, S.D. and D.D.B.; project administration, S.D. and D.D.B.; funding acquisition, S.D. and D.D.B. All authors have read and agreed to the published version of the manuscript.

Funding: This research was funded by the Vingroup Innovation Foundation, grant number VI-NIF.2019.DA17 and the UK Natural Environment Research Council (NE/S002847/1).

Institutional Review Board Statement: The authors certify that all data collected during the study are presented in this manuscript; no data has been or will be published separately. The researcher team has an obligation to conduct this research with integrity and transparency. The researcher team has protected and respected all personal data provided by participants through rigorous and appropriate procedures for confidentiality and anonymisation. The research was approved by the Hanoi University of Natural Resources and Environment with approval code 1873/TDDHHN.

Informed Consent Statement: Informed consent was obtained from all subjects involved in the study.

Data Availability Statement: The data presented in this study are available on request from the corresponding author.

Acknowledgments: The authors would like to express our gratitude to Lucia Wright Contreras at URBANgrad, Technical University of Darmstadt, Germany for her excellent comments; to Hoang Nam Nguyen at the Faculty of Urban and Environmental Natural Resource Economics & Management, National Economics University for his guidance; and to the students of class DH_7QM_1 (Nga, Nguyen, Tinh, Hoan) at the Faculty of Environment, Hanoi University of Natural Resources and Environment, Vietnam, in conducting the field survey.

Conflicts of Interest: The authors declare that they have no conflict of interest.

Appendix A

HANOI UNIVERSITY OF NATURAL RESOURCES & ENVIRONMENT	**SOCIALIST REPUBLIC OF VIETNAM** **Independence—Freedom—Happiness**

QUESTIONNAIRE FOR HANOI URBAN COMMUNITIES TO IMPROVE THE DOMESTIC WATER SUPPLY SERVICE
(Translated from the Vietnamese version)

Safe drinking water in both quantity and quality is an essential need of all communities. With a densely populated population like the capital Hanoi, ensuring a high quality source of drinking water for the communities becomes more urgent than ever. In order to improve the domestic water supply service for the capital, we are a research team from Hanoi University of Natural Resources and Environment. We would like to conduct a survey to have a better understanding of the current situation of Hanoi urban domestic water supply service. Filling out this survey will take about 15 min from Sirs/Madams. The research team commits that all information Sirs/Madams provide by filling this questionnaire will be treated as confidential and will be used for scientific purposes only.

Thank you very much for your kind cooperation!

(Please put an **X** mark in the bank square in front of your choices)

Appendix A.1. Questions about Water Resources Used for Domestic Purposes

Appendix A.1.1. What Is the Main Source of Water You Use for Cooking and Eating?

Water supply from city water company
Well water
River and lake water
Rain water

Appendix A.1.2. What Is the Main Source of Water That You Use for Washing and Sanitation?

Water supply from city water company
Well water
River and lake water

Rain water

Appendix A.1.3. What Is the Main Source of Water You Use for Gardening and Car Washing?

Water supply from city water company
Well water
River and lake water
Rain water

Appendix A.2. Questions for Those Who Use the Hanoi Urban Domestic Water Supply

Appendix A.2.1. Could You Please Tell Me, How Much in % the Amount of Supply Water Meets Your Family's Water Need?

Less than 50% 50–60% 60–80% 80–100% Over 100%

Appendix A.2.2. How Much Is Your Family's Monthly Water Consumption?

Less than 10 m^3 10–20 m^3 20–30 m^3 Over 30 m^3

Appendix A.2.3. How Much Is Your Monthly Water Bill?

Less than 200,000 dong 200,000–500,000 dong
500,000–1,000,000 dong Over 1,000,000 dong

Appendix A.2.4. How Many Days Does Your Family Have No Domestic Water Supply in a Month?

No water cut-off
1 day/month
2 day/month Over
3 day/month

Appendix A.2.5. During the Water Cut Off Day, How Long Is the Period of Water Cut-Off?

Within 6 h
Within 12 h

Appendix A.2.6. Did You Receive Notice before Water Cut-Off Day? If Yes, What Is the Method of Notification?

No notification
Receive from water supply company (HAWACO) website
Receive from the community's radio
Receive from the community's bulletin
Receive from HAWACO documents

Appendix A.2.7. How Do You Feel about the Current Water Quality?

Clean water (colorless, odorless, and tasteless)
Water is cloudy, scum
Water sometimes has color/smell/taste
Water often has a strange color/smell/taste

Appendix A.2.8. How Do You Treat Water before Drinking/Cooking?

Advanced water purifiers
Just boiling
No treatment, just use directly

Appendix A.2.9. The Following Are 06 Water Borne Diseases. Have You or Anyone in Your Family Suffered from Any of the Water Borne Diseases?

Not yet
Slightly
Diarrhea
Skin disease
Gynecological diseases
Dengue fever, Japanese encephalopathy
Helminthic infection

Appendix A.2.10. For How Long the Water Supply System You Are Using Has Been Installed?

Less than 5 years   5–10 years   10–15 years   Over 15 years

Appendix A.2.11. Has Your Water Supply System Been Regularly Maintained?

Don't know
No maintenance service
Regularly checked and maintained

Appendix A.2.12. What Is Your Family's Water Monthly Payment Method?

Via home-visiting staff
Via supermarket/Post office
Via apps on smart- phone/computer

Appendix A.3. Questions for Those Who Do Not Use Hanoi Water Supply

Appendix A.3.1. What Are Your Main Reasons for Not Using The Water Supply?

Expensive installation cost
Too high water price compared to our water affordability
The water quality is not high enough
Unstable and dripping water
Unreliable water payment methods

Appendix A.3.2. Under the Difficulty of the Current Polluted Natural Water Sources in and nearby Hanoi, Do You Use One of the Following Treatment Methods before Using Water for Drinking Purposes?

Advanced water purifiers
A sand gravel filter
Just boiling

Appendix A.3.3. Do You Plan to Use the Water Supply for Domestic Purpose in the Near Future?

No
Yes, we will

Appendix A.4. Questions of Which Aspects of Hanoi Water Supply System the Communities Expect to Be Improved

Appendix A.4.1. Please Number in Descending Order of Priority (3—The Highest Priority; 2—Igh Priority; 1—No Priority) the Things to Do to Improve Hanoi Water Supply System Service?

Supporting the system installment costs
Reducing the water price
Improving the water supply quality

Improving the stable water supply quantity
Providing the reliable water payment methods
Training the professional water payment collection staffs

Appendix A.4.2. To Improve the Performance of Hanoi Urban Domestic Water Supply System as You Expected, Are You Willing to Support the Water Price?

Yes
No, my family's income is low
No, I'm afraid our support will not be used properly

Appendix A.4.3. The Current Price of Water Supply Is 7000 dong/m^3, How Much Do You Think That This Current Price Could Be Increased to Have a Better Budget for HUDWSS Improvement?

… … … … … … … … … … …..

What is the maximum amount which you are willing to support to improve HUDWSS's performance?

… … … … … … … … … … …..

Appendix A.5. Questions of Personal Information

a. Living area (District):… … … … … … … … … … … … … … … … … …
b. Gender:… … … … … … … … … … … … … … … … … …...
c. Age:

Less than 18 years old | 18–24 years old
24–55 years old | Over 55 years old

d. Current job:

Student | Worker
Employees and officer | Work at home
Business

e. Number of people in your family:… … … … … … people
f. Total family income:

Less than 3 million dong | 3–5 million dong
5–10 million dong | 10–30 million dong
Over 30 million dong

Thank you very much for your kind assistance!

References

1. UN-Water. *Water Scarcity*; United Nations Water: San Francisco, CA, USA, 2014.
2. Vairavamoorthy, K.; Gorantiwar, S.D.; Pathirana, A. Managing urban water supplies in developing countries—Climate change and water scarcity scenarios. *Phys. Chem. Earth* **2008**, *33*, 330–339. [CrossRef]
3. He, C.; Liu, Z.; Wu, J.; Pan, X.; Fang, Z.; Li, J.; Bryan, B.A. Future global urban water scarcity and potential solutions. *Nat. Commun.* **2021**, *12*, 4667. [CrossRef] [PubMed]
4. FAO. *Coping with Water Scarcity: Challenge of the Twenty-First Century*; Food and Agriculture Organization of the United Nations, FAO Regional Office for the Near East: Cairo, Egypt, 2007.
5. Suzenet, G.; Tal, A.; Boymanns, D. Sustainable Water Management for the City: Technologies for Improving Domestic Water Supply. *Built Environ.* **2002**, *28*, 138–151. Available online: http://www.jstor.org/stable/23288797 (accessed on 19 September 2021).
6. Hutton, G.; Haller, L.; Bartram, J. Global cost-benefit analysis of water supply and sanitation interventions. *J. Water Health* **2007**, *5*, 481–502. [CrossRef]
7. Briscoe, J. The Financing of Hydropower, Irrigation and Water Supply Infrastructure in Developing Countries. *Int. J. Water Resour. Dev.* **1999**, *15*, 459–491. [CrossRef]
8. Varian, H.R. *Microeconomic Analysis*; W. W. Norton & Co Inc.: New York, NY, USA, 1992.
9. Platt, J.; Piper, S. Willingness and Ability to Pay for Rural Water. In *Water Policy and Management: Solving the Problems*; ASCE: Reston, VA, USA, 1994.
10. Al-Ghuraiz, Y.; Enshassi, A. Ability and willingness to pay for water supply service in the Gaza Strip. *Build. Environ.* **2005**, *40*, 1093–1102. [CrossRef]

11. Gunatilake, H.; Yang, J.C.; Pattanayak, S.; Berg, C. *Willingness-to-Pay and Design of Water Supply and Sanitation Projects: A Case Study*; Asian Development Bank: Mandaluyong, Philippines, 2006. Available online: http://hdl.handle.net/11540/2351 (accessed on 19 September 2021).
12. Carson, R.T. *Contingent Valuation: A User's Guide; University of California at San Diego, Economics Working Paper Series qt2mw607q7*; Department of Economics, UC San Diego: San Diego, CA, USA, 1999.
13. Meunier, S.; Manning, D.T.; Quéval, L.; Cherni, J.A.; Dessante, P.; Zimmerle, D. Determinants of the marginal willingness to pay for improved domestic water and irrigation in partially electrified Rwandan villages. *Int. J. Sustain. Dev. World Ecol.* **2019**, *26*, 547–559. [CrossRef]
14. Kaliba, A.R.M.; Norman, D.W.; Chang, Y.-M. Willingness to pay to improve domestic water supply in rural areas of Central Tanzania: Implications for policy. *Int. J. Sustain. Dev. World Ecol.* **2003**, *10*, 119–132. [CrossRef]
15. Ayenew, B.; Belay, A.; Tesfay, Y. Economic Value of Wondo Genet Catchment Forest in Domestic Water Supply Services, Southern Ethiopia. *J. Econ. Sustain. Dev.* **2015**, *6*, 9.
16. Mezgebo, G.K.; Ewnetu, Z. Households willingness to pay for improved water services in urban areas: A case study from Nebelet town, Ethiopia. *JDAE* **2015**, *7*, 12–19. [CrossRef]
17. Makwinja, R.; Kosamu, I.B.M.; Kaonga, C.C. Determinants and Values of Willingness to Pay for Water Quality Improvement: Insights from Chia Lagoon, Malawi. *Sustainability* **2019**, *11*, 4690. [CrossRef]
18. Omole, D.O.; Okunowo, O.S. People Perception of Domestic Water Supply Situation in Ogun State, Nigeria. *Res. J. Appl. Sci. Eng. Technol.* **2016**, *12*, 94–99. [CrossRef]
19. So, Y.K.; Seung, H.Y.; Chang, S.K. Measuring the Willingness to Pay for Tap Water Quality Improvements: Results of a Contingent Valuation Survey in Pusan. *Water* **2013**, *5*, 1638–1652. [CrossRef]
20. Reis, N. *Tracing and Making the State: Policy Practices and Domestic Water Supply in the Mekong Delta, Vietnam*; LIT Verlag Münster: Berlin, Germany, 2012.
21. MONRE. *National Environment Report 2016—Urban Environment*; Ministry of Natural Resources and Environment: Hanoi, Vietnam, 2016.
22. Khoi, D.D. Hanoi Metropolitan Area. In *Urban Development in Asia and Africa*; 2017 edition; Springer: New York, NY, USA, 2017; pp. 131–150.
23. MONRE. *National Environment Report 2018—Water Environment in River Basins*; Ministry of Natural Resources and Environment: Hanoi, Vietnam, 2018.
24. HAWACO. *Annual Report*; Hanoi Water Limited Company: Hanoi, Vietnam, 2015.
25. Whittington, D.; Briscoe, J.; Mu, X.; Barron, W. Estimating the Willingness to Pay for Water Services in Developing Countries: A Case Study of the Use of Contingent Valuation Surveys in Southern Haiti. *Econ. Dev. Cult. Chang.* **1990**, *38*, 293–311. [CrossRef]
26. VIWASE. *The Report of Drainage Planning of Hanoi Capital up to 2030, Vision to 2050*; Vietnam Water, Sanitation and Environment Joint Stock Company: Hanoi, Vietnam, 2013. (In Vietnamese)
27. Wright-Contreras, L.; March, H.; Schramm, S. Fragmented landscapes of water supply in suburban Hanoi. *Habitat Int.* **2017**, *61*, 64–74. [CrossRef]
28. GSO. *Statistical Summary Book of Vietnam*; General Statistic Office of Vietnam, Statistical Publishing House: Hanoi, Vietnam, 2017; p. 49.
29. Wright-Contreras, L. A transnational urban political ecology of water infrastructures: Global water policies and water management in Hanoi. Public Works Manag. *Policy* **2019**, *24*, 195–212.
30. HAWACO. *Temporary Water Shut-Off Schedule*; Hanoi Water Limited Company: Hanoi, Vietnam, 2016. Available online: http://hawacom.vn/?cat=67 (accessed on 16 October 2020).
31. HAWACO. Hanoi water supply system. In Proceedings of the Part of the Guided Visit to HAWACO's Hoa Binh WTP at the 37th WEDC International Conference: Sustainable Water and Sanitation Services for all in a Fast Changing World, Co-Hosted by Loughborough University and the National University of Civil Engineering (NUCE), Hanoi, Vietnam, 15–19 September 2014.
32. UNDP. Urban Poverty Assessment in Hanoi and Ho Chi Minh City; United Nations Development Programme. 2010. Available online: http://www.vn.undp.org/content/vietnam/en/home/library/poverty/urban-poverty-assessment-in-ha-noi-and-ho-chi-minh-city.html (accessed on 14 September 2017).
33. Du Bui, D.; Kawamura, A.; Tong, T.N.; Amaguchi, H.; Nakagawa, N. Spatio-temporal analysis of recent groundwater-level trends in the Red River Delta, Vietnam. *Hydrogeol. J.* **2012**, *20*, 1635–1650. [CrossRef]
34. Du Bui, D.; Kawamura, A.; Tong, T.N.; Amaguchi, H.; Trinh, T.M. Aquifer system for potential groundwater resources in Hanoi, Vietnam. *Hydrol. Process.* **2012**, *26*, 932–946. [CrossRef]
35. Bui, N.T.; Kawamura, A.; Bui, D.D.; Amaguchi, H.; Bui, D.D.; Truong, N.T.; Do, H.H.T.; Nguyen, C.T. Groundwater sustainability assessment framework: A demonstration of environmental sustainability index for Hanoi, Vietnam. *J. Environ. Manag.* **2019**, *241*, 479–487. [CrossRef] [PubMed]
36. Jiang, Y.; Rohendi, A. Domestic water supply, residential water use behaviour, and household willingness to pay: The case of Banda Aceh, Indonesia after ten years since the 2004 Indian Ocean Tsunami. *Environ. Sci. Policy* **2018**, *89*, 10–22. [CrossRef]
37. Rananga, H.T.; Gumbo, J.R. Willingness to Pay for Water Services in Two Communities of Mutale Local Municipality, South Africa: A Case Study. *J. Hum. Ecol.* **2015**, *49*, 231–243. [CrossRef]

38. Akeju, T.J.; Oladehinde, G.J.; Abubakar, K. An Analysis of Willingness to Pay (WTP) for Improved Water Supply in Owo Local Government, Ondo State, Nigeria. *Asian Res. J. Soc. Sci.* **2018**, *5*, 1–15. [CrossRef]
39. Rodríguez-Tapia, L.; Revollo-Fernández, D.A.; Morales-Novelo, J.A. Household's Perception of Water Quality and Willingness to Pay for Clean Water in Mexico City. *Economies* **2017**, *5*, 12. [CrossRef]
40. Wahid, N.A.; Hooi, C.K. Factors Determining Household Consumer's Willingness to Pay for Water Consumption in Malaysia. *Asian Soc. Sci.* **2015**, *11*, 26. [CrossRef]
41. Khan, H.; Iqbal, F.; Saeed, I. Estimating willingness to pay for improvements in drinking water quality: Evidence from Peshawar, Northern Pakistan. *Environ. Econ.* **2010**, *1*, 38–43.
42. Saker, M.A.R.; Alam, K. Willingness to Pay for Improved Water Services in Rajshahi City, Bangladesh. *Asian J. Water Environ. Pollut.* **2013**, *10*, 41–49.
43. Salazar, S.S.; Gómez, F.G.; Guardiola, J. Willingness to pay to improve urban water supply: The case of Sucre, Bolivia. *Water Policy* **2015**, *17*, 112–125. [CrossRef]
44. Parveen, S.; Ahmad, J.; Rahman, M.U. Estimating Willingness to Pay for Drinking Water Quality in Nowshera Pakistan: A Domestic Study for Public Health. *JAAS* **2016**, *19*, 48–56.
45. Venkatachalam, L. Factors influencing household willingness to pay (WTP) for drinking water in peri-urban areas: A case study in the Indian context. *Water Policy* **2006**, *8*, 461–473. [CrossRef]
46. Krejcie, R.V.; Morgan, D.W. Determining Sample Size for Research Activities. *Educ. Psychol. Meas.* **1970**, *30*, 607–610. [CrossRef]
47. Eridadi, H.M.; Yoshihiko, I.; Alemayehu, E.; Kiwanuka, M. Evaluation of willingness to pay toward improving water supply services in Sebeta town, Ethiopia. *J. Water Sanit. Hyg. Dev.* **2021**, *11*, 282–294. [CrossRef]
48. Ogunniyi, L.T.; Sanusi, W.A.; Ezekiel, A.A. Determinants of rural household willingness to pay for safe water in Kwara State, Nigeria. *AACL Bioflux* **2011**, *4*, 660–669.
49. Awad, I.; Holländer, R. Applying Contingent Valuation Method to Measure the Total Economic Value of Domestic Water Services: A Case Study in Ramallah Governorate, Palestine. *Eur. J. Econ. Financ. Adm. Sci.* **2010**, *20*, 76–93.
50. Groothuis, P.A.; Cockerill, K.; Mohr, T.M. Water does not flow up hill: Determinants of willingness to pay for water conservation measures in the mountains of western North Carolina. *J. Behav. Exp. Econ.* **2015**, *59*, 88–95. [CrossRef]
51. Beyene, F.; Lema, Z. Willingness to Pay for Improved Rural Water Supply in Goro-Gutu District of Eastern Ethiopia: An Application of Contingent Valuation. *J. Econ. Sustain. Dev.* **2012**, *3*, 145–159.
52. Pham, K.N.; Tran, V.H.S. Household Demand for Improved Water Services in Ho Chi Minh City: A Comparison of Contin-gent Valuation and Choice Modeling Estimates. In *Economy and Environment Program for Southeast Asia (EEPSEA) 2005*; EEPSEA Research Report rr2005063; EEPSEA: Ho Chi Minh City, Vietnam, 2005.
53. Pattanayak, S.K.; Jui, C.Y.; Whittington, D.; Bal Kumar, K.C. Coping with unreliable public water supplies: Averting expenditures by households in Kathmandu, Nepal. *Water Resour. Res.* **2005**, *41*, W0201. [CrossRef]
54. Byambadorj, A.; Han, S.L. Household Willingness to Pay for Wastewater Treatment and Water Supply System Improvement in a Ger Area in Ulaanbaatar City, Mongolia. *Water* **2019**, *11*, 1856. [CrossRef]
55. Fujita, Y.; Fujii, A.; Furukawa, S.; Ogawa, T. Estimation of Willingness-to-Pay (WTP) for Water and Sanitation Services through Contingent Valuation Method (CVM): A Case Study in Iquitos City, The Republic of Peru. *BICI Rev.* **2005**, *11*, 59–87.
56. Hernández, J.R.; Salazar, S.S. Estimating the non-market benefits of water quality improvement for a case study in Spain: A contingent valuation approach. *Environ. Sci. Policy* **2012**, *22*, 47–59. [CrossRef]
57. Tussupova, K.; Berndtsson, R.; Bramryd, T.; Beisenova, R. Investigating Willingness to Pay to Improve Water Supply Services: Application of Contingent Valuation Method. *Water* **2015**, *7*, 3024–3039. [CrossRef]
58. Brox, J.A.; Kumar, R.C.; Stollery, K.R. Willingness to pay for water quality and supply enhancements in the Grand River Watershed. Can. *Water Resour. J.* **1996**, *21*, 275–288. [CrossRef]
59. Chatterjee, C.; Triplett, R.; Johnson, C.K.; Ahmed, P. Willingness to pay for safe drinking water: A contingent valuation study in Jacksonville, FL. *J. Environ. Manag.* **2017**, *203*, 413–421. [CrossRef] [PubMed]
60. Majumdar, C.; Gupta, G. Willingness to pay and municipal water pricing in transition: A case study. *J. Integr. Environ. Sci.* **2009**, *6*, 247–260. [CrossRef]
61. Casey, J.F.; Kahn, R.K.; Rivas, A. Willingness to pay for improved water service in Manaus, Amazonas. *Brazil. Ecol. Econ.* **2006**, *58*, 365–372. [CrossRef]
62. Sule, B.F.; Okeola, O.G. Measuring willingness to pay for improved urban water supply in Offa City, Kwara State, Nigeria. Water Sci. Technol. *Water Supply* **2010**, *10*, 933–941. [CrossRef]
63. Wang, H.; Shi, Y.; Kim, Y.; Kamata, T. Valuing water quality improvement in China: A case study of Lake Puzhehei in Yun-nan Province. *Ecol. Econ.* **2013**, *94*, 56–65. [CrossRef]
64. Ahmad, I.; Haq, M.; Sattar, A. *Factors Determining Public Demand for Safe Drinking Water (A Case Study of District Peshawar)*; Pakistan Institute of Development Economics Islamabad: Islamabad, Pakistan, 2010.
65. Hundie, S.K.; Abdisa, L.T. Households' Willingness to Pay for Improved Water Supply: Application of the Contingent Valuation Method: Evidence from Jigjiga Town, Ethiopia. *Rom. Econ. J.* **2016**, *19*, 191–214.
66. Dhungana, A.R.; Baral, B. Factors Affecting Willingness to Pay for Improved Water Supply System in Rural Tanahu, Nepal. *Janapriya J. Interdiscip. Stud.* **2017**, *5*, 1–13. [CrossRef]

67. Haq, M.; Mustafa, U.; Ahmad, I. Household's Willingness to Pay for Safe Drinking Water: A Case Study of Abbottabad District. *Pak. Dev. Rev.* **2007**, *46*, 1137–1153. [CrossRef]
68. Odwori, E.O. Factors Determining Households' Willingness to Pay for Improved Water Supply Services in Nzoia River Basin, Kenya. *Int. J. Innov. Res. Adv. Stud. IJIRAS* **2020**, *7*, 165–176. Available online: http://www.ijiras.com/2020/Vol_7-Issue_7/paper_22.pdf (accessed on 19 September 2021).
69. Vásquez, W.F.; Raheem, N.; Quiroga, D.; Ochoa-Herrera, V. Household preferences for improved water services in the Galápagos Islands. *Water Resour. Econ.* **2021**, *34*, 100180. [CrossRef]
70. Wondimu, S.; Bekele, W. Determinants of individual willingness to pay for quality water supply: The case of Wonji Shoa Sugar Estate, Ethiopa. *WIT Trans. Ecol. Environ.* **2011**, *153*, 59–72.
71. Danh, V.T.; Khai, H.V. Estimating residents' willingness to pay for groundwater protection in the Vietnamese Mekong Delta. *Appl. Water Sci.* **2017**, *7*, 421–431.
72. Rahman, M.M.; Alam, K.; Karim, R.; Islam, M.K. Willingness to Pay for Improved Water Supply: A Policy Implications for Future Water Security. *Am. J. Environ. Resour. Econ.* **2017**, *2*, 116–122. [CrossRef]
73. Jianjun, J.; Wenyu, W.; Ying, F.; Xiaomin, W. Measuring the willingness to pay for drinking water quality improvements: Results of a contingent valuation survey in Songzi, China. *J. Water Health* **2016**, *14*, 504–512. [CrossRef]
74. Akhtar, S.; Sohail, A.; Tariq, U.; Khan, F.A.; Asghar, S. Willingness to pay for improved drinking water facility in Samsani Khui, Johar Town, Lahore, Pakistan. *AAES* **2017**, *2*, 187–193.
75. Guilfoos, T.; Hayden, S.; Uchida, E.; Oyanedel-Craver, V. WTP for water filters and water quality testing services in Guatemala. *Water Resour. Econ.* **2019**, *31*, 100139. [CrossRef]
76. Orgill, J.; Shaheed, A.; Brown, J.; Jeuland, M. Water quality perceptions and willingness to pay for clean water in peri-urban Cambodian communities. *J. Water Health* **2013**, *11*, 489–506. [CrossRef]
77. Olajuyigbe, A.E.; Fasakin, J.O. Citizens' Willingness to Pay for Improved Sustainable Water Supply in a Medium-Sized City in South Western Nigeria. *Curr. Res. J. Soc. Sci.* **2010**, *2*, 41–50.
78. Adenike, A.A.; Titus, O.B. Determinants of Willingness to Pay for Improved Water Supply in Osogbo Metropolis; Osun State, Nigeria. *Res. J. Soc. Sci.* **2009**, *4*, 1–6.
79. Vintró, C.; Sanmiquel, L.; Freijo, M. Environmental sustainability in the mining sector: Evidence from Catalan companies. *J. Clean Prod.* **2014**, *84*, 155–163. [CrossRef]
80. Ayanshola, A.M.; Sule, B.F.; Salami, A.W. Evaluation of Willingness to Pay for Reliable and Sustainable household Water Use in Ilorin, Nigeria. *Ethiop. J. Environ. Stud. Manag.* **2013**, *6*, 754. [CrossRef]

water

MDPI

Article

An Expanded Interpretive Structural Modeling Analysis of the Barriers to Integrated Flood Risk Management Adaptation in Metro Manila

Jean Margaret Mercado [1,*], Akira Kawamura [2,*] and Reynaldo Medina [3]

1 Department of Civil Engineering, University of Santo Tomas, España Blvd, Sampaloc, Manila 1008, Metro Manila, Philippines
2 Department of Civil and Environmental Engineering, Tokyo Metropolitan University, 1-1 Minami-Osawa, Hachioji, Tokyo 192-0397, Japan
3 Woodfields Consultants, Inc., 153 Kamias Road Extension, Kamias 1102, Kamias Rd, Diliman, Quezon City 1102, Metro Manila, Philippines; reynaldo.medina@wci.com.ph
* Correspondence: jrmercado@ust.ust.edu.ph (J.M.M.); kawamura@tmu.ac.jp (A.K.)

Abstract: The implementation of integrated flood risk management (IFRM) is still in its infancy in both developed and developing countries, yet some countries have already encountered barriers to IFRM adaptation. The interrelationships between these barriers need to be determined and analyzed systematically, as such an analysis is the groundwork for decision-making when devising solutions to overcome the barriers. Interpretive Structural Modeling (ISM) is a popular and systematic method for analyzing the interrelationship between variables in broad study areas. This study applies the proposed expanded ISM (Ex-ISM) approach to comprehensively analyze the interrelationships between the barriers to IFRM in Metro Manila. Ex-ISM enhances conventional ISM in that the symbolism is modified to explicitly show the contextual interrelationships, the step for hierarchy assignment is simplified, and the diagram shows all of the interrelationships that allow a comprehensive analysis. The results obtained using the Ex-ISM method do not deviate from those yielded by the conventional ISM method, but the Ex-ISM method allows an easy assignment of hierarchy, and it shows not only the direct but also the indirect interrelationships to provide a comprehensive analysis of the relationships between the barriers.

Keywords: barriers; integrated flood risk management; interrelationships; interpretive structural modeling; expanded ISM; Metro Manila

Citation: Mercado, J.M.; Kawamura, A.; Medina, R. An Expanded Interpretive Structural Modeling Analysis of the Barriers to Integrated Flood Risk Management Adaptation in Metro Manila. *Water* **2023**, *15*, 1029. https://doi.org/10.3390/w15061029

Academic Editor: Renato Morbidelli

Received: 27 November 2022
Revised: 20 February 2023
Accepted: 2 March 2023
Published: 8 March 2023

1. Introduction

Integrated flood risk management (IFRM) has been increasingly implemented in many developed and developing countries because the traditional approach to flood control and prevention using structural measures or hard engineering interventions fails to cope with the residual risks brought on by extreme weather events [1]. IFRM is a relatively modern approach that includes non-structural measures, not only structural measures, the non-structural measures being those that do not require any physical construction but use policy, knowledge, and practice to reduce flood risks and impacts, in particular through policies and laws, public awareness-raising, training, and education [2]. IFRM prioritizes more non-structural measures as it aims to proactively manage flood risks by "keeping people away from water" rather than "keeping water away from people" [3]. So far, most research on IFRM heavily concentrates on the hydrological and hydraulic processes.

Among the megacities in Asia, Metro Manila, the Philippines' center of political and economic activities, is considered to be the most at risk of climate impacts, mainly due to its exposure to tropical cyclones [4]. Flooding has been the most frequent natural disaster and a major cause of destruction in Metro Manila, in which the most disastrous flooding in the last decades was brought by Typhoon Ketsana in September 2009 [5]. In the Philippines, an IFRM Master Plan was established in 2012 after the onslaught of Typhoon Ketsana. The transition to IFRM from traditional flood protection, however, had been difficult due to the critical issues or obstacles identified as "barriers" that hindered a smooth adaption to this approach. Therefore, identifying the barriers to IFRM adaptation is an essential task so that decision-makers and practitioners can devise an appropriate, realistic course of action and propose the required policy changes to overcome them.

In developed countries, the identified barriers are mostly related to governance and include issues such as political opposition, economic tensions, fragmented governance structures, and weak enforcement of building restrictions [1,6–10]. Meanwhile, the research on barriers to IFRM adaptation in developing countries remains limited to almost no studies except for authors' previous study [5]. We have identified barriers to IFRM in Metro Manila, and it is found that they are multifaceted and numerous, and some are unique and are expected to be more severe and alarming than those found in developed countries [5].

After identifying the barriers to IFRM adaptation, understanding their interrelationships is also crucial, since barriers are often interrelated with one another as they can alleviate, augment, reinforce, or trigger one another [11]. A systematic analysis of barrier interrelationships is imperative so that decision-makers can make a rational assessment rather than an intuitive judgment when devising a plan to overcome the barriers. However, the interrelationships between the barriers to IFRM adaptation have not been analyzed yet, and there is no universally accepted framework within which such barriers are analyzed, as far as the authors know.

One systematic approach that can be used for barrier analysis is the Interpretive Structural Modeling (ISM) method. The ISM method is popular for analyzing the interrelationships among issues, concerns, or variables in a complex problem. It offers to translate ill-articulated variables in a problem into a structural diagram that shows the contextual, direct interrelationships and the hierarchy among the variables. The advantages of this method compared to other methods such as structural equation modeling (SEM), the Delphi method, or the Analytic Hierarchy Process (AHP) include to following: (1) no requirement for large or statistical datasets, whereas SEM requires a priori statistical data [12], (2) fewer required experts, as even an individual can apply this method, whereas SEM and the Delphi method gather data from a large number of experts/respondents [13], and (3) the display of both the interrelation and the hierarchy (ranking) of the variables, whereas the Delphi Method and AHP only shows the ranking [14,15]. These advantages have contributed to its popularity in broad areas of study, such as systems engineering [16], waste management [11,17], supplier selection [18,19], supply chain flexibility [20,21], and knowledge management [22], among others. However, the ISM method has not yet been applied to the barriers related to natural hazards and disaster risk reduction management, such as IFRM. Consequently, we tried to apply the ISM method in order to analyze the interrelationships between the barriers to IFRM adaptation in Metro Manila for the first time, using the conventional ISM method with some modifications [5].

Since the development of ISM in the 1970s, there have been only minor modifications or improvements to this method, and most studies have only applied the original ISM method. Sushil [23,24] proposed three modifications in ISM, and the first two modifications led to the modified approach called Total ISM (TISM). Sushil [23,24] proposed (1) the interpretation of the links (arrows) in the ISM diagram using an interpretative matrix to explain "why" or "how" the interrelationship exists, (2) the inclusion of significant indirect interrelationships in the ISM diagram, and (3) the use of a simultaneous pairwise comparison and transitivity check to lessen the efforts for these repetitive tasks. Despite these modifications, the numerical representations of the four types of contextual relationships in ISM do not explicitly represent the active and passive interrelations because of the use of binary values (0 and 1), which may intuitively indicate the non-existence or existence of a relationship. In addition, identifying the hierarchy among the variables is a very tedious task, especially when the number of variables considered becomes higher due to repetitive tasks in a step. For these reasons, we believe there is a need to modify some of the steps in ISM so that analyzing the results of the interrelationships between the barriers to IFRM adaptation, especially for developing countries, can become intuitive, straightforward, and comprehensive. Hence, this study proposes an expanded ISM (Ex-ISM) approach that enhances some of the steps in conventional ISM.

The primary objective of this study is to intuitively and easily determine the comprehensive interrelationships among the barriers to IFRM adaptation in Metro Manila using the proposed Ex-ISM method. The following sections describe the barriers to IFRM adaptation in Metro Manila and the proposed Ex-ISM method; elaborate on the application of the proposed method to the barriers to IFRM adaptation in Metro Manila; present the results and compare them with the results given by the conventional method; and give the conclusions.

2. Materials and Methods

2.1. Barriers to IFRM Adaptation in Metro Manila

We conducted a literature review to identify the barriers to IFRM adaptation in Metro Manila. Journal articles, books, and reports published in the last two decades that discuss issues in flood management in Metro Manila were used to identify the barriers to IFRM adaptation in Metro Manila. The reviewed articles related to Metro Manila's flood problem include 15 internationally published papers; 2 locally published papers in the Philippines; 3 project reports from the Department of Public Works and Highways (DPWH) completed in 2000, 2004, and 2013; and 2 books that feature case studies from Metro Manila. The barriers to IFRM adaptation in Metro Manila were noted if they are recurring issues concerning flood management or are cited at least once in the literature.

As for the investigation's results, 12 barriers to IFRM adaptation in Metro Manila were identified and are shown in Table 1 with their descriptions, and these were categorized into three categories: governance, social, and technological resources. The barriers identified are relatively numerous, and some are unique to developed countries because of the socio-economic conditions of a developing country.

The barriers presented in Table 1 are time-dependent and may vary for each location. However, in this study, the identified barriers in Table 1 are given condition as materials for both conventional and expanded ISM. Both approaches only deal with the interrelationships among the given barriers, so they do not deal with the temporal and spatial issues among them. The temporal and spatial issues will be considered in detail before applying in the ISM methods. In this study, under the same given condition for both conventional and expanded ISM, we attempted to compare how Ex-ISM is different from conventional ISM.

Table 1. Barriers to IFRM adaptation in Metro Manila, Philippines. (See Reference [5] for Details).

Category	Barrier		Description
Governance	G1	Lack of a sole organizing body	Fragmented governance in flood management in Metro Manila because the current institutional framework does not have a clear demarcation of tasks among government agencies
	G2	Lack of communication	Weak inter-agency communication and the lack of information exchange and communication on the local level
	G3	Lack of funding	Low funding for flood mitigation and control projects
	G4	Lack of flood control measures	Inadequate existing flood control infrastructures
Social	S1	Informal settlers	Encroachment of marginalized communities in flood-prone areas where they can live cheaply
	S2	Poor solid waste management	Clogged waterways and drainages due to solid wastes
	S3	Poor social planning	Services provided by the government of the Philippines are inadequate to address issues related to the community

Table 1. *Cont.*

Category		Barrier	Description
Technological Resources	T1	Lack of technological capabilities	Absence of real-time flood forecasts, water level, and rainfall depth updates
	T2	Sparse data and limited access	Available hydro-meteorological information is thinly distributed, is not automated, and is measured on a large time interval only
	T3	Lack of experts	Lack of experts from government agencies and local government units
	T4	Lack of data processing systems	No data processing systems, which resulted in hydro-meteorological information just being stored and not used for analysis
	T5	Deterioration of flood control structures	Deterioration of existing flood control structures (e.g., pumping stations, drainage systems, hydraulic control structures) due to poor maintenance and poor solid waste management

2.2. Expanded Interpretive Structural Modeling (Ex-ISM)

Warfield developed ISM in the 1970s [25,26]. ISM is a structural modeling method that determines the interrelations between variables in complex issues or systems. ISM transforms unclear and poorly articulated variables into a structural model or diagram by employing discrete mathematics and elementary graph theory so that theoretical, conceptual, and computational leverage is efficiently exploited [20]. This method's output is a diagram showing interrelationships through directed links or arrows and the hierarchy between variables. This ISM diagram aids users, decision-makers, or practitioners in visualizing, interpreting, and understanding the variables, which can help them devise solutions to the problem [27,28].

This study proposes an expanded ISM (Ex-ISM) method, which enhances the numerical representation of the four contextual relations by expanding them from binary to trinary values (0, 1, and −1). The trinary values can express active and passive interrelations using 1 and −1, respectively. Expanded Boolean multiplication and addition operations are introduced so that ISM can accommodate the calculation of trinary values. The proposed Ex-ISM method further simplifies the tedious, repetitive tasks involved in assigning the hierarchy (levels), and it aims to present comprehensive interrelationships among the barriers by showing the indirect interrelationships, not only the direct ones, in the ISM diagram. Indirect interrelationships indicates that, between two barriers with a direct interrelationship, there may be another barrier that can be influenced by the two.

Fundamentally, conventional ISM has five steps [29], as shown in Figure 1. In the proposed Ex-ISM approach, Step 1 and Step 2 are expanded, and Step 3 is simplified, as depicted in Figure 1. The succeeding paragraphs discuss in detail the expansion and simplification proposed in this study.

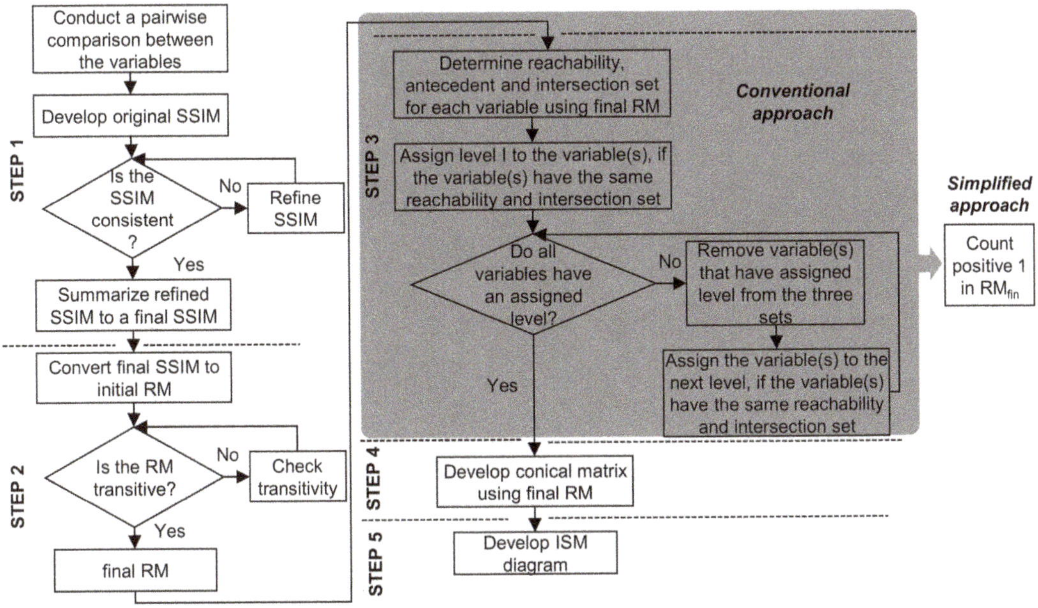

Figure 1. Flow chart of the ISM method with the simplified approach.

2.2.1. Step 1—Developing a Structural Self-Interaction Matrix (SSIM)

Pairwise comparisons between the identified variables are conducted first to develop the SSIM. The pairwise comparisons can be conducted individually or in a group, so ISM is often considered a group learning tool [30]. A contextual relationship of the "leads to" or "influences" type must be chosen for the pairwise comparison [29]. In this study, only five experts were asked to establish the SSIM because it is extremely difficult to find experts with overarching experience in flood control and management in Metro Manila. The 5 identified experts have more than 20 years of experience and are members of the mandated national government agency in charge of flood control planning and implementation in the Philippines.

There are four types of interrelations based on the chosen contextual relationship for the pairwise comparison, and four symbols represent these interrelations. In the conventional ISM approach, the four symbols are V, A, X, and O, and these symbols are used to fill up only the upper triangular half of the SSIM.

In this study, we changed the conventional symbolisms to provide meaningful representations in the SSIM, and we also propose to fill up all of the cells of the SSIM (except the diagonal, which is kept blank) with the new symbols in the Ex-ISM method. The following are the new symbols used in Ex-ISM, which are based on the "influencing" type of contextual relation:

1. The letter "V" is replaced with the symbol "+", which denotes that variable *i* influences variable *j* and *j* does not influence *i*.
2. The letter "A" is replaced with "-", which denotes that variable *i* is influenced by variable *j* and *j* is not influenced by *i*.
3. The letter "X" is replaced with the symbol "±" or "∓", which means that variable *i* and variable *j* influence each other.
4. The letter "O" is replaced with the symbol "0", which means that variable *i* and variable *j* are independent of one another.

5. To fill up the lower triangular half of the SSIM, the corresponding passive pairwise comparison SSIM(j, i) of the active SSIM(i, j) is filled with the corresponding passive symbolism, i.e., if SSIM(i, j) = + or −, then SSIM(j, i)= − or +.

After eliciting the input of the five experts, an original SSIM is established and checked for consistency, wherein consistency is considered a tally with a majority (three of the more common responses in this study). For any inconsistent tally in the original SSIM, we asked the experts to reconsider the relation in the pairwise comparison. After checking the consistency, the resulting SSIM is referred to as the refined SSIM. The refined SSIM is further summarized as the final SSIM by considering the majority response. This approach addresses the handling of inconsistent data due to the experts' inference. The final SSIM is the only input for the following ISM method.

2.2.2. Step 2—Developing a Reachability Matrix (RM)

In this step, the final SSIM in Step 1 is transformed into an initial reachability matrix (RM_{init}). The RM_{init} in conventional ISM is a binary-valued matrix in which the values consist only of "1" and "0" to represent "an influence" and "no influence" relations between variables i and j, respectively. The rules to convert the SSIM to the RM_{init} in the conventional approach are as follows:

1. If SSIM(i, j) = V, then $RM_{init}(i, j) = 1$ and $RM_{init}(j, i) = 0$.
2. If SSIM(i, j) = A, then $RM_{init}(i, j) = 0$ and $RM_{init}(j, i) = 1$.
3. If SSIM(i, j) = X, then $RM_{init}(i, j) = 1$ and $RM_{init}(j, i) = 1$.
4. If SSIM(i, j) = O, then $RM_{init}(i, j) = 0$ and $RM_{init}(j, i) = 0$.
5. For the diagonal element SSIM(i, i), $RM_{init}(i, i) = 1$.

On the other hand, the proposed Ex-ISM method utilizes a trinary-valued matrix (1, −1, and 0) to explicitly represent the four types of interrelations between variables i and j. The following rules are introduced to convert the SSIM to RM_{init}:

1. If the SSIM(i, j) is "+", then $RM_{init}(i, j)$ is 1 and $RM_{init}(j, i)$ is −1.
2. If the SSIM(i, j) is "−", then $RM_{init}(i, j)$ is −1 and $RM_{init}(j, i)$ is 1.
3. If the SSIM(i, j) is "±" or "∓", then $RM_{init}(i, j)$ is ±1 or ∓1.
4. If the SSIM(i, j) is "0", then $RM_{init}(i, j)$ is 0.
5. For the blank diagonal element SSIM(i, i), $RM_{init}(i, i) = \pm 1$.

This new rule not only simplifies the transformation of the SSIM to RM_{init} but also retains the interrelations from the SSIM and explicitly represents the active and passive interrelations between variables i and j using 1 and −1 values, respectively. Subsequently, a transitivity check is conducted iteratively in the RM_{init} to derive the final RM (RM_{fin}). The transitive relations in the RM_{fin} suggest an indirect relation between variables i and j. The conventional way to conduct the transitivity check is expressed by the following equation [31]:

$$RM_{fin} = {RM_{init}}^{k} = {RM_{init}}^{k+1},\ k > 1 \quad (1)$$

where k denotes the powers. Equation (1) is calculated using matrix Boolean multiplication and addition operations. All transitive relations, i.e., 0 in $RM_{init}(i, j)$ that changed into 1 in $RM_{fin}(i, j)$, are conventionally denoted by one "*" in the RM_{fin}.

On the other hand, the RM_{init} in the Ex-ISM approach is a trinary-valued matrix, so the conventional approach to the transitivity check is not applicable. To cope with this, we propose an expanded Boolean multiplication and addition for Ex-ISM, shown in Table 2, to calculate the −1 values. Firstly, the RM_{init} is deconstructed into two matrices, $+RM_{init}$ and $-RM_{init}$. The following are the rules to derive these matrices:

Table 2. Expanded Boolean multiplication and addition for Ex-ISM.

Multiplication				Addition			
×	**0**	**1**	**−1**	**+**	**0**	**1**	**−1**
0	0	0	0	0	0	1	−1
1	0	1	B	1	1	1	±1/∓1 [a]
−1	0	B	−1	−1	−1	±1/∓1 [a]	−1

B—blank; [a]—upper sign corresponds to the original relation in RM_{init} while the bottom sign corresponds to transitive or indirect relation in RM_{fin}.

1. If $RM_{init}(i, j)$ has a positive sign, then $+RM_{init}(i, j) = 1$; otherwise $+RM_{init}(i, j) = 0$.
2. If $RM_{init}(i, j)$ has a negative sign, then $-RM_{init}(i, j) = -1$; otherwise $-RM_{init}(i, j) = 0$.

Then, the transitivity check for $+RM_{init}$ and $-RM_{init}$ is expressed using the following equations:

$$+RM_{fin} = +{RM_{init}}^{k} = +{RM_{init}}^{k+1},\ k > 1 \tag{2}$$

$$-RM_{fin} = -{RM_{init}}^{k} = -{RM_{init}}^{k+1},\ k > 1 \tag{3}$$

These two matrices are calculated using the expanded Boolean multiplication and addition in Table 2. The matrix operation for $+RM_{init}$ in Equation (2) is the same as that in Equation 1, because it is a binary matrix. Thus, $+RM_{fin}$ is the same RM_{fin} as the conventional approach. In the Ex-ISM approach, the transitive or indirect relations are represented by multiple "*" to distinctly indicate the number of iterations for the transitive relations, whereas only one "*" is indicated in the conventional approach. Finally, in this step, the obtained $+RM_{fin}$ and $-RM_{fin}$ are combined using the expanded Boolean addition operation in Table 2 to derive the RM_{fin}. For multi-signed 1 (±1 or ∓1) in Table 2, the upper sign corresponds to the original relation in the RM_{init}, while the lower sign corresponds to a transitive or indirect relation in the RM_{fin}.

2.2.3. Step 3—Assigning Levels for Each Variable

The conventional approach to assigning levels to each variable is one of the most tedious steps in ISM, and it is incredibly taxing when the number of variables becomes larger. The shaded part of Figure 1 shows the conventional ISM flowchart for Step 3. Following is the description of the flowchart. Three sets for each variable i are first determined using the RM_{fin}: the "reachability set", which consists of those variables that it influences and itself (when variable $j = 1$ within its row i); the "antecedent set", which consists of those variables that influence it and itself (when variable $i = 1$ within its column j); and the "intersection set", which consists of the intersection between the reachability and antecedent sets. To assign the levels, variables with the same reachability and intersection sets are assigned to level I, and these level I variables are eliminated from the three sets to result in new reachability, antecedent, and intersection sets. Then, this process is carried out recursively until all variables are assigned to a corresponding level.

In the Ex-ISM method, we propose to simplify this step. Using the RM_{fin} given by the Ex-ISM method, the positive values, i.e., 1, are counted at each row of the variable i. The counted value responds to the level for variable i. The highest level assigned to variable i is the total number of variables considered for the analysis, while the lowest is 1. The levels derived from the proposed Ex-ISM method have the same order as in the conventional ISM method, but the magnitude of the levels is also considered.

2.2.4. Step 4—Developing a Conical Matrix (CM)

Using the RM_{fin} and the assigned levels, a CM is derived by rearranging RM_{fin} according to the levels across the rows and columns. The CM is conventionally arranged in an ascending order of the levels across the rows and columns, but we changed the arrangement into a descending order in the Ex-ISM method so that the ISM diagram can be constructed according to the modified schematic described in the next step.

2.2.5. Step 5—Developing the ISM Diagram

The ISM diagram is drawn using the CM. The ISM diagram is a kind of directed graph (or digraph) that shows a set of interconnected variables representing an interrelation while also showing its hierarchy. In the conventional approach, only the direct interrelations between the variables are shown in the ISM diagram, while indirect interrelations are disregarded. The direct interrelations are represented by 1 in CM(i, j), while the indirect ones have "*" attached to the 1. To construct the ISM diagram conventionally, an arrow is drawn from variable i to j if CM(i, j) = 1, and the variables are arranged in ascending levels from top to bottom.

By contrast, we also show the indirect interrelations in the ISM diagram of the Ex-ISM method, because indirect interrelations may cause an underlying ripple effect or chains of reactions to other variables [32]. A separate diagram for this is constructed similarly to the diagram for the direct relations. In addition, descending levels are adopted to arrange the ISM diagram in the Ex-ISM approach. This rearrangement depicts the ISM diagram in a pyramid schematic, where the highest-level variables (most influential) are placed at the top, while the lowest-level variables (least influential) are placed at the bottom. In this way, the hierarchy among the variables is consistently depicted in the ISM diagram.

The ISM diagram in the Ex-ISM method shows the direct interrelationships as those having a solid line arrow drawn from variable i to variable j if CM(i, j) has a positive sign. As for the ISM diagram that shows the indirect interrelationships, an arrow is drawn from variable i to variable j if CM(i, j) has a positive sign with "*". The arrows are formatted into varying broken lines or line types for "*", "**", "***", and so on to distinguish them, which was not done in the conventional approach.

3. Results and Discussion

The direct and indirect interrelationships between identified barriers to IFRM adaptation in Metro Manila were established using the Ex-ISM method. The results given by the Ex-ISM method are presented in the succeeding paragraphs, while the conventional ISM results are shown in Appendix A.

For Step 1, the pairwise comparison between the 12 barriers was derived from the 5 experts engaged, and this is shown in Table 3 and called the refined SSIM. There are five symbols in the refined SSIM, which correspond to each expert's inference regarding the pairwise comparison. For example, in refined SSIM(1, 2), there are four "+" symbols, indicating that four experts inferred that barrier G1 influences barrier G2, while only a single "-" symbol indicates that one expert inferred that G1 is influenced by G2. This refined SSIM was already checked for any inconsistency, so a majority response from the five experts can be derived to come up with the final SSIM shown in Table 4. Contrary to the conventional SSIM in Appendix A, Table A1, the final SSIM of the Ex-ISM method in Table 4 is completely filled with the new symbols (+, −, ±, and ∓) that show the active and passive interrelations explicitly. For example, the SSIM(1,2) in Table 4 is "+", so SSIM(2, 1) is "−", which means that barrier G1 influences barrier G2 and barrier G2 is influenced by barrier G1, respectively.

Table 3. Refined Structural Self-Interaction Matrix (refined SSIM).

i	Category	Barrier	1 G1	2 G2	3 G3	4 G4	5 S1	6 S2	7 S3	8 T1	9 T2	10 T3	11 T4	12 T5
			Governance				Social			Technological Resources				
1	Governance	G1		+++ +-	+++ ++	+++ +0	+00 00	++0 00	++0 00	+++ 00	+++ ++	+++ +0	+++ +±	+++ +±
2		G2	- - - -∓		+++ +±	++0 00	+++ +0	++0 0+	+++ 0±	+00 0±	+++ ±±	-00 00	-00 00	+00 00
3		G3	- - - - -	— -∓		+++ -±	-00 0±	000 ±±	+00 0±	+++ -±	+++ 0±	+++ 0±	+++ ± ±	+++ ±±
4		G4	- - - -0	–0 00	— +∓		+00 0±	+00 00	+++ 00	+++ +±	+++ +±	++± ±±	+0± ±±	+++ ++
5	Social	S1	-00 00	— -0	+00 0∓	-00 0∓		+++ +±	+– -±	000 00	000 00	000 00	000 00	000 0±
6		S2	–0 00	–0 0-	000 ∓∓	-00 00	— -∓		— 00	-00 00	000 00	-00 00	000 00	000 0±
7		S3	–0 00	— 0∓	-00 0∓	— 00	-++ +∓	+++ 00		-00 00	-00 0±	— 00	000 ±±	000 0±
8	Technological Resources	T1	— 00	-00 0∓	— +∓	— -∓	000 00	+00 00	+00 00		+++ +±	++- -	+±± ±±	+++ 0±
9		T2	— -	— ∓∓	— 0∓	— -∓	000 00	000 00	+00 0∓	— -∓		+– -	+– -±	+00 0±
10		T3	— -0	+00 00	— 0∓	–∓ ∓∓	000 00	+00 00	+++ 00	–+ ++	-++ ++		+++ 00	+++ 00
11		T4	— -∓	+00 00	— ∓∓	-0∓ ∓∓	000 00	000 00	000 ∓∓	-∓∓ ∓∓	-++ +∓	— 00		0+0 ++
12		T5	— -∓	-00 00	— ∓∓	— -	000 0∓	000 0∓	000 0∓	— 0∓	-00 0∓	— ∓∓	0-0 -	

Table 4. Final Structural Self-Interaction Matrix (final SSIM).

i	j	1	2	3	4	5	6	7	8	9	10	11	12
	Barrier	G1	G2	G3	G4	S1	S2	S3	T1	T2	T3	T4	T5
1	G1		+	+	+	0	0	0	+	+	+	+	+
2	G2	–		+	0	+	+	+	0	+	0	0	0
3	G3	–	–		+	0	0	0	+	+	+	+	+
4	G4	–	0	–		0	0	+	+	+	±	±	+
5	S1	0	–	0	0		+	–	0	0	0	0	0
6	S2	0	–	0	0	–		–	0	0	0	0	0
7	S3	0	–	0	–	+	+		0	0	-	0	0
8	T1	–	0	–	–	0	0	0		+	–	±	+
9	T2	–	–	–	–	0	0	0	–		–	–	0
10	T3	–	0	–	∓	0	0	+	+	+		+	+
11	T4	–	0	–	∓	0	0	0	∓	+	–		+
12	T5	–	0	–	-	0	0	0	–	0	–	–	

For Step 2, the final SSIM is transformed to RM_{init}, as shown in Table 5. The transformation of the final SSIM to RM_{init} was more effortless with Ex-ISM than with the conventional ISM method, because the SSIM of Ex-ISM is already filled with the new symbols. The trinary values in Table 5 also retain the active and passive relations from the SSIM, e.g., $RM_{init}(1, 2) = 1$ indicates that barrier G1 influences barrier G2, while $RM_{init}(2, 1) = -1$ indicates that barrier G2 is influenced by barrier G1. In contrast, the binary values used in the conventional RM_{init}, shown in Appendix A, Table A2, represent $RM_{init}(2, 1)$ as "0", which does not distinguish whether there is an "influenced by" or a "no" relation. Then, the RM_{init} of the Ex-ISM method is first deconstructed into positive and negative initial reachability matrices ($+RM_{init}$ and $-RM_{init}$) before the transitivity check.

Table 5. Initial Reachability Matrix (RM_{init}).

i	*j*	1	2	3	4	5	6	7	8	9	10	11	12
	Barrier	**G1**	**G2**	**G3**	**G4**	**S1**	**S2**	**S3**	**T1**	**T2**	**T3**	**T4**	**T5**
1	**G1**	±1	1	1	1	0	0	0	1	1	1	1	1
2	**G2**	−1	±1	1	0	1	1	1	0	1	0	0	0
3	**G3**	−1	−1	±1	1	0	0	0	1	1	1	1	1
4	**G4**	−1	0	−1	±1	0	0	1	1	1	±1	±1	1
5	**S1**	0	−1	0	0	±1	1	−1	0	0	0	0	0
6	**S2**	0	−1	0	0	−1	±1	−1	0	0	0	0	0
7	**S3**	0	−1	0	−1	1	1	±1	0	0	−1	0	0
8	**T1**	−1	0	−1	−1	0	0	0	±1	1	−1	±1	1
9	**T2**	−1	−1	−1	−1	0	0	0	−1	±1	−1	−1	0
10	**T3**	−1	0	−1	∓1	0	0	1	1	1	±1	1	1
11	**T4**	−1	0	−1	∓1	0	0	0	∓1	1	−1	±1	1
12	**T5**	−1	0	−1	−1	0	0	0	−1	0	−1	−1	±1

The $+RM_{init}$ and $-RM_{init}$ are just the same as the positive and negative parts in Table 5 when deconstructed, and these are operated on using the proposed expanded Boolean multiplication operation for the Ex-ISM method in Table 2 to check the transitivity and derive the final positive and negative reachability matrices ($+RM_{fin}$ and $-RM_{fin}$) in Tables 6 and 7, respectively. The $+RM_{fin}$ (Table 6) and $-RM_{fin}$ (Table 7) were combined according to the proposed expanded Boolean addition in Table 2, which resulted in the RM_{fin} shown in Table 8. The number of "*" indicates the iterations for the transitive or indirect relations, whereas only one "*" is indicated in the conventional approach, as shown in Table A3.

Table 6. Final Positive Reachability Matrix ($+RM_{fin}$).

	j	1	2	3	4	5	6	7	8	9	10	11	12
i	**Barrier**	**G1**	**G2**	**G3**	**G4**	**S1**	**S2**	**S3**	**T1**	**T2**	**T3**	**T4**	**T5**
1	**G1**	1	1	1	1	1^{*}	1^{*}	1^{*}	1	1	1	1	1
2	**G2**	0	1	1	1^{*}	1	1	1	1^{*}	1	1^{*}	1^{*}	1^{*}
3	**G3**	0	0	1	1	1^{**}	1^{**}	1^{*}	1	1	1	1	1
4	**G4**	0	0	0	1	1^{*}	1^{*}	1	1	1	1	1	1
5	**S1**	0	0	0	0	1	1	0	0	0	0	0	0
6	**S2**	0	0	0	0	0	1	0	0	0	0	0	0
7	**S3**	0	0	0	0	1	1	1	0	0	0	0	0
8	**T1**	0	0	0	1^{*}	1^{***}	1^{***}	1^{**}	1	1	1^{**}	1	1
9	**T2**	0	0	0	0	0	0	0	0	1	0	0	0
10	**T3**	0	0	0	1	1^{*}	1^{*}	1	1	1	1	1	1
11	**T4**	0	0	0	1	1^{**}	1^{**}	1^{*}	1	1	1^{*}	1	1
12	**T5**	0	0	0	0	0	0	0	0	0	0	0	1

Table 7. Final Negative Reachability Matrix ($-RM_{fin}$).

	j	1	2	3	4	5	6	7	8	9	10	11	12
i	Barrier	G1	G2	G3	G4	S1	S2	S3	T1	T2	T3	T4	T5
1	G1	−1	0	0	0	0	0	0	0	0	0	0	0
2	G2	−1	−1	0	0	0	0	0	0	0	0	0	0
3	G3	−1	−1	−1	0	0	0	0	0	0	0	0	0
4	G4	−1	−1 *	−1	−1	0	0	0	−1 *	0	−1	−1	0
5	S1	−1 *	−1	−1 **	−1 *	−1	0	−1	−1 ***	0	−1 *	−1 **	0
6	S2	−1 *	−1	−1 **	−1 *	−1	−1	−1	−1 ***	0	−1 *	−1 **	0
7	S3	−1 *	−1	−1 *	−1	0	0	−1	−1 **	0	−1	−1 *	0
8	T1	−1	−1 *	−1	−1	0	0	0	−1	0	−1	−1	0
9	T2	−1	−1	−1	−1	0	0	0	−1	−1	−1	−1	0
10	T3	−1	−1 *	−1	−1	0	0	0	−1 **	0	−1	−1 *	0
11	T4	−1	−1 *	−1	−1	0	0	0	−1	0	−1	−1	0
12	T5	−1	−1 *	−1	−1	0	0	0	−1	0	−1	−1	−1

Table 8. Final Reachability Matrix (RM_{fin}) and Levels of each barrier.

	j	1	2	3	4	5	6	7	8	9	10	11	12	Level
i	Barrier	G1	G2	G3	G4	S1	S2	S3	T1	T2	T3	T4	T5	
1	G1	±1	1	1	1	1 *	1 *	1 *	1	1	1	1	1	12
2	G2	−1	±1	1	1 *	1	1	1	1 *	1	1 *	1 *	1 *	11
3	G3	−1	−1	±1	1	1 **	1 **	1 *	1	1	1	1	1	10
4	G4	−1	−1 *	−1	±1	1 *	1 *	1	±1 *	1	±1	±1	1	9
5	S1	−1 *	−1	−1 **	−1 *	±1	1	−1	−1 ***	0	−1 *	−1 **	0	2
6	S2	−1 *	−1	−1 **	−1 *	−1	±1	−1	−1 ***	0	−1 *	−1 **	0	1
7	S3	−1 *	−1	−1 *	−1	1	1	±1	−1 **	0	−1	−1 *	0	3
8	T1	−1	−1 *	−1	∓1 *	1 ***	1 ***	1 **	±1	1	∓1 **	±1	1	9
9	T2	−1	−1	−1	−1	0	0	0	−1	±1	−1	−1	0	1
10	T3	−1	−1 *	−1	∓1	1 *	1 *	1	±1 **	1	±1	±1 *	1	9
11	T4	−1	−1 *	−1	∓1	1 **	1 **	1 *	∓1	1	∓1 *	±1	1	9
12	T5	−1	−1 *	−1	−1	0	0	0	−1	0	−1	−1	±1	1

The results show that some barriers to IFRM adaptation have indirect interrelations up to the third degree. Those values in Table 8 with "*", "**", and "***" signifies indirect relations to the first, second, and third degree for barriers *i* and *j*, respectively. For multi-signed (±1 or ∓1) indirect relations, the upper sign signifies the original sign in the RM_{init}, while the lower sign signifies a transitive or indirect interrelation in the RM_{fin}. In addition, there are three barrier interrelations (RM_{fin} = (4, 8), (8, 10) (10, 11) in Table 8) that become multi-signed after the transitivity check. Such details are not specified in the conventional ISM shown in Appendix A, Table A3.

For Step 3, the assignment of the levels for each barrier is much simpler in the Ex-ISM method than in the conventional one, where a recursive process is necessary. We used the RM_{fin} in Table 8 to determine the levels by just counting the positive 1s in the rows for each barrier. The counted values are assigned as the levels in the Ex-ISM method, which are shown in the last column of Table 8. The tedious conventional process, in which nine iterations were performed to assign a level for each barrier, is presented in Appendix A, Tables A4–A8.

For Step 4, the levels and RM_{fin} in Table 8 are rearranged in descending order across the columns and rows to derive the CM in Table 9, whereas an ascending order is adopted in the conventional approach shown in Appendix A, Table A9.

Table 9. Conical matrix (CM).

Level		12	11	10	9	9	9	9	3	2	1	1	1
	Barrier	**G1**	**G2**	**G3**	**G4**	**T1**	**T3**	**T4**	**S3**	**S1**	**S2**	**T2**	**T5**
12	**G1**	±1	1	1	1	1	1	1	1 *	1 *	1 *	1	1
11	**G2**	−1	±1	1	1 *	1 *	1 *	1 *	1	1	1	1	1 *
10	**G3**	−1	−1	±1	1	1	1	1	1 *	1 **	1 **	1	1
9	**G4**	−1	−1 *	−1	±1	±1 *	±1	±1	1	1 *	1 *	1	1
9	**T1**	−1	−1 *	−1	∓1 *	±1	∓1 **	±1	1 **	1 ***	1 ***	1	1
9	**T3**	−1	−1 *	−1	∓1	±1	±1	±1	1	1 *	1 *	1	1
9	**T4**	−1	−1 *	−1	∓1	∓1	∓1 *	±1	1 *	1 **	1 **	1	1
3	**S3**	−1 *	−1	−1 *	−1	−1 **	−1	−1 *	±1	1	1	0	0
2	**S1**	−1 *	−1	−1 **	−1 *	−1 ***	−1 *	−1 **	−1	±1	1	0	0
1	**S2**	−1 *	−1	−1 **	−1 *	−1 ***	−1 *	−1 **	−1	−1	±1	0	0
1	**T2**	−1	−1	−1	−1	−1	−1	−1	0	0	0	±1	0
1	**T5**	−1	−1 *	−1	−1	−1	−1	−1	0	0	0	0	±1

Finally, for Step 5, the CM in Table 9 is used to draw the ISM diagram. Figure 2 shows two ISM diagrams, i.e., Figure 2a,b shows the direct and indirect interrelationships, respectively. The conventional ISM diagram disregards the indirect interrelationships, but showing indirect relations may provide a comprehensive analysis of the variables' underlying chains of influence. The pyramid schematic in the Ex-ISM method is the opposite of the conventional one. The ISM diagrams in Figure 2 show a noticeable gap between the upper (Level 9~12) and lower (Level 1~3) level barriers, which is not possible in the conventional ISM diagram because it only shows the order of the levels, not the magnitude. The gap in the hierarchy shown by the redefined magnitude of the levels in the Ex-ISM method may imply that the barriers at the upper level have an unforeseen impact or influence on the lower-level barriers.

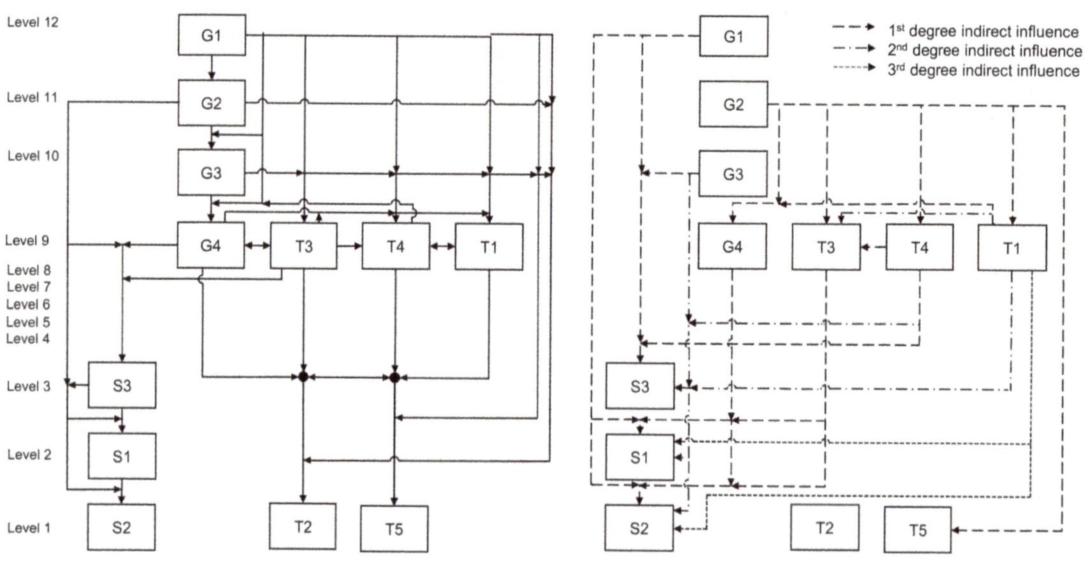

Figure 2. ISM diagram for the (**a**) direct and (**b**) indirect influence interrelationships between the barriers to IFRM adaptation in Metro Manila.

From Figure 2, the most influential barrier to IFRM adaptation in Metro Manila is G1 (Lack of sole organizing body). This may imply that the establishment or assignment of a lead agency in IFRM that supports planning, implementation, operations, and maintenance has to be carried out, since there are currently too many key players in flood risk management in Metro Manila. The second most influential barrier is G2 (Lack of coordination), which directly influences G4 (Lack of flood control measures) and all of the social barriers. The ISM diagrams in Figure 2 also reveal that G4 (Lack of flood control measures), T1 (Lack of data processing systems), T3 (Lack of experts), and T4 (Lack of data processing system) are directly influencing each other, and that these are directly influenced by G3 (Lack of funding) and G1. On the other hand, the least influential barriers are those related to the social category (S2—poor solid waste management) and the technological category (T2—Sparse data and limited access and T5—Deterioration of flood control structures).

In Figure 2a, we recognize that the social barriers (S1~S3) are mainly influenced by barrier G2, which we have regarded as the key to overcoming the social barriers. However, the indirect interrelationships in Figure 2b show that the barriers in the social category are further influenced indirectly to the first and second degrees by the other barriers in the governance (G1, G3, and G4) category, and to the first–third degrees by level 9 technological resource issues (T1, T3, and T4).

Studies in developed countries [33–35] have suggested that collaborative governance and technology-based tools can help enhance social learning by understanding people's interests and values in regard to IFRM. However, overcoming the social barriers in Metro Manila may be more complicated due to several interrelations of indirect influence. Overcoming them will require strong political will, a large amount of financial support, and science-based evidence and technological capabilities, which are inherently lacking in many developing countries. These indirect interrelations may imply that overcoming the social barriers necessitates improving other barriers in the governance and technological resource categories, not only barrier G2.

In the same way, barriers in the technological category are mainly influenced by G1 and G3. The barriers in this category are also indirectly influenced in the first degree by barrier G2. The results suggest that barrier G2 also plays a significant role in overcoming the technological resource barriers in Metro Manila, but the context may differ in developed countries. Effective risk communication in many developed countries emphasizes the translation of risks from scientists to practitioners, decision-makers, and laypeople to articulate risks, forecasts, warnings, or even uncertainties that can aid decision-making and social learning [36–38]. However, Metro Manila's technological capabilities are far behind those of developed countries, as is the case in many developing countries, so there is no operational flood forecasting and warning system in Metro Manila [39]. The role of communication in overcoming the technological barriers in Metro Manila may suggest that solutions must come from the practitioners, the decision-makers, or those who hold government positions, since they can stimulate an increase in funding for technological improvement in flood management. Thus, overcoming technological barriers necessitates political will, communication, and funding, as expressed by the direct and indirect interrelationships of barriers G1, G2, and G3 in Figure 2.

Uncertainty is a crucial issue, and this is inherent in the data and modeling techniques in flood risk management [40], flood defense engineering [41], and decision-making [41,42]. However, the ISM approach (including both the conventional and the expanded ISM methods) does not deal with any uncertainty once the input matrix, i.e., the final SSIM that is automatically constructed from the Refined SSIM, is determined. In other words, once the Refined SSIM is constructed, ISM is a deterministic and systematic method that can only show the interrelationships between the barriers. In this paper, the Refined SSIM is treated as the given condition for both conventional and expanded ISM, and we focus on showing the usefulness of the Ex-ISM approach compared to conventional ISM.

On the other hand, we agree that the data gathered for both the Ex-ISM and conventional ISM approaches, i.e., the identified barriers to IFRM adaptation and the contextual interrelationships in the SSIM based on the experts' judgment, are inherently uncertain. The experts' judgments in the SSIM has inherent uncertainties because they are based on how experts perceive flood risk, so, in this paper, the experts' judgments in the SSIM were iteratively checked in order to construct a Refined SSIM with less uncertainty. Nevertheless, if data inputs with less uncertainty are available, the Ex-ISM method can be used to systematically determine the barriers' interrelationships and their hierarchy through the ISM diagrams for the direct and indirect relationships.

The Ex-ISM and the conventional ISM approach do not deal with systems failure and temporal variability among the barriers, as they only show the interrelationships between variables. Nevertheless, the results of the Ex-ISM method provide a comprehensive analysis by showing both the direct and indirect interrelationships between the barriers in the ISM diagrams. Those barriers recognized as being most influential have been further emphasized through the redefined magnitudes of the hierarchical levels. This may give additional direction to the decision-makers regarding which barriers must be eliminated first and foremost. The diagram showing the direct and indirect interrelationships can help decision-makers and stakeholders visualize and understand the complex interrelationships among the barriers, since the barriers are only abstract concepts. Thus, the ISM diagram can be used as a tool when devising a plan to overcome the barriers. This analysis of the barriers' interrelationships may lead to an effective plan for transitioning to IFRM, especially for Metro Manila, since the identified barriers have impeded the transition [43].

4. Conclusions

Identifying and analyzing the interrelationships between the barriers to IFRM adaption is crucial before and during the implementation of IFRM in both developed and developing countries. These tasks are necessary but are not widely performed. Therefore, this study attempts to conduct a comprehensive analysis of the barriers to IFRM adaptation in Metro Manila, Philippines, by applying a proposed Ex-ISM method for barrier interpretation.

The proposed Ex-ISM method is an enhancement of conventional ISM, and it aims to (1) utilize trinary values to explicitly represent the active and passive interrelations between the variables by introducing both 1 and -1 in the reachability matrices; (2) introduce an expanded Boolean multiplication and addition to accommodate the calculation of the trinary values; (3) simplify the repetitive and tedious task of assigning the levels for each variable; and (4) provide comprehensive interrelations in the ISM diagram by showing not only the direct interrelations but also the indirect interrelations.

Because Ex-ISM contains conventional ISM, the limitations of conventional ISM are still inherent in Ex-ISM. Both methods heavily depend on experts' judgment to create the final SSIM as the input for any ISM approach. The handling of experts' inferences that may have inconsistencies has still not been standardized for the ISM approach. In this study, however, inconsistencies were dealt with by having multiple pairwise comparison iterations with the experts engaged. Once the final SSIM is created, the following procedures are systematically and automatically calculated according to the ISM methods of interest.

The Ex-ISM method was applied to 12 identified barriers to IFRM adaptation in Metro Manila. The results from the Ex-ISM method did not deviate from the results obtained using the conventional ISM approach, but the proposed Ex-ISM method provided enhanced results through the trinary values in the reachability matrices. Furthermore, the ISM diagrams that show both the direct and indirect influence interrelationships provided a deeper understanding of the barriers. Therefore, the Ex-ISM method, which shows the indirect interrelationships, presents a comprehensive analysis of the interrelationships between the barriers to IFRM. The hierarchy is also emphasized by the magnitude of the levels, not only the order, which may imply the underlying impacts of upper-level barriers on the lower-level barriers.

The most influential barriers to IFRM that were identified belong to the governance category, and their interrelations with the other barriers were comprehensively analyzed in this study. The indirect relationships are also interpreted, emphasizing that their influence on the other barriers needs to be considered as well when devising a strategic plan to overcome said barriers and implement IFRM. Indirect relations are equally important as direct relations, as the former are the ripple effect of the latter.

Overall, the study's results presented comprehensive interrelations between the barriers to IFRM. Also, the proposed Ex-ISM application is less tedious, and the results were enhanced and were more comprehensive compared to the conventional ISM method.

Author Contributions: Conceptualization: J.M.M. and A.K.; Methodology: J.M.M. and A.K.; formal analysis and investigation: J.M.M. and A.K.; Data curation: J.M.M. and R.M.; Writing—original draft preparation: J.M.M. and A.K.; Writing—review and editing J.M.M., A.K. and R.M.; Supervision, A.K. and R.M. All authors have read and agreed to the published version of the manuscript.

Funding: This research was conducted through the support of "Tokyo Human Resources Fund for City Diplomacy" and "Advanced Research Network Enhancement Program" by the Tokyo Metropolitan Government.

Institutional Review Board Statement: Not applicable.

Informed Consent Statement: Not applicable.

Data Availability Statement: Not applicable.

Acknowledgments: The authors wish to express their sincerest gratitude to Yuuto Yokota of Nippon Shuppan Hanbai, Inc. and Hideo Amaguchi of Tokyo Metropolitan University for their utmost support and assistance in developing the proposed Ex-ISM approach.

Conflicts of Interest: The authors declare no conflict of interest.

Appendix A

The results obtained using conventional ISM for the SSIM, Initial RM, and Final RM, the level partitioning iterations, and the CM are presented in the succeeding tables.

Table A1. Structural Self-Interaction Matrix (SSIM) (Conventional Method).

i	*j*	1	2	3	4	5	6	7	8	9	10	11	12
	Barrier	**G1**	**G2**	**G3**	**G4**	**S1**	**S2**	**S3**	**T1**	**T2**	**T3**	**T4**	**T5**
1	**G1**		V	V	V	O	O	O	V	V	V	V	V
2	**G2**			V	O	V	V	V	O	V	O	O	O
3	**G3**				V	O	O	O	V	V	V	V	V
4	**G4**					O	O	V	V	V	X	X	V
5	**S1**						V	A	O	O	O	O	O
6	**S2**							A	O	O	O	O	O
7	**S3**								O	O	A	O	O
8	**T1**									V	A	X	V
9	**T2**										A	A	O
10	**T3**											V	V
11	**T4**												V
12	**T5**												

Table A2. Initial Reachability Matrix (RM_{init}) (Conventional Method).

i	*j*	1	2	3	4	5	6	7	8	9	10	11	12
	Barrier	**G1**	**G2**	**G3**	**G4**	**S1**	**S2**	**S3**	**T1**	**T2**	**T3**	**T4**	**T5**
1	**G1**	1	1	1	1	0	0	0	1	1	1	1	1
2	**G2**	0	1	1	0	1	1	1	0	1	0	0	0
3	**G3**	0	0	1	1	0	0	0	1	1	1	1	1
4	**G4**	0	0	0	1	0	0	1	1	1	1	1	1
5	**S1**	0	0	0	0	1	1	0	0	0	0	0	0
6	**S2**	0	0	0	0	0	1	0	0	0	0	0	0
7	**S3**	0	0	0	0	1	1	1	0	0	0	0	0
8	**T1**	0	0	0	0	0	0	0	1	1	0	1	1
9	**T2**	0	0	0	0	0	0	0	0	1	0	0	0
10	**T3**	0	0	0	1	0	0	1	1	1	1	1	1
11	**T4**	0	0	0	1	0	0	0	1	1	0	1	1
12	**T5**	0	0	0	0	0	0	0	0	0	0	0	1

Table A3. Final Reachability Matrix (RM_{fin}) (Conventional Method).

i	*j*	1	2	3	4	5	6	7	8	9	10	11	12
	Barrier	**G1**	**G2**	**G3**	**G4**	**S1**	**S2**	**S3**	**T1**	**T2**	**T3**	**T4**	**T5**
1	**G1**	1	1	1	1	1 *	1 *	1 *	1	1	1	1	1
2	**G2**	0	1	1	1 *	1	1	1	1 *	1	1 *	1 *	1 *
3	**G3**	0	0	1	1	1 *	1 *	1 *	1	1	1	1	1
4	**G4**	0	0	0	1	1 *	1 *	1	1	1	1	1	1
5	**S1**	0	0	0	0	1	1	0	0	0	0	0	0
6	**S2**	0	0	0	0	0	1	0	0	0	0	0	0
7	**S3**	0	0	0	0	1	1	1	0	0	0	0	0
8	**T1**	0	0	0	1 *	1 *	1 *	1 *	1	1	1 *	1	1
9	**T2**	0	0	0	0	0	0	0	0	1	0	0	0
10	**T3**	0	0	0	1	1 *	1 *	1	1	1	1	1	1
11	**T4**	0	0	0	1	1 *	1 *	1 *	1	1	1 *	1	1
12	**T5**	0	0	0	0	0	0	0	0	0	0	0	1

Table A4. Level Partitioning (Conventional Method)—1st Iteration.

Barrier	Reachability Set	Antecedent Set	Intersection Set	Level
1	1.2.3.4.5.6.7.8.9.10.11.12	1	1	
2	2.3.4.5.6.7.8.9.10.11.12	1.2	2	
3	3.4.5.6.7.8.9.10.11.12	1.2.3	3	
4	4.5.6.7.8.9.10.11.12	1.2.3.4.8.10.11	4.8.10.11	
5	5.6	1.2.3.4.5.7.8.10.11	5	
6	6	1.2.3.4.5.6.7.8.10.11	6	I
7	5.6.7	1.2.3.4.7.8.10.11	7	
8	4.5.6.7.8.9.10.11.12	1.2.3.4.8.10.11	4.8.10.11	
9	9	1.2.3.4.8.9.10.11	9	I
10	4.5.6.7.8.9.10.11.12	1.2.3.4.8.10.11	4.8.10.11	
11	4.5.6.7.8.9.10.11.12	1.2.3.4.8.10.11	4.8.10.11	
12	12	1.2.3.4.8.10.11.12	12	I

Table A5. Level Partitioning (Conventional Method)—2nd Iteration.

Barrier	Reachability Set	Antecedent Set	Intersection Set	Level
1	1.2.3.4.5.7.8.10.11	1	1	
2	2.3.4.5.7.8.10.11	1.2	2	
3	3.4.5.7.8.10.11	1.2.3	3	
4	4.5.7.8.10.11	1.2.3.4.8.10.11	4.8.10.11	
5	5	1.2.3.4.5.7.8.10.11	5	II
7	5.7	1.2.3.4.7.8.10.11	7	
8	4.5.7.8.10.11	1.2.3.4.8.10.11	4.8.10.11	
10	4.5.7.8.10.11	1.2.3.4.8.10.11	4.8.10.11	
11	4.5.7.8.10.11	1.2.3.4.8.10.11	4.8.10.11	

Table A6. Level Partitioning—3rd Iteration.

Barrier	Reachability Set	Antecedent Set	Intersection Set	Level
1	1.2.3.4.7.8.10.11	1	1	
2	2.3.4.7.8.10.11	1.2	2	
3	3.4.7.8.10.11	1.2.3	3	
4	4.7.8.10.11	1.2.3.4.8.10.11	4.8.10.11	
7	7	1.2.3.4.7.8.10.11	7	III
8	4.7.8.10.11	1.2.3.4.8.10.11	4.8.10.11	
10	4.7.8.10.11	1.2.3.4.8.10.11	4.8.10.11	
11	4.7.8.10.11	1.2.3.4.8.10.11	4.8.10.11	

Table A7. Level Partitioning (Conventional Method)—4th Iteration.

Barrier	Reachability Set	Antecedent Set	Intersection	Level
1	1.2.3.4.8.10.11	1	1	
2	2.3.4.8.10.11	1.2	2	
3	3.4.8.10.11	1.2.3	3	
4	4.8.10.11	1.2.3.4.8.10.11	4.8.10.11	IV
8	4.8.10.11	1.2.3.4.8.10.11	4.8.10.11	IV
10	4.8.10.11	1.2.3.4.8.10.11	4.8.10.11	IV
11	4.8.10.11	1.2.3.4.8.10.11	4.8.10.11	IV

Table A8. Level Partitioning (Conventional Method)—5th and 6th Iteration.

Barrier	Reachability Set	Antecedent Set	Intersection Set	Level
1	1.2.3.	1	1	VII
2	2.3.	1.2	2	VI
3	3.	1.2.3	3	V

Table A9. Conical Matrix.

Level	VII	VI	V	IV	IV	IV	IV	III	II	I	I	I	Level
Barrier	**G1**	**G2**	**G3**	**G4**	**T1**	**T3**	**T4**	**S3**	**S1**	**S2**	**T2**	**T5**	
G1	1	1	1	1	1	1	1	1 *	1 *	1 *	1	1	VII
G2	0	1	1	1 *	1 *	1 *	1 *	1	1	1	1	1 *	VI
G3	0	0	1	1	1	1	1	1 *	1 *	1 *	1	1	V
G4	0	0	0	1	1	1	1	1	1 *	1 *	1	1	IV
T1	0	0	0	1 *	1	1 *	1	1 *	1 *	1 *	1	1	IV
T3	0	0	0	1	1	1	1	1	1 *	1 *	1	1	IV
T4	0	0	0	1	1	1 *	1	1 *	1 *	1 *	1	1	IV
S3	0	0	0	0	0	0	0	1	1	1	0	0	III
S1	0	0	0	0	0	0	0	0	1	1	0	0	II
S2	0	0	0	0	0	0	0	0	0	1	0	0	I
T2	0	0	0	0	0	0	0	0	0	0	1	0	I
T5	0	0	0	0	0	0	0	0	0	0	0	1	I

References

1. Van Herk, S.; Rijke, J.; Zevenbergen, C.; Ashley, R. Understanding the Transition to Integrated Flood Risk Management in the Netherlands. *Environ. Innov. Soc. Transit.* **2015**, *15*, 84–100. [CrossRef]
2. United Nations International Strategy for Disaster Reduction. *UNISDR Terminology on Disaster Risk Reduction*; United Nations: Geneva, Switzerland, 2009.
3. Scott, M.; Kuhlicke, C.; Steinführer, A.; Sultana, P.; Thompson, P.; Minnery, J.; O'Neall, E.; Cooper, J.; Adamson, M.; Russell, E. Living with Flood Risk/The More We Know, the More We Know We Don't Know: Reflections on a Decade of Planning, Flood Risk Management and False Precision/Searching for Resilience or Building Social Capacities for Flood Risks?/Participatory Floodplain Manage. *Plan. Theory Pract.* **2013**, *14*, 114–120. [CrossRef]
4. Worldwide Fund for Nature, Mega-Stress for Megacities, A Climate Vulnerability Ranking of Major Coastal Cities in Asia. 2009. Available online: https://www.wwf.org.uk/sites/default/files/2009-13/mega_stress_cities_report.pdf (accessed on 15 January 2018).
5. Mercado, J.M.R.; Kawamura, A.; Amaguchi, H. Interrelationships of the Barriers to Integrated Flood Risk Management Adaptation in Metro Manila, Philippines. *Int. J. Disaster Risk Reduct.* **2020**, *49*, 101683. [CrossRef]
6. Vinke-de Kruijf, J.; Kuks, S.M.M.; Augustijn, D.C.M. Governance in Support of Integrated Flood Risk Management? The Case of Romania. *Environ. Dev.* **2015**, *16*, 104–118. [CrossRef]
7. Mysiak, J.; Testella, F.; Bonaiuto, M.; Carrus, G.; De Dominicis, S.; Cancellieri, U.G.; Firus, K.; Grifoni, P. Flood Risk Management in Italy: Challenges and Opportunities for the Implementation of the EU Floods Directive (2007/60/EC). *Nat. Hazards Earth Syst. Sci.* **2013**, *13*, 2883–2890. [CrossRef]
8. Samuels, P.; Klijn, F.; Dijkman, J. An Analysis of the Current Practice of Policies on River Flood Risk Management in Different Countries. *Irrig. Drain.* **2006**, *55*, S141–S150. [CrossRef]
9. Xia, C.; Pahl-Wostl, C. Understanding the Development of Flood Management in the Middle Yangtze River. *Environ. Innov. Soc. Transit.* **2012**, *5*, 60–75. [CrossRef]
10. Dieperink, C.; Hegger, D.L.T.; Bakker, M.H.N.; Kundzewicz, Z.W.; Green, C.; Driessen, P.P.J. Recurrent Governance Challenges in the Implementation and Alignment of Flood Risk Management Strategies: A Review. *Water Resour. Manag.* **2016**, *30*, 4467–4481. [CrossRef]
11. Dos Muchangos, L.S.; Tokai, A.; Hanashima, A. Analyzing the Structure of Barriers to Municipal Solid Waste Management Policy Planning in Maputo City, Mozambique. *Environ. Dev.* **2015**, *16*, 76–89. [CrossRef]
12. Barve, A.; Kanda, A.; Shankar, R. Analysis of Interaction among the Barriers of Third Party Logistics. *Int. J. Agil. Syst. Manag.* **2007**, *2*, 109–129. [CrossRef]
13. Yadav, D.K.; Barve, A. International Journal of Disaster Risk Reduction Analysis of Critical Success Factors of Humanitarian Supply Chain: An Application of Interpretive Structural Modeling. *Int. J. Disaster Risk Reduct.* **2015**, *12*, 213–225. [CrossRef]
14. Alora, A.; Barua, M.K. Barrier Analysis of Supply Chain Finance Adoption in Manufacturing Companies. *Benchmarking* **2019**, *26*, 2122–2145. [CrossRef]
15. Bui, T.D.; Tsai, F.M.; Tseng, M.L.; Ali, M.D.H. Identifying Sustainable Solid Waste Management Barriers in Practice Using the Fuzzy Delphi Method. *Resour. Conserv. Recycl.* **2020**, *154*, 104625. [CrossRef]
16. Ertas, A.; Smith, M.W.; Tate, D.; Lawson, W.D.; Baturalp, T.B. Complexity of System Maintainability Analysis Based on the Interpretive Structural Modeling Methodology: Transdisciplinary Approach. *J. Syst. Sci. Syst. Eng.* **2016**, *25*, 254–268. [CrossRef]
17. Chandramowli, S.; Transue, M.; Felder, F.A. Analysis of Barriers to Development in Landfill Communities Using Interpretive Structural Modeling. *Habitat Int.* **2011**, *35*, 246–253. [CrossRef]

18. Beikkhakhian, Y.; Javanmardi, M.; Karbasian, M.; Khayambashi, B. The Application of ISM Model in Evaluating Agile Suppliers Selection Criteria and Ranking Suppliers Using Fuzzy TOPSIS-AHP Methods. *Expert Syst. Appl.* **2015**, *42*, 6224–6236. [CrossRef]
19. Thakkar, J.; Deshmukh, S.G.; Gupta, A.D.; Shankar, R. Selection of Third-Party Logistics (3PL): A Hybrid Approach Using Interpretive Structural Modeling (ISM) and Analytic Network Process (ANP). *Supply Chain Forum Int. J.* **2018**, *6*, 32–46. [CrossRef]
20. Agi, M.A.N.; Nishant, R. Understanding Influential Factors on Implementing Green Supply Chain Management Practices: An Interpretive Structural Modelling Analysis. *J. Environ. Manag.* **2017**, *188*, 351–363. [CrossRef]
21. Mudgal, R.K.; Shankar, R.; Talib, P.; Raj, T. Modelling the Barriers of Green Supply Chain Practices: An Indian Perspective. *Int. J. Logist. Syst. Manag.* **2010**, *7*, 81–107. [CrossRef]
22. Singh, M.D.; Shankar, R.; Narain, R.; Agarwal, A. An Interpretive Structural Modeling of Knowledge Management in Engineering Industries. *J. Adv. Manag. Res.* **2003**, *1*, 28–40. [CrossRef]
23. Sushil. Modified ISM/TISM Process with Simultaneous Transitivity Checks for Reducing Direct Pair Comparisons. *Glob. J. Flex. Syst. Manag.* **2017**, *18*, 331–351. [CrossRef]
24. Sushil. Interpreting the Interpretive Structural Model. *Glob. J. Flex. Syst. Manag.* **2012**, *13*, 87–106. [CrossRef]
25. Warfield, J.N. Binary Matrices in System Modeling. *IEEE Trans. Syst. Man. Cybern.* **1973**, *5*, 441–449. [CrossRef]
26. Warfield, J.N. Societal Systems Planning, Policy, and Complexity. *Cybern. Syst.* **1978**, *8*, 113–115.
27. Farris, D.R.; Sage, A.P. On the Use of Interpretive Structural Modeling for Worth Assessment. *Comput. Electr. Eng.* **1975**, *2*, 149–174. [CrossRef]
28. Watson, R.H. Interpretive Structural Modeling—A Useful Tool for Technology Assessment? *Technol. Forecast. Soc. Change* **1978**, *11*, 165–185. [CrossRef]
29. Attri, R.; Dev, N.; Sharma, V. Interpretive Structural Modelling (ISM) Approach: An Overview. *Res. J. Manag. Sci.* **2013**, *2*, 3–8. [CrossRef]
30. Agarwal, A.; Shankar, R.; Tiwari, M.K. Modeling Agility of Supply Chain. *Ind. Mark. Manag.* **2007**, *36*, 443–457. [CrossRef]
31. Huang, J.; Tzeng, G. Multidimensional Data in Multidimensional Scaling Using the Analytic Network Process. *Pattern Recognit. Lett.* **2005**, *26*, 755–767. [CrossRef]
32. Saxena, J.P.; Sushil; Vrat, P. Impact of Indirect Relationships in Classification of Variables—A Micmac Analysis for Energy Conservation. *Syst. Res.* **1990**, *7*, 245–253. [CrossRef]
33. Evers, M.; Jonoski, A.; Almoradie, A.; Lange, L. Collaborative Decision Making in Sustainable Flood Risk Management: A Socio-Technical Approach and Tools for Participatory Governance. *Environ. Sci. Policy* **2016**, *55*, 335–344. [CrossRef]
34. Van Herk, S.; Zevenbergen, C.; Rijke, J.; Ashley, R.; Van Herk, S.; Zevenbergen, C.; Rijke, J.; Ashley, R. Collaborative Research to Support Transition towards Integrating Flood Risk Management in Urban Development. *J. Flood Risk Manag.* **2011**, *4*, 306–317. [CrossRef]
35. Nye, M.; Tapsell, S.; Twigger-Ross, C. New Social Directions in UK Flood Risk Management: Moving towards Flood Risk Citizenship? *J. Flood Risk Manag.* **2011**, *4*, 288–297. [CrossRef]
36. Faulkner, H.; Parker, D.; Green, C.; Beven, K. Developing a Translational Discourse to Communicate Uncertainty in Flood Risk between Science and the Practitioner. *Ambio* **2007**, *36*, 692–703. [CrossRef] [PubMed]
37. Demeritt, D.; Nobert, S. Models of Best Practice in Flood Risk Communication and Management. *Environ. Hazards* **2014**, *13*, 313–328. [CrossRef]
38. Rollason, E.; Bracken, L.J.; Hardy, R.J.; Large, A.R.G. Rethinking Flood Risk Communication. *Nat. Hazards* **2018**, *92*, 1665–1686. [CrossRef]
39. Gilbuena, R.; Kawamura, A.; Medina, R.; Amaguchi, H.; Nakagawa, N. Gap Analysis of the Flood Management System in Metro Manila, Philippines: A Case Study of the Aftermath of Typhoon Ondoy. *IAHS Publ.* **2013**, *357*, 32–40.
40. Hall, J.W.; Meadowcroft, I.C.; Sayers, P.B.; Bramley, M.E. Integrated Flood Risk Management in England and Wales. *Nat. Hazards Rev.* **2003**, *4*, 126–135. [CrossRef]
41. Hall, J.W. Flood Risk Management: Decision Making under Uncertainty. In *Applied Uncertainty Analysis for Flood Risk Management*; Imperial College Press: London, UK, 2014; pp. 3–24. [CrossRef]
42. Kekez, T.; Knezic, S.; Andricevic, R. Incorporating Uncertainty of the System Behavior in Flood Risk Assessment–Sava River Case Study. *Water* **2020**, *12*, 2676. [CrossRef]
43. Mercado, J.M.R.; Kawamura, A.; Amaguchi, H.; Rubio, C.J.P. Fuzzy Based Multi-Criteria M&E of the Integrated Flood Risk Management Performance Using Priority Ranking Methodology: A Case Study in Metro Manila, Philippines. *Int. J. Disaster Risk Reduct.* **2021**, *64*, 102498. [CrossRef]

water

MDPI

Article

Study on a Water-Level-Forecast Method Based on a Time Series Analysis of Urban River Basins—A Case Study of Shibuya River Basin in Tokyo

Naoki Koyama [1,*], Mizuki Sakai [2] and Tadashi Yamada [1]

[1] Research and Development Initiative, Chuo University, 1-13-27, Kasuga, Bunkyo-ku, Tokyo 112-8551, Japan
[2] Civil, Human and Environmental Engineering Course, Graduate School of Science and Engineering, Chuo University, 1-13-27, Kasuga, Bunkyo-ku, Tokyo 112-8551, Japan
* Correspondence: knaoki002@g.chuo-u.ac.jp; Tel.: +81-3-3817-1621

Abstract: In urban basins, localized torrential rain increases the water level of rivers in an extremely short time, thereby leading to flooding within an hour. Therefore, to achieve early evacuation, the water level should be accurately forecasted. The outflow process in urban areas employs the sewer system to discharge the water back to rivers. However, the data for the sewer system are not freely available, and it requires much work and time to design a physical model based on such data. Thus, a vector autoregressive model to develop a water level forecast system that uses observed rainfall and water level is being used. Additionally, this model was used to ensure information conducive to evacuation approximately 20 min in advance and to assess its forecast accuracy, despite the very limited data—water levels at one point and average rainfall at another—without the need to build a physical model such as that which is used in sewer pipe calculations. Compared to the observed water level, the calculated water level increased faster; and thus, the forecast leaned toward safety in evacuation. Furthermore, the data from past five torrential rainfall events to achieve a stable forecast; this method can be applied to basins with limited observation data. Therefore, these results indicate that this method can be applied as a water level forecast method for basins with an extremely fast flood arrival time.

Keywords: urban floods; forecast method; water level; vector auto regressive model; time series analysis

Citation: Koyama, N.; Sakai, M.; Yamada, T. Study on a Water-Level-Forecast Method Based on a Time Series Analysis of Urban River Basins—A Case Study of Shibuya River Basin in Tokyo. *Water* **2023**, *15*, 161. https://doi.org/10.3390/w15010161

Academic Editors: Akira Kawamura and Kei Nakagawa

Received: 27 November 2022
Revised: 22 December 2022
Accepted: 30 December 2022
Published: 31 December 2022

1. Introduction

Floods are the most frequently occurring natural disaster, causing damage every year worldwide. Specifically, the damage from "urban floods," which target cities, is substantial [1,2], and rapid urbanization is changing the scale and frequency of urban floods [3–7]. Rainwater that used to infiltrate the ground or was stored on the ground surface is now discharged into rivers as surface water due to urbanization, thereby increasing the flood damage in urban areas. Rainwater in urban areas is usually discharged through the sewer system, but when its discharge capacity is exceeded, rainwater that cannot be discharged to the sewer system remains on the ground surface, or it flows out of manholes, leading to inland floods. Furthermore, the increasingly extreme nature of weather phenomena due to climate change is a global issue, putting cities at risk [6–9]. In Japan, localized bursts of short torrential rain called "guerrilla rainstorms" are becoming more frequent in urban areas, and not only rainfall of 50 mm or more per hour, but also rainfall of 80 mm or more per hour has been observed an increasing number of times [8].

To manage these urban floods and protect urban functions, flood measures that consider an increase in external forces must be shifted due to climate change. Flood measures are generally divided as "structure measures" and "non-structure measures". According to Itsukushima et al. [10], structure measures against urban floods can be classified into three types: "rivers", "the sewer system", and "basins". Measures for "rivers" include embankment, channel excavation, and setting back of levees [11]. However, in Japanese cities, dense structures such as buildings and homes have been built along urban rivers, thereby complicating the construction of embankments and channel excavation. Furthermore, it is difficult to secure the land for constructing levees, while the project costs are massive, which also makes the measure difficult to implement. Measures for "the sewer system" include larger diameters for sewer pipes, underground storage facilities for sewage water, and so on, but enlarging the pipes in massive sewer systems is expensive and time consuming. Measures for "basins" could include the installation of a rainwater storage facility in each home to control discharge, the installation of a dry well or water storage facility for each city block, and so on. Kenji Kawaike et al. [12] assessed the flood control effect of these discharge control facilities, showing that they reduce flood damage. In addition, there are many studies that showed that green infrastructure measures, such as permeable pavements, rooftop greening, and green gardens are useful in flood mitigation [13–15]. However, in the case of mitigation measures where the unit is a building, such as discharge control facilities and the green infrastructure measures mentioned above, the flood control effect of each measure is limited; and thus, to achieve success in minimizing urban floods, measures at a scale of the entire basin are necessary. Thus, to achieve structure measures to reduce flood damage in urban areas, a massive amount of money and time is necessary. Therefore, measures at the level of each home that are easier to incorporate cannot truly solve the problems.

However, non-structure measures could include the provision of information during floods for rapid evacuation, and the provision of flood risk information to guide planned town building. As a manner of providing flood risk information, some institutions and countries provide a "hazard map" for each city [16–19]. However, while this "hazard map" is useful in informing individuals about the flood risk of the areas in which they live, it does not provide real-time information to determine evacuation. To that end, water level forecast information for urban rivers is necessary [20]. In Japan, the river water level is used to determine the issuance of evacuation information during floods. If the water level for rivers can be accurately forecast with a several-hour lead or for cities with a lead of less than an hour, evacuation can be appropriately guided. With accurate and fast forecasting of water levels, lead time can be ensured for evacuation, minimizing the damage from floods.

Water level forecast methods can be broadly divided into physical models and statistical models [21]. In this study, we focused on a statistical model, a vector autoregressive model of time series analysis, and hypothesized that applying this model to water level forecasts would lead to more accurate and fast forecasts.

Among statistical models, time series analysis and machine learning are the common flood forecast methods. Generally, flood forecast methods based on machine learning can consider nonlinearity of phenomena and have enjoyed rapid development in recent years [22–24]. However, statistical models, represented by autoregressive model of time series analysis, attempt to establish the relationship between multiple phenomena without internally describing physical process related to the past; and thus, it is called system theoretical transfer function models. Since the vector autoregressive model, a time series analysis method, employed in this study is a linear model, some might point out that it is unable to consider nonlinearity. However, urban river basins in this study are mostly paved by concrete, where the outflow process is relatively simple compared to mountain basins. Therefore, the effect of nonlinearity is likely minimal. Furthermore, multivariate autoregressive models have been used in various fields in the past and are being studied with the aim of applying them to water level forecasts in the field of hydrology [25–29]. For example, Niedzielski et al. forecasted the water level of Oder River in southern Poland using a multivariate autoregressive model [28,29]. However, many past studies on water level forecast focused on the time axis of hours, days, and months. There is no study that focused on urban rivers with an extremely short flood arrival time of less than an hour, and the validity of each forecast method remains unclear, the objective is to assess the accuracy of water level forecasts using this method.

Meanwhile, the outflow process of urban basins discharges rain to rivers through the sewer system; and thus, physical models that consider this outflow process and calculation methods that follow the flow of water through the sewer system have been extensively studied [30]. Specifically, commercial software is often used to forecast floods including the sewer system in urban areas, such as MIKE URBAN [31], Infoworks CS [32], and SWMM [33]. Many studies used these software programs to study flood forecast [34–36]. However, the sewer system pipe data and water level data are not easy to observe since these structures are underground, where the data on the network of pipes are not freely accessible, often being difficult to obtain [37,38]. In addition, the amount of data for pipes and manholes is massive even for a small urban basin of several 10 s of km^2. It takes massive work and time to design a water level forecast system based on physical models that include pipe calculations. In contrast, in the case of multivariate autoregressive model employed for this study, data required for the forecast are limited to observation data of rainfall and water levels. Thus, it is highly suited as a water level forecast system for urban rivers.

Therefore, in this study, a water level forecast method is developed which conductive to accommodates floods caused by localized torrential rains in order to reduce casualty loss of urban floods, which have become more frequent in recent years, with an aim of supporting more rapid issuance of evacuation information.

2. Study Area and Data

The target basin was Shibuya River Basin, which flows through Shibuya Ward, Tokyo Metropolis (Figure 1). Shibuya River is an urban river that flows along the site of the Tokyo Olympics. The basin area is approximately 12.5 km^2 with a length of approximately 2.6 km. Figure 1a shows Shibuya River Basin, where the red line shows the sewer system network, and the blue line shows Shibuya River. Other than the green areas that indicate Meiji Jingu Grand Shrine and Shinjuku Gyoen, a park, Shibuya River Basin is covered by structures. Rainwater passes through the sewer system and flows into Shibuya River. Figure 1b shows the sewer system pipeline network prepared from the sewer system ledger data managed by Bureau of Sewerage Tokyo Metropolitan Government [39]. The total length of the sewer system pipes is 243 km, with 8800 manholes. These numbers show how difficult it is to forecast the water level for the 2.6 km along the river upon plotting the entire pipeline network and manholes.

Next, the focal point is on torrential rainfall events caused by localized torrential rain. Figure 2 shows the localized torrential rain in Shibuya River Basin that occurred on May 18, 2017. Figure 2a shows Shibuya River under normal conditions. These photographs show that the river carries little water under normal conditions, and the buildings fill the area along the river right up to the embankment. Figure 2b shows the river during a flood. At the bottom of Figure 2 are the hyetograph and hydrograph at that time. These diagrams show that the time between the rainfall peak and the water level peak was ~30 min, showing the extremely fast flood arrival time. Therefore, urban basins with a small area have a short flood arrival time, resulting in an extremely short lead time for residents to evacuate.

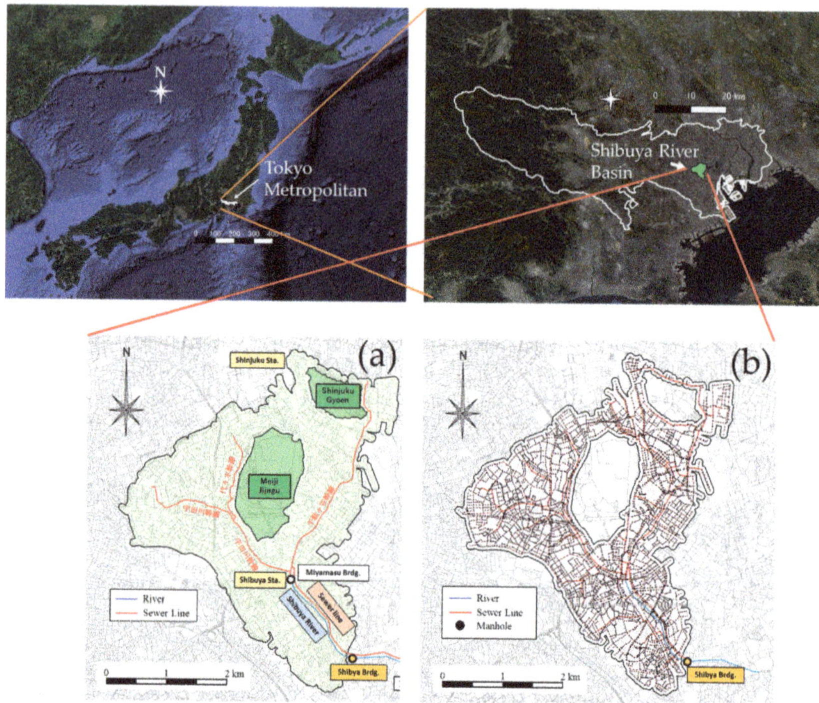

Figure 1. Location of the Study Area (a): Shibuya River Basin and (b): Pipeline Network.

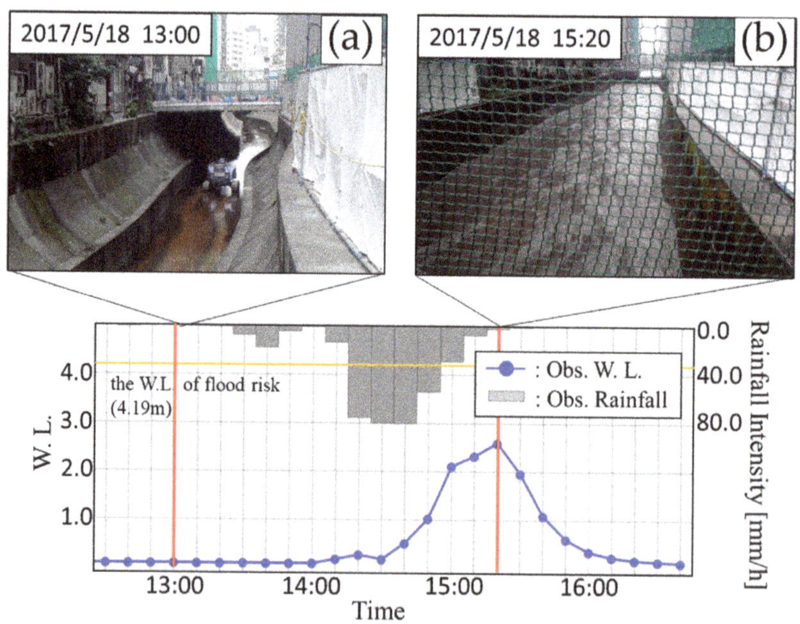

Figure 2. An example of localized torrential rain in Shibuya River Basin (18 May 2017). (a) State of the river during low conditions. (b) State of the river during of flooding.

There were nine torrential rainfall events were pinpointed. Figure 3 shows the rainfall and water level during each torrential event as a graph. Meanwhile, Figure 4 shows the difference in the rainfall peak and water level peak for each torrential rainfall event. Then, Table 1 shows the data figures used in Figure 3, with the maximum rainfall and maximum water level for each event. In most torrential rainfall events, the difference between the rainfall peak and water level peak was extremely short: 10–20 min. Next, the explanation about the observation data of rainfall and water level will be broken down. In Shibuya River Basin, there is one water level observation station [40] managed by Tokyo Metropolis (Figure 1b, orange circle), where observations are made every 10 min. As for the rainfall data, the rainfall observed data by the Xband-MP radar which installed by the Ministry of Land, Infrastructure, Transport and Tourism [41] were applied. Observations were taken every minute, and the 10 min average was taken to match the time resolution of the water level meter.

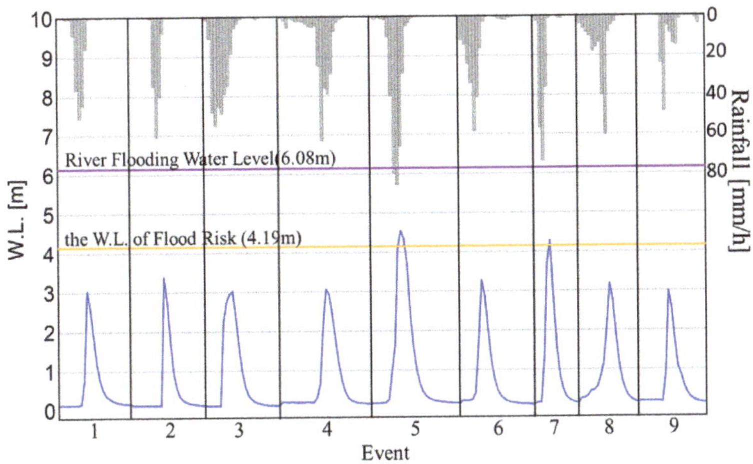

Figure 3. Localized torrential rainfall events since 2015 (9 Events).

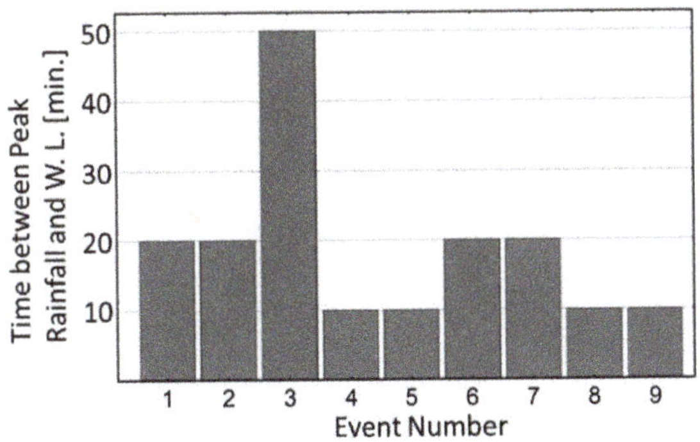

Figure 4. Difference in the time of the rainfall peak and water level peak for each torrential rainfall event.

Table 1. Observed numerical data on rainfall and water levels at each event, with colored maximum rainfall intensity and water level for each event.

Time [min.]	Rain 1 [mm/h]	W.L. 1 [m]	Rain 2 [mm/h]	W.L. 2 [m]	Rain 3 [mm/h]	W.L. 3 [m]	Rain 4 [mm/h]	W.L. 4 [m]	Rain 5 [mm/h]	W.L. 5 [m]	Rain 6 [mm/h]	W.L. 6 [m]	Rain 7 [mm/h]	W.L. 7 [m]	Rain 8 [mm/h]	W.L. 8 [m]	Rain 9 [mm/h]	W.L. 9 [m]
0	0	0.14	0	0.12	0	0.12	0	0.1	0	0.15	0	0.23	0	0.2	0	0.18	0	0.19
10	0	0.14	0	0.12	0	0.12	0	0.1	0	0.15	0	0.23	0	0.2	0	0.18	0	0.19
20	0	0.13	0	0.12	0	0.12	0	0.1	0	0.15	0	0.24	0	0.2	0	0.18	0	0.19
30	0	0.13	0	0.12	0	0.12	0	0.1	0	0.16	0	0.24	0	0.2	0	0.18	0	0.2
40	0	0.13	0	0.12	0	0.11	0	0.1	0	0.16	0	0.24	0	0.2	0	0.18	0	0.19
50	0	0.13	0	0.12	0	0.11	0	0.1	0	0.15	0	0.24	0	0.2	0	0.18	0	0.19
60	0	0.13	0	0.12	0	0.11	0	0.1	0	0.15	0	0.24	0	0.2	0.47	0.18	0	0.19
70	0	0.13	0	0.12	0	0.11	0.88	0.1	0	0.15	0	0.24	0	0.21	1.79	0.18	0	0.19
80	0	0.13	0	0.12	0	0.11	5.3	0.1	0	0.15	0	0.24	0	0.21	0.94	0.18	0	0.19
90	0	0.13	0	0.12	0	0.11	0.76	0.1	0.61	0.15	0	0.25	0	0.21	0.07	0.18	0	0.19
100	0	0.13	0	0.12	0	0.11	0.16	0.1	1.3	0.15	0	0.25	0	0.21	0.09	0.18	0	0.19
110	0	0.13	0	0.12	0	0.11	0	0.1	0.5	0.15	0	0.25	0	0.21	0.16	0.18	0	0.19
120	0	0.13	0	0.12	0	0.11	0	0.1	1.04	0.16	0	0.25	0	0.21	1.44	0.19	0	0.19
130	0	0.13	0	0.12	0	0.1	0	0.1	1.54	0.17	0	0.25	0	0.21	6.93	0.18	0	0.19
140	0	0.13	0	0.12	0	0.1	0	0.1	10.32	0.25	0	0.25	0	0.21	2.25	0.18	0	0.19
150	0	0.13	0	0.12	0	0.1	0	0.1	40.22	0.98	0	0.25	0	0.21	3.23	0.29	0	0.19
160	0	0.13	0	0.12	0	0.1	0	0.1	79.8	1.65	0	0.25	0	0.21	5.07	0.28	0	0.19
170	0	0.13	0	0.12	0	0.1	0	0.1	85.69	4.07	0	0.25	0	0.21	1.22	0.27	0	0.19
180	0	0.13	0	0.12	0	0.11	0	0.1	66.18	4.56	0.03	0.25	0	0.2	0	0.25	0	0.19
190	0	0.13	0	0.12	0	0.1	0	0.1	28.87	4.36	0.14	0.25	0	0.21	0.01	0.21	0	0.19
200	0	0.13	0	0.12	0	0.11	0	0.1	4.82	3.76	0.65	0.25	0	0.21	0.29	0.19	0	0.2
210	0	0.13	0	0.12	0	0.11	0	0.1	3.1	2.43	0	0.25	0	0.21	0.07	0.18	0	0.19
220	0.26	0.13	0	0.12	0	0.11	0	0.1	0.87	1.59	0	0.25	0	0.2	0.11	0.18	0	0.19
230	0.66	0.13	0	0.12	0	0.11	0	0.1	0.15	1.03	0	0.25	0	0.2	1.03	0.18	0	0.19
240	0.59	0.13	0	0.12	0	0.1	0	0.19	0	0.68	0	0.25	0	0.2	0.08	0.18	0	0.19
250	8.87	0.13	0	0.12	0	0.11	0	0.19	0	0.46	0	0.25	0	0.2	0.01	0.18	0	0.19
260	36.82	0.13	0.74	0.12	0	0.11	0	0.19	0	0.35	0	0.25	0	0.2	0.81	0.17	0	0.19
270	51.2	0.15	34.96	0.12	0	0.11	0	0.19	0	0.28	0	0.25	0	0.2	1.07	0.3	0	0.19
280	44.8	0.78	61	0.12	0	0.11	0	0.19	0	0.24	0	0.25	0	0.2	0.07	0.22	0	0.19
290	15.89	3.04	40.29	0.12	0	0.11	0	0.19	0	0.21	0	0.25	0	0.2	0	0.18	0	0.19
300	0.55	2.57	8.27	3.4	0	0.11	0	0.19	0	0.2	0	0.25	0	0.2	0.34	0.16	0	0.19
310	0	1.88	0.01	2.77	0	0.11	0	0.19	0	0.18	0	0.25	0	0.2	0.2	0.16	0	0.19
320	0	1.21	0	1.76	0	0.11	0	0.18	0	0.18	0	0.25	0	0.2	0.27	0.16	0	0.19

Table 1. *Cont.*

Time [min.]	Rain 1 [mm/h]	W.L. 1 [m]	Rain 2 [mm/h]	W.L. 2 [m]	Rain 3 [mm/h]	W.L. 3 [m]	Rain 4 [mm/h]	W.L. 4 [m]	Rain 5 [mm/h]	W.L. 5 [m]	Rain 6 [mm/h]	W.L. 6 [m]	Rain 7 [mm/h]	W.L. 7 [m]	Rain 8 [mm/h]	W.L. 8 [m]	Rain 9 [mm/h]	W.L. 9 [m]
330	0.44	0.78	0	0.95	0.04	0.11	0	0.18	0	0.16	0.04	0.25	0	0.2	0.17	0.16	0	0.19
340	0.22	0.53	0	0.56	1.42	0.11	0	0.19	0	0.16	0.19	0.25	1.03	0.2	0.62	0.16	0	0.19
350	0.02	0.38	0	0.35	0.41	0.11	0.12	0.19	0	0.15	0.09	0.25	0.02	0.19	2.25	0.18	0	0.19
360	0.17	0.31	0	0.25	9.97	0.11	1.96	0.19	0	0.15	0.15	0.25	0	0.18	0	0.18	0.12	0.19
370	0.03	0.25	0	0.2	8.7	0.1	5.63	0.18	0	0.15	0.32	0.25	0.21	0.17	0	0.17	23.95	0.18
380	0	0.22	0	0.17	1.98	0.1	0.78	0.19	0	0.14	0.57	0.25	24.12	0.16	0	0.16	48.66	0.88
390	0	0.19	0	0.15	0.55	0.1	1.29	0.18	0	0.14	0.5	0.24	9	0.16	0	0.16	5.26	3
400	0	0.18	0	0.14	0.04	0.1	2.22	0.18	0	0.14	12.07	0.22	3.18	0.16	0	0.16	8.91	2.57
410	0	0.16	0	0.13	0	0.1	2.02	0.18	0	0.14	14.48	0.21	1.28	0.16	0	0.16	14.06	1.75
420	0	0.16	0	0.12	0	0.1	2.5	0.18	0.01	0.13	24.17	0.21	0.51	0.16	0.02	0.16	14.92	1.07
430	0	0.15	0	0.12	0	0.1	3.5	0.19	0.03	0.13	34.01	0.24	0	0.16	0.06	0.16	1.57	0.87
440	0	0.14	0	0.11	0	0.1	4.86	0.18	0	0.12	58.56	0.43	0	0.16	0.3	0.16	0.04	0.59
450	0	0.14	0	0.11	0	0.1	5.21	0.18	0	0.12	41.32	1.76	0	0.16	0.18	0.16	0.18	0.4
460	0.06	0.14	0	0.11	0	0.1	4.56	0.31	0	0.12	16.17	3.29	0	0.15	0.04	0.16	0.11	0.29
470	0	0.13	0	0.11	0	0.1	24.93	0.67	0	0.12	2.92	2.87	0	0.16	0	0.16	0.04	0.23
480	0	0.13	0	0.11	0	0.1	63.09	2.14	0	0.11	0.35	2	0	0.16	0.12	0.16	1.26	0.19
490	0	0.13	0	0.11	0.01	0.1	36.2	3.07	0	0.11	0	1.21	0	0.16	0	0.16	3.92	0.17
500	0	0.13	0	0.1	1.42	0.1	39.04	2.93	5.98	0.1	0	0.75	0	0.16	0	0.16	0.5	0.15
510	0	0.13	0	0.1	10.87	0.1	28.98	2.4	10.68	0.1	0.55	0.47	0	0.16	0	0.16	0.1	0.14
520	0	0.13	0	0.1	48.34	0.1	7.19	1.76	3.04	0.1	1.31	0.33	0	0.16	0	0.16	0.58	0.13
530	0	0.13	0	0.1	55.49	0.1	2.61	1.17	2.65	0.1	1.7	0.25	0	0.16	0	0.16	0.1	0.12
540	0	0.13	0	0.1	45.55	0.1	1.24	0.75	0.32	0.09	1.93	0.2	0	0.16	0	0.16	0.04	0.12
550	0	0.13	0	0.1	48.86	2.22	0.43	0.46	0.5	0.09	0.89	0.17	0	0.16	0	0.16	0.42	0.11
560	0	0.13	0	0.1	38.94	2.74	0.01	0.3	0.19	0.09	0.65	0.15	0	0.16	0	0.16	0.29	0.11
570	0	0.13	0	0.1	34.7	2.95	0	0.22	0.01	0.09	0.23	0.14	0	0.16	1.63	0.16	1.13	0.11
580	0	0.13	0	0.1	18.4	3.02	0	0.19	0	0.09	0.14	0.13	0	0.16	3.47	0.16	1.46	0.1
590	0	0.13	0	0.1	5.68	2.37	0	0.17	0	0.09	0.04	0.12	0	0.16	5.06	0.16	0.91	0.1
600	0	0.13	0	0.1	3.21	1.6	0	0.16	0	0.09	0	0.12	2.5	0.16	2.61	0.17	0.22	0.1
610	0	0.12	0	0.1	1.13	0.97	0	0.15	0.26	0.1	0	0.12	57.05	0.16	14.52	0.17	0	0.1
620	0	0.12	0	0.1	0.05	0.6	0	0.14	0.64	0.1	0	0.11	73.98	1.31	7.66	0.2	0	0.1
630	0	0.12	0	0.1	0	0.38	0.01	0.14	3.77	0.1	0	0.11	36.22	3.71	3.91	0.26	0	0.1
640	0	0.12	0	0.1	0.03	0.27	0.13	0.14	5.34	0.16	0.11	0.11	1.25	4.3	6.31	0.28	0	0.1

Table 1. *Cont.*

Time [min.]	Rain 1 [mm/h]	W.L. 1 [m]	Rain 2 [mm/h]	W.L. 2 [m]	Rain 3 [mm/h]	W.L. 3 [m]	Rain 4 [mm/h]	W.L. 4 [m]	Rain 5 [mm/h]	W.L. 5 [m]	Rain 6 [mm/h]	W.L. 6 [m]	Rain 7 [mm/h]	W.L. 7 [m]	Rain 8 [mm/h]	W.L. 8 [m]	Rain 9 [mm/h]	W.L. 9 [m]
650	0	0.12	0	0.1	0.15	0.2	0.06	0.14	3.67	0.2	2.91	0.11	0.06	2.82	8.52	0.33	0	0.1
660	0	0.12	0	0.1	1.15	0.17	0	0.14	2.88	0.23	4.51	0.11	0.05	1.6	10.21	0.45	0	0.1
670	0	0.13	0	0.1	0.12	0.15	0	0.14	0.17	0.19	4.14	0.11	1	0.87	17.9	0.47	0	0.14
680	0	0.13	0	0.18	0	0.13	0	0.14	0	0.16	3.4	0.11	1.37	0.51	15.95	0.56	0	0.17
690	0	0.13	0	0.19	0	0.12	0	0.14	0	0.13	2.43	0.13	0.15	0.32	13.43	0.82	0	0.17
700	0	0.13	0	0.19	0	0.11	0	0.14	0	0.12	5.73	0.18	0	0.24	40.35	1.16	0	0.17
710	0	0.13	0	0.19	0	0.11	0	0.14	0	0.1	7.65	0.18	0	0.19	60.6	2.1	0	0.17
720	0	0.13	0	0.19	0	0.1	0	0.14	0	0.1	4.68	0.16	0	0.16	8.27	3.2	0	0.17
730	0	0.12	0	0.19	0	0.1	0	0.14	0	0.1	0.5	0.2	0	0.14	4.82	2.74	0	0.18
740	0	0.12	0	0.19	0	0.1	0	0.14	0	0.1	0	0.19	0	0.13	2.96	1.88	0	0.17
750	0	0.12	0	0.19	0	0.1	0	0.14	0	0.09	0	0.16	0	0.12	0.54	1.15	0	0.17
760	0	0.12	0	0.19	0	0.09	0	0.14	0	0.09	0	0.13	0	0.11	0.36	0.73	0	0.17
770	0	0.12	0	0.19	0	0.09	0	0.14	0	0.09	0	0.12	0	0.1	0.28	0.47	0	0.17
780	0	0.12	0	0.19	0	0.09	0	0.14	0	0.09	0	0.11	0	0.1	0.07	0.32	0	0.17
790	0	0.15	0	0.19	0	0.1	0	0.14	0	0.1	0	0.11	0	0.1	0	0.24	0	0.17
800	0	0.18	0	0.19	0	0.1	0	0.14	0	0.09	0	0.19	0	0.1	0	0.19	0	0.17
810	0	0.2	0	0.19	0	0.1	0	0.14	0	0.1	0	0.19	0	0.09	0	0.16	0	0.17
820	0	0.2	0	0.19	0	0.1	0	0.14	0	0.1	0	0.19	0	0.09	0	0.15	0	0.17
830	0	0.21	0	0.19	0	0.09	0	0.14	0	0.1	0	0.19	0.29	0.09	0	0.13	0	0.17

3. Analytical Method for Water Level Forecast with Time Series Analysis

3.1. About Vsector Autoregressive Model

The time series analysis is a statistical model, which is used to relate phenomena with past data and forecast/control phenomena [42]. First, let us show the univariate autoregressive model equation used for the most basic time series analysis in Equation (1).

$$h_n = \sum_{i=1}^{N} a_i h_{n-i} + \varepsilon_n \tag{1}$$

where h is an datum (water level), n is an time, a is a parameter, N is a degree, and ε is white noise.

Equation (1) expresses the water level at an arbitrary time h_n as a product of the past water level time series h_{n-i} and parameter a_i in a linear sum with white noise ε_i. N is a degree, which is the amount of past data considered. Since the parameter a_i is the weight of the past observation values h_{n-I}, the size of a_i expresses the impact of past observations made for h_n.

When considering real rivers, the flow rate of rivers gathers from distributaries upstream to downstream, and flow rate at each point interacts. Thus, the autoregressive model were expanded and expressed in Equation (2) the impact of each tributary included.

$$\begin{bmatrix} h_n^1 \\ h_n^2 \\ \vdots \\ h_n^p \end{bmatrix} = \sum_{i=1}^{N} \begin{bmatrix} a_{11}^i & a_{12}^i & \cdots & a_{1l}^i \\ a_{21}^i & a_{22}^i & & a_{2l}^i \\ \vdots & & \ddots & \vdots \\ a_{p1}^i & a_{p2}^i & \cdots & a_{pl}^i \end{bmatrix} \begin{bmatrix} h_{n-i}^1 \\ h_{n-i}^2 \\ \vdots \\ h_{n-i}^p \end{bmatrix} + \begin{bmatrix} \varepsilon_n^1 \\ \varepsilon_n^2 \\ \vdots \\ \varepsilon_n^p \end{bmatrix} \tag{2}$$

Equation (2) is a generalization of Equation (1), which is called multivariate autoregressive model or vector autoregressive model, where N is the degree of the model, P is the number of water levels used, l is the order of the water level, h_n^p is the water level at the point p at time n, and a_{pl}^i is the impact of the water level at the point l on the water level at the point p before the time i.

Furthermore, in the rainfall outflow process, rain falls on a basin, which infiltrates and outflows from the ground surface, flows into rivers and the sewer system, and becomes a water level at a certain point. Water level information at an arbitrary observation point is the final information in the rainfall outflow process, and the first start is the rainfall information. Thus, instead of only using the water level information, by adding the rainfall information, we can add the information prior to the water level information to the model, thereby considering the rapid rise in water levels in urban basins. Therefore, rainfall data were added into Equation (2) and hence Equation (3) is obtained.

$$\begin{bmatrix} h_n^1 \\ \vdots \\ h_n^p \\ r_n^1 \\ \vdots \\ r_n^s \end{bmatrix} = \sum_{i=1}^{N} \begin{bmatrix} a_{11}^i & \cdots & a_{1p}^i b_{11}^i & \cdots & b_{1s}^i \\ \vdots & \ddots & \vdots \ \ \vdots & \ddots & \vdots \\ a_{p1}^i & \cdots & a_{pl}^i b_{11}^i & \cdots & b_{sm}^i \end{bmatrix} \begin{bmatrix} h_{n-i}^1 \\ \vdots \\ h_{n-i}^p \\ r_{n-i}^1 \\ \vdots \\ r_{n-i}^s \end{bmatrix} + \begin{bmatrix} \varepsilon_{qn}^1 \\ \vdots \\ \varepsilon_{qn}^p \\ \varepsilon_{rn}^1 \\ \vdots \\ \varepsilon_{rn}^s \end{bmatrix} \tag{3}$$

where r is rainfall data and s is the number of rainfall data, m is the order of the rainfall, and b_{sm}^i is the impact of the rainfall at the point m on the rainfall at the point s before the time i. With this Equation, we can see that water level and rainfall are used to obtain the water level at an arbitrary point. Next, let us explain the forecast method for water level. With Equation (3), the current water level and rainfall data are obtained from the past data; thus, by including the current data in this Equation, we can forecast one step ahead, which is expressed in Equation (4). Here, for simplification, we will only examine the water level in this Equation.

$$\hat{\mathbf{h}}^p_{n+1} = \sum_{i=1}^{N}\sum_{l=1}^{P} \mathbf{a}^i_{pl}\mathbf{h}^l_{n+1-i} + \varepsilon^p_n \tag{4}$$

where $\hat{\mathbf{h}}^p_{n+1}$ is the water level forecast. The forecast result is used recursively to forecast the next step, and when the forecast is performed after the time x, it can be expressed with Equation (5).

$$\hat{\mathbf{h}}^p_{n+x} = \sum_{j=1}^{x-1}\sum_{l=1}^{P} \mathbf{a}^i_{pl}\hat{\mathbf{h}}^l_{n+x-j} + \sum_{i=1}^{N-x}\sum_{l=1}^{P} \mathbf{a}^i_{pl}\mathbf{h}^l_{n+1-i} + \varepsilon^p_n \tag{5}$$

With the same line of thinking, a forecast can be made with Equation (5) that includes rainfall.

In this study, rainfall and water level were used to forecast the water level. In the case of the Shibuya River basin, the water level data used were limited to one point. For rainfall, since the area of the basin is small at 10 s of km^2, the fact that spatial distribution of rainfall is limited, hence the average rainfall for the basin. Since our goal was to provide information for evacuation during floods, forecasts up to 30 min ahead were calculated. The reason for predicting up to 30 min is that the flood arrival time in the target basin is several tens of minutes and the time from the start of the hydrograph to the peak can be predicted.

The characteristic of this model Is that forecasts can be made from the observation data up to the present point, which makes it applicable for areas without forecast rainfall data.

3.2. Methods to Design Water Level Forecast Model and Verify Its Accuracy

When designing the water level forecast model of Equation (5) used in this study, the number of water level data P and the number of rainfall data S will be selected, while determining the degree N. As for the number of water level data, since there was only one water level observation station in the target basin, P = 1. For rainfall, assuming that the impact of the spatial distribution of rainfall was limited due to the small size of the basin (10 s of km^2). Thus, the model used the average rainfall for the basin with s = 1.

Next, for the degree N, Akaike Information Criterion (AIC) proposed by Akaike [43] is applied. AIC is expressed with Equation (6).

$$AIC = -2\log(L) + 2K \tag{6}$$

where L is the maximum log-likelihood and K is the number of model parameters. Figure 5 shows the relationship between AIC and the degree in this calculation. The smallest value in the AIC, N = 5, was selected.

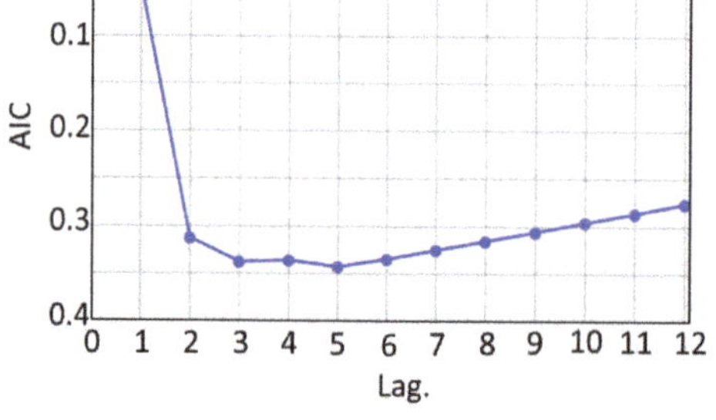

Figure 5. Relation of Lagged time and AIC.

The data were used to estimate parameters targeted nine localized torrential rainfalls since 2015 (Figure 3). Since the amount of data used to estimate the parameters for the water level forecast model was limited, we performed fold cross-validation that uses all limited data [44]. The steps of the fold cross-validation are as follows.

1. Target torrential rainfall event is divided into nine data sets as shown in Figure 3.
2. Data from one torrential rainfall event are used as the verification data.
3. Data from the remaining seven torrential rainfall events are used as the data set for parameter estimates.
4. With the set data, parameters are estimated and the accuracy is verified.
5. Steps 2, 3, and 4 are repeated to assess the accuracy of the forecast.

4. Results and Discussions

4.1. Verifying the Accuracy of the Forecast

The perspective important to verify the accuracy of the water level forecast model is whether the accuracy of the rising flood is ensured. This is especially important for urban basins with a short flood arrival time as in the present study since the time used to evacuate is also short. In the case of urban rivers, basically, embankments are not soil but concrete. Thus, it is rare for the embankment to be breached, and most floods are caused by overtopping. Thus, the judgment criterion in such cases is whether the water would top the height of the embankment or not; thus, the accuracy of peak water level forecast is crucial.

We present the water level forecast results of the vector autoregressive model which uses rainfall and water level data. Figure 6 is the water level forecast result for all nine torrential rainfall events. First, let us look at the torrential rainfall event 5 with the highest peak water level. Figure shows that the during the water level rise, its increase is captured, but it is underestimated compared to the measured water level. This trend was observed in all 10 min, 20 min, and 30 min forecasts. Specifically, at the start of the event when only weak rainfall was observed, the water level forecast was extremely low. On the other hand, when strong rainfall of over 50 mm/h was observed, the difference with the observed water level became small.

Next, let us look at the torrential rainfall event 2. With this torrential rainfall event, compared to the other torrential rainfall events, it takes $\geq$10 min to reach the peak water level. However, the flood arrival time for this torrential rainfall event 2 was the shortest at 10 min. In the case of such flood, the sign for the water level rise cannot be detected; and thus, it is considered the most dangerous urban flood. Figure 6 (event 2) shows that the rising trend of water level was forecast 20 min before the peak water level. The forecast was possible despite the fact that the observed water level had not risen at all. Therefore, using the proposed water level forecast model, forecasts are on the safe side from the viewpoint of evacuation.

In other flood events, forecast water levels were generally underestimated compared to the observed values; however, a rise in water level can be captured. Therefore, this method contributes to the decision to issue the evacuation information.

Next, Figure 7 shows the result of verifying the accuracy of peak water level forecast. This figure shows that generally, peak water level is underestimated in the forecast. However, other than the event 2, 10 min forecasts had an accuracy of 1 m or less compared to the observed values.

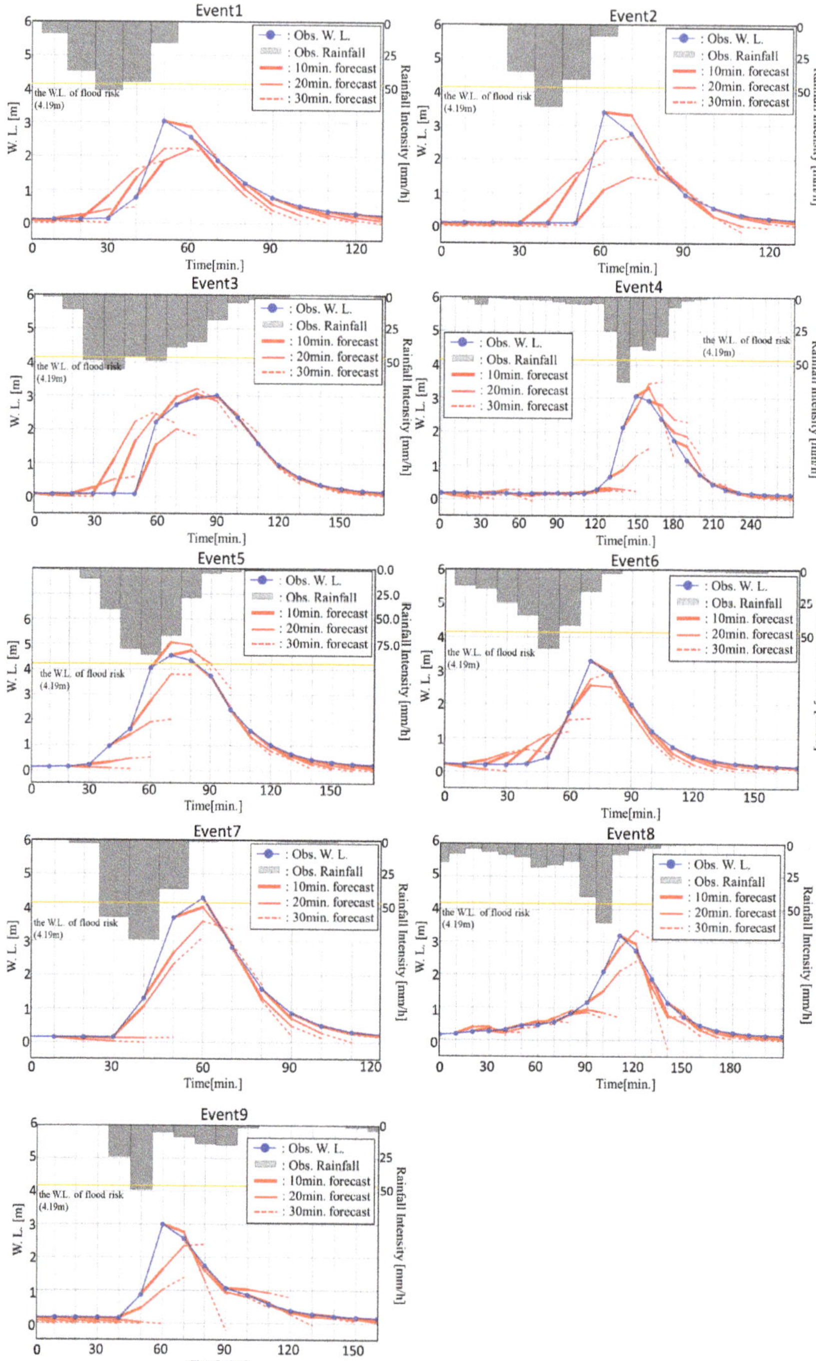

Figure 6. Forecasted water level results up to 30 min ahead for all events.

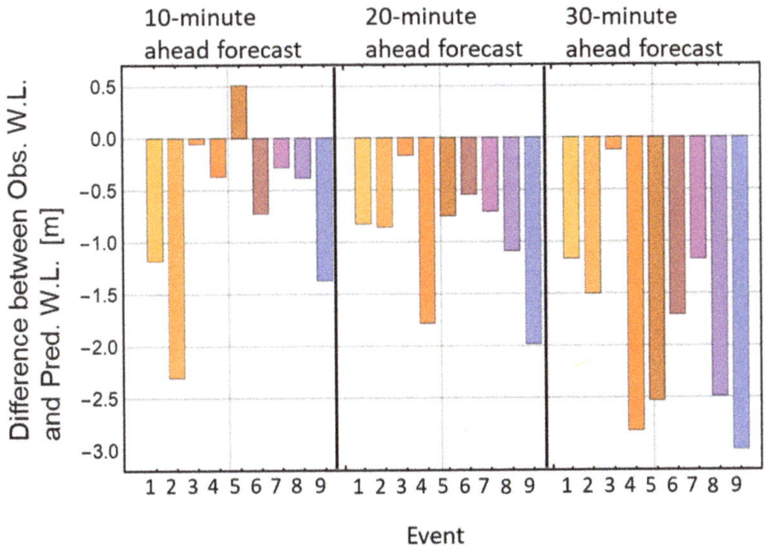

Figure 7. Difference in the observed water level and water level at the peak for each forecast time.

The overall accuracy of the 20 min forecast was lower, where the difference in the water level is approximatery1.5 m lower except for the torrential rainfall event 3.

The accuracy of the forecast for each torrential rainfall event, in the case of the above-mentioned event 2, was about 2 m lower in the 10 min forecast. This was likely since the heavy rain exceeding 50 mm/h already ended by the time for the 10 min forecast. In contrast, 20 min forecast had the peak difference with the observed value of less than 1 m. This was likely since heavy rainfall exceeding 50 mm/h was falling at the time of forecast. In the case of torrential rainfall event 3, the forecast was highly accurate for all 10 min, 20 min, and 30 min forecasts. Figure 6 shows that the time difference in the rainfall peak and the water level peak of torrential rainfall event 3 was longer than other torrential rainfall events. These results show that though the accuracy of the water level forecast is not favorable at the rise of the water level, the forecast was adequate once the water level rises close to the peak. This means that the peak water level can be forecast even when strong rainfall continues.

Figure 8 superimposes 10 min, 20 min, and 30 min forecasts with the observed water levels for all time periods from the rise to the fall of water level in the hydrograph for each torrential rainfall event. The red solid line is 45°, while the red dashed line is ±1 m of the red solid line. Figure 8a assesses observed and forecast values for 10 min ahead, which shows that most forecasts are within 1 m. In some cases, the forecast water levels rose during the normal condition (0.15 to 0.2 m) before the observed water levels rose. This shows that, as mentioned earlier, when heavy rain exceeding 50 mm/h was observed in each torrential rainfall event, the forecast was on the safe side. In the falling period of the hydrograph, the forecast was highly accurate. In addition, Figure 8 shows that 20 min and 30 min forecasts had lower water levels compared to the observed level as the forecast time increased.

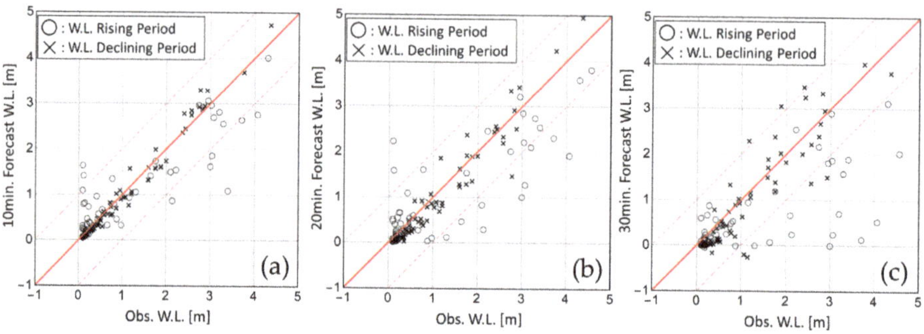

Figure 8. Comparison of the calculated and forecast values for the entire period in the hydrograph. (**a**): 10 min forecast, (**b**): 20 min forecast, and (**c**): 30 min forecast. Circles in figure are values during the rising period in the hydrograph, while cross are values during the falling period.

4.2. The Number of Torrential Rainfall Events Necessary for Stable Water Level Forecast

To verify the accuracy of water level forecasts by the present water level forecast method, it was calculated using all torrential rainfall events since 2015 that could be used in the target basin. However, in the basin of urban rivers, to which we want to introduce this method, the amount of data at the time of flood at the observation points is likely limited. Thus, we assessed the stability of forecast values when the number of torrential rainfall events used for parameter estimate was changed. For the assessment, the variations in the forecast values were examined by comparing them to the results of changing the number of torrential rainfall events used for parameter estimates between 1 and 7.

Table 2 shows the number of torrential rainfall events that can be prepared with the number of torrential events used to estimate parameters. The number of data sets that can be prepared varies based on the number of torrential rainfall events used to estimate parameters. The average of the dataset was taken for each forecast time of each torrential rainfall event and assessed the stability. *RMSE* was also used to validate the accuracy of the forecast. For the verification, the result of the hydrograph from the rise of the water level to the peak water level were applied, which is crucial during evacuation. *RMSE* is shown in Equation (7).

$$RMSE = \sqrt{\sum_{n=1}^{N_{peak}} \left(h^i_{ob} - h^i_{pre}\right)^2 / N_{peak}} \quad (7)$$

where h^i_{ob} is the observed values, h^i_{pre} is the forecast values, and N_{peak} is the amount of data.

Table 2. Number of events used for parameter estimation.

Number of Flood Events Used	Number of Datasets
1	${}_8C_1 = 8$
2	${}_8C_2 = 28$
3	${}_8C_3 = 56$
4	${}_8C_4 = 70$
5	${}_8C_5 = 56$
6	${}_8C_6 = 28$
7	${}_8C_1 = 8$

The torrential rainfall event that was the target of forecast in this verification was torrential rainfall event 5, where the water level rose the most (Figure 3). The result is shown in Figure 9. Frames 1 to 7 in Figure 9 show the number of torrential rainfall events used to estimate parameters (the number of torrential rainfall events used was changed between 1 and 7).

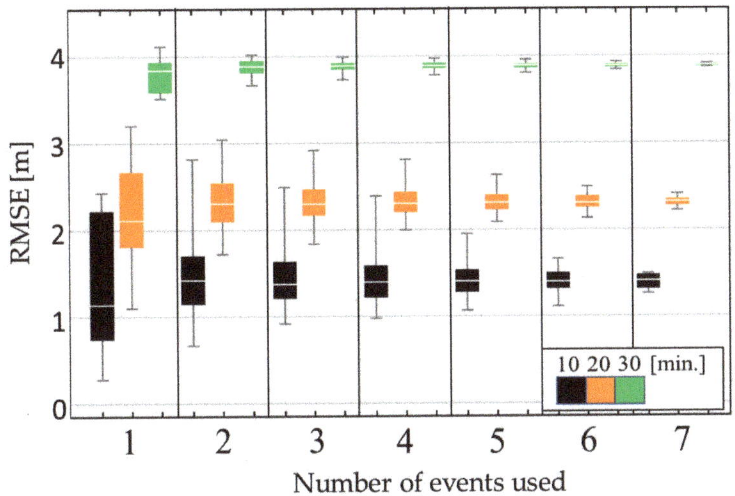

Figure 9. Differences in forecast accuracy for different numbers of flood events.

This result shows that the variations in the forecast values decreased when the number of torrential rainfall events used to estimate parameters was higher. Specifically, for the 10 min forecast (black bar), the range of variations in forecast values became notable based on the number of events. However, when we look at the average values, even for 10, 20, and 30 min forecasts, there was little difference in the average as long as at least two torrential rainfall events were used to estimate parameters. If there were about six torrential rainfall events, variations became small. Therefore, we assume that as long as there are data from about six large flood events, parameters become stable along with forecast values.

Data used for the water level forecast in this study included one of water level and one of rainfall. Since the water level data comprised the observation data from the station installed at the downstream side of the target, Shibuya River Basin, this is the final information in the rainfall outflow process, not allowing the use of upstream information (observation data). Therefore, the rainfall data at the start of the rainfall outflow process, and the information at the end of the process was limited to the water level information. Therefore, there was no information on the tributaries, which is the middle of the outflow process (in the case of this study, this would be the sewer system pipeline network). Thus, it was difficult to ensure the accuracy of the present forecast. If we could use the information from the upstream of the forecast point (water level observation station), we could design a forecast model with higher precision.

5. Conclusions

In this study, we designed a water level forecast method with the goal of providing early evacuation information in the basin of an urban river where the flood arrival time was extremely short at much less than one hour. Assessment of its accuracy is performed. For the water level forecast method, the vector autoregressive model was applied. For forecasting, the model only applied using real-time data of rainfall and water level.

The result showed that in the rising period of the hydrograph, forecast values were lower, but its difference was 1 m or less. Thus, it was sufficient information to encourage evacuation. In the case of localized torrential rainfall, it has been reported that evacuation times of a few minutes can be fatal [45]. The model's ability to forecast these tens of minutes is extremely important because the reality of localized torrential rainfall is that the time from the start of the rainfall to the peak water level is only a few tens of minutes, and these localized torrential rains frequently cause flooding damage in major cities in Japan. By incorporating the rainfall data into the forecast method, the rise in the water level was forecasted from the point when there was no rise in the real-time water level yet. Therefore, the present method can be used as a method to forecast water levels in basins without forecast rainfall data by only using observation data. The analysis did not take into account physical factors such as drainage systems by sewers, which means that the results obtained by using very limited data show that the method could be applied to many basins.

By changing the number of torrential rainfall events used to estimate the parameters of the water level forecast model, the numbers of torrential rainfall events were examined to calculate stable forecast values. The result showed that since five or more torrential rainfall events reduced the variations in the forecast values, as long as there are flood data from about five events, this model can be applied to urban river basins with limited observation data.

Finally, water level data from one point and rainfall data from one point to design this method was used. This was because there was only one water level observation station in the present target basin, but by using the upstream information, the accuracy of forecasts can be improved dramatically. However, there is hardly any location in Japan where the water level within the sewer system is being observed. To make more accurate forecasts to reduce the flood damage from short urban rivers, many rivers present similar challenges. The present method that allows forecasting of water levels even with a limited amount of data is useful in forecasting the water levels of urban rivers. Thus, to improve the accuracy of the present forecast method, it is necessary to increase the number of cases in other urban rivers for further verification.

Author Contributions: Conceptualization, N.K. and T.Y.; methodology, N.K.; software, N.K.; validation, N.K. and M.S.; formal analysis, N.K.; investigation, M.S.; resources, N.K.; data curation, M.S.; writing—original draft preparation, N.K.; writing—review and editing, N.K. and T.Y.; visualization, M.S.; supervision, T.Y.; project administration, T.Y.; funding acquisition, T.Y. All authors have read and agreed to the published version of the manuscript.

Funding: This research was supported by Unit for Research and Application Solution of Water-Related Disaster Science and Information of Research and Development Initiative (RDI), Chuo University.

Data Availability Statement: Not applicable.

Acknowledgments: We would like to thank Tokyo Metropolitan Government and Tokyu Construction Co., Ltd. for assisting with data collection.

Conflicts of Interest: The authors declare no conflict of interest.

References

1. Jha, A.K.; Bloch, R.; Lamond, J. *Cities and Flooding: A Guide to Integrated Urban Flood Risk Management for the 21st Century*; The World Bank Press: Washington, DC, USA, 2012.
2. Galloway, G.; Brody, S.; Reilly, A.; Highfield, W.E. *The Growing Threat of Urban Flooding: A National Challenge*; University of Maryland: College Park, MD, USA; Texas A&M University: Galveston, TX, USA, 2018.
3. Hollis, G.E. The effects of urbanization on floods of different recurrence interval. *Water Resour. Res.* **1975**, *11*, 431–435. [CrossRef]
4. Morita, S.; Izumi, K. Quantitative analysis, the difference of flood runoff depth associated with urbanization. *Proc. Jpn. Conf. Hydraul.* **1985**, *29*, 19–24.
5. Muis, S.; Güneralp, B.; Jongman, B.; Aerts, J.C.; Ward, P.J. Flood risk and adaptation strategies under climate change and urban expansion: A probabilistic analysis using global data. *Sci. Total Environ.* **2015**, *538*, 445–457. [CrossRef] [PubMed]
6. Mahmoud, S.H.; Gan, T.Y. Urbanization and climate change implications in flood risk management: Developing an efficient decision support system for flood susceptibility mapping. *Sci. Total Environ.* **2018**, *636*, 152–167. [CrossRef] [PubMed]
7. Wang, S.; McGrath, R.; Hanfin, J.; Lynch, P.; Semmler, T.; Nolan, P. The impact of climate change on storm surges over Irish waters. *Ocean Model.* **2008**, *25*, 83–94. [CrossRef]
8. Suzuki, T. Japan's inundation risk map of storm surge under global warming and inundation damage function of coastal area. *Annu. J. Civ. Eng. Ocean.* **2012**, *68*, I-870–I-875. [CrossRef]

9. Ye, S.; Li, H.Y.; Leung, L.R.; Guo, J.; Ran, Q.; Demissie, Y.; Sivapalan, M. Understanding flood seasonality and its temporal shifts within the contiguous united states. *J. Hydrol. Meteorol.* **2017**, *18*, 1997–2009. [CrossRef]
10. Iwanaga, Y.; Ideta, k.; Sato, T.; Shimatani, Y. Study on outflow effect of on-site storage and soil improvement for small watershed. *J. Jpn. Soc. Civ. Eng. Ser B1 (Hydraul. Eng.)* **2016**, *72*, 49–58.
11. Tokyo Metropolitan Government: Tokyo Metropolitan Government Basic Policy on Torrential Rainfall Countermeasures (Revised). 2014. Available online: https://www.toshiseibi.metro.tokyo.lg.jp/kiban/gouu_houshin/ (accessed on 28 November 2022).
12. Kawaike, K.; Nakagawa, H. Mitigation Effect of ON-SITE Storage Facilities on Pluvial Inundation in Highly Urbanized Area. *J. Jpn. Soc. Civ. Eng. Ser. B1 (Hydraul. Eng.)* **2018**, *74*, I-1537–I-1542. [CrossRef]
13. Hou, J.M.; Liu, F.F.; Tong, Y.; Guo, K.H.; Ma, L.; Li, D. Numerical simulation for runoff regulation in rain garden using 2D hydrodynamic model. *Ecol. Eng.* **2020**, *153*, 105794. [CrossRef]
14. Gagliano, A.; Nocera, F.; Detommaso, M.; Evola, G. Thermal behavior of an extensive green roof: Numerical simulations and experimental investigations. *Int. J. Heat Mass Transf.* **2016**, *34*, 226–234. [CrossRef]
15. Liu, X.; Chui, T.F.M. Evaluation of green roof performance in mitigating the impact of extreme storms. *Water* **2019**, *11*, 815. [CrossRef]
16. CIMA Foundation. Input Paper Prepared for the Global Assessment Report on Disaster Risk Reduction 2015. 2015. Available online: https://www.preventionweb.net/english/hyogo/gar/2015/en/gar-pdf/GAR2015_EN.pdf?_gl=1*q6zrfx*_ga*MTI4MzUyNzU0OS4xNjY3NTY0NTQ3*_ga_D8G5WXP6YM*MTY2NzU2NDU0Ni4xLjEuMTY2NzU2NDYzMS4wLjAuMA (accessed on 28 November 2022).
17. JRC. River Flood Hazard Maps at European and Global Scale. Available online: https://data.jrc.ec.europa.eu/collection/id-0054 (accessed on 4 November 2022).
18. Winsemius, H.C.; Van Beek, L.P.H.; Jongman, B.; Ward, P.J.; Bouwman, A. A frame work for global river flood risk assessments. *Hydrol. Earth Syst. Sci.* **2013**, *17*, 1871–1892. [CrossRef]
19. Ministry of Land, Infrastructure, Transport and Tourism. Hazard Map Portal Site. Available online: https://disaportal.gsi.go.jp/ (accessed on 4 November 2022).
20. Morooka, Y.; Yamada, T. Study on the characteristics of rainfall runoff in the Kinugawa river basin and the evacuation behavior of the residents at the time of Kanto and Tohoku flood disaster in September, 2015. *J. Disaster Res.* **2017**, *12*, 176–186. [CrossRef]
21. Beven, K.J. *Rainfall-Runoff Modeling*; Wiley: Hoboken, NJ, USA, 2001.
22. Fotovatikhah, F.; Herrera, M.; Shamshirband, S.; Chau, K.; Faizollahzadeh Ardabili, S.; Piran, M.J. Survey of computational intelligence as basis to big flood management: Challenges, research directions and future work. *Eng. Appl. Comp. Fluid Mech.* **2018**, *12*, 411–437. [CrossRef]
23. Yaseen, Z.M.; Sulaiman, S.O.; Deo, R.C.; Chau, K.W. An enhanced extreme learning machine model for river flow forecasting: State-of-the-art, practical applications in water resource engineering area and future research direction. *J. Hydrol.* **2019**, *569*, 387–408. [CrossRef]
24. Cho, M.; Kim, C.; Jung, K.; Jung, H. Water level prediction model applying a long short-term memory (lstm)–gated recurrent unit (gru) method for flood prediction. *Water* **2022**, *14*, 2221. [CrossRef]
25. Makoto, H.; Makoto, S.; Hironobu, M. Application of multivariate autoregressive model to groundwater behavior. *J. Hydraul. Eng.* **1995**, *1995*, 93–102.
26. Hartini, S.; Hadi, M.P.; Sudibyakto, S.; Poniman, A. Application of vector auto regression model for rainfall-river discharge analysis. *Forum Geogr.* **2015**, *29*, 1–10. [CrossRef]
27. Miedzielski, T. A data-based regional scale autoregressive rainfall-runoff model: A study from the odra river. *Stoch. Environ. Res. Risk Assess.* **2007**, *21*, 649–664. [CrossRef]
28. Niedzielski, T.; Miziński, B. Real-time hydrograph modelling in the upper nysa klodzka river basin (sw poland): A Two-Model hydrologic ensemble prediction approach. *Stoch. Environ. Res. Risk Assess.* **2017**, *31*, 1555–1576. [CrossRef]
29. Ali, G. Cointegration VAR and VECM and ARIMAX econometric approaches for water quality variates. *J. Stat Econ. Methods* **2015**, *4*, 1–38.
30. Mignot, E.; Li, X.; Dewals, B. Experimental modelling of urban flooding: A review. *J. Hydrol.* **2019**, *568*, 334–342. [CrossRef]
31. Danish, D. Hydraulic Institutek: Mike Urban, Mike Urban. 2017. Available online: https://www.mikepoweredbydhi.com/products/mike-urban (accessed on 28 November 2022).
32. Mwh: InfoWorks CS, Innovyze. 2010. Available online: http://www.innovyze.com/products/infoworks (accessed on 28 November 2022).
33. Rossman, L. *Storm Water Management Model Users' Manual, Version 5.1*; Environmental Protection Agency: Washington, DC, USA, 2015.
34. Henonin, J.; Russo, B.; Mark, O.; Gourbesville, P. Real-time urban flood forecasting and modelling—A state of the art. *J. Hydroinf.* **2013**, *15*, 717–736. [CrossRef]
35. Wang, H.; Hu, Y.; Guo, Y.; Wu, Z.; Yan, D. Urban flood forecasting based on the coupling of numerical weather model and stormwater model: A case study of zhengzhou city. *J. Hydrol. Reg. Stud.* **2022**, *39*, 100985. [CrossRef]
36. Sidek, L.M.; Jaafar, A.S.; Majid, W.H.A.W.A.; Basri, H.; Marufuzzaman, M.; Fared, M.M.; Moon, W.C. High-resolution hydrological-hydraulic modeling of urban floods using infoworks ICM. *Sustainability* **2021**, *13*, 10259. [CrossRef]

37. Mair, M.; Zischg, J.; Rauch, W.; Sitzenfrei, R. Where to find water pipes and sewers?—On the correlation of infrastructure networks in the urban environment. *Water* **2017**, *9*, 146. [CrossRef]
38. Qi, W.; Ma, C.; Xu, H.; Chen, Z.; Zhao, K.; Han, H. A review on applications of urban flood models in flood mitigation strategies. *Nat. Hazards* **2021**, *108*, 31–62. [CrossRef]
39. Tokyo Metropolitan Government. Sewerage Registers System. Available online: http://www.gesuijoho.metro.tokyo.jp/semiswebsystem/TspAgreementWeb.aspx (accessed on 4 November 2022).
40. Tokyo Metropolitan Government. Comprehensive Information System for Flood and Disaster Prevention. Available online: https://www.kasen-suibo.metro.tokyo.lg.jp/im/uryosuii/tsim0102g.html (accessed on 4 November 2022).
41. National Institute for Land and Infrastructure Management Ministry of Land, Infrastructure, Transport and Tourism (NILIM), Japan. *Technical Documentation on the Practical Application of XRAIN (X-Band Polarimetric (Multi Parameter) Radar Information Network) Rainfall Observation*; Technical Note of NIKIM; NILIM: Tsukuba, Japan, 2016; Volume 909.
42. Hyndman, R.J.; Athanasopoulos, G. *Forecasting: Principles and Practice*; OTexts: Heathmont, VIC, Australia, 2014.
43. Akaike, H. A new look at the statistical model identification. *IEEE Trans. Autom. Control* **1974**, *6*, 716–723. [CrossRef]
44. Lachenbruch, P.A.; Mickey, M.R. Estimation of Error Rates in Discriminant Analysis. *Technometrics* **1968**, *10*, 1–11. [CrossRef]
45. Eiichi, N.; Hiroto, S.; Ryuta, N.; Hiroyuki, Y.; Kosei, Y. Early Detection of Baby-Rain-Cell Aloft in a Severe Storm and Risk Projection for Urban Flash Flood. *Adv. Meteorol.* **2017**, *2017*, 5962356. [CrossRef]

MDPI
St. Alban-Anlage 66
4052 Basel
Switzerland
www.mdpi.com

Water Editorial Office
E-mail: water@mdpi.com
www.mdpi.com/journal/water

www.ingramcontent.com/pod-product-compliance
Lightning Source LLC
Chambersburg PA
CBHW040804250926
51850CB00007BA/1082
* 9 7 8 3 0 3 6 5 9 2 4 2 8 *